Innovations in Green Chemistry
and Green Engineering

This volume collects selected topical entries from the *Encyclopedia of Sustainability Science and Technology* (ESST). ESST addresses the grand challenges for science and engineering today. It provides unprecedented, peer-reviewed coverage of sustainability science and technology with contributions from nearly 1,000 of the world's leading scientists and engineers, who write on more than 600 separate topics in 38 sections. ESST establishes a foundation for the research, engineering, and economics supporting the many sustainability and policy evaluations being performed in institutions worldwide.

Paul T. Anastas • Julie B. Zimmerman
Editors

Innovations in Green Chemistry and Green Engineering

Selected Entries from the Encyclopedia of Sustainability Science and Technology

 Springer

Editors
Paul T. Anastas
Center for Green Chemistry
  and Green Engineering
Yale University
New Haven, CT 06520, USA

Julie B. Zimmerman
Department of Chemical Engineering
Environmental Engineering Program
Yale University
New Haven, CT 06520-8286, USA

This book consists of selections from the Encyclopedia of Sustainability Science and Technology edited by Robert A. Meyers, originally published by Springer Science+Business Media New York in 2012.

ISBN 978-1-4939-0138-8   ISBN 978-1-4614-5817-3 (eBook)
DOI 10.1007/978-1-4614-5817-3
Springer New York Heidelberg Dordrecht London

Contents

Chapter 1
Green Chemistry and Chemical Engineering, Introduction

Robert A. Meyers, Paul T. Anastas, and Julie B. Zimmerman

The goal of Green Chemistry and Chemical Engineering is to minimize waste, totally eliminate the toxicity of waste, minimize energy use, and utilize green energy (solar thermal, solar electric, wind, geothermal, etc.) – that is, non fossil fuel. Clearly, fossil fuels have their own waste and toxicity problems even though usually remote from the site of chemical production.

The objective in preparing this section is to provide a significant sampling of the scientific and engineering basis of green chemistry and engineering with specific processes as examples. There are nine detailed entries written at a level for use by university students through practicing professionals. For ease of use by students, each entry begins with a glossary of terms, while at an average length of 20 print pages each, sufficient detail is presented for utilization by professionals in government, universities, and industry. The reader is also directed to closely related sections in our *Encyclopedia*: *Solar Thermal Energy* (see the entry, Solar Energy in Thermo-chemical Processing); *Hydrogen Production Science and Technology* (see the entry, Photo-catalytic Hydrogen Production and also Hydrogen via Direct Solar Production); *Solar Radiation* (see the entry on Photosynthetically Active Radiation: Measurement and Modeling); *Geothermal Power Stations*; and the section on *Batteries*.

This chapter was originally published as part of the Encyclopedia of Sustainability Science and Technology edited by Robert A. Meyers. DOI:10.1007/978-1-4419-0851-3

R.A. Meyers
RAMTECH LIMITED, Laekspur,
CA, USA

P.T. Anastas
Center for Green Chemistry and Green Engineering, Yale University,
225 Prospect Street, New Haven, CT 06520, USA
e-mail: paul.anastas@yale.edu

J.B. Zimmerman
Department of Chemical Engineering Environmental Engineering Program,
Yale University, New Haven, CT 06520-8286, USA
e-mail: julie.zimmerman@yale.edu

P.T. Anastas and J.B. Zimmerman (eds.), *Innovations in Green Chemistry and Green Engineering*, DOI 10.1007/978-1-4614-5817-3_1,
© Springer Science+Business Media New York 2013

Each of the entries is summarized below.

Gas Expanded Liquids for Sustainable Catalysis – A gas-expanded liquid (GXL) phase is generated by dissolving a compressible gas such as CO_2 or a light olefin into the traditional liquid phase at mild pressures (tens of bar) (). When CO_2 is used as the expansion gas, the resulting liquid phase is termed a CO_2-expanded liquid or CXL. GXLs combine the advantages of compressed gases such as CO_2 and of traditional solvents in an optimal manner. GXLs retain the beneficial attributes of the conventional solvent (polarity, catalyst/reactant solubility) but with higher miscibility of permanent gases (O_2, H_2, CO, etc.) compared to organic solvents at ambient conditions and enhanced transport rates compared to liquid solvents (ii, iii, iv, v). The enhanced gas solubility's in GXLs have been exploited to alleviate gas starvation (often encountered in homogeneous catalysis with conventional solvents), resulting in a 1–2 orders of magnitude greater rates than in neat organic solvent or $scCO_2$. *Environmental advantages* include substantial replacement of organic solvents with environmentally benign CO_2. *Process advantages* include reduced flammability due to CO_2 presence in the vapor phase and milder process pressures (tens of bar) compared to $scCO_2$ (hundreds of bar). GXLs thus satisfy many of the attributes of an ideal alternative solvent.

Green Catalytic Transformations – A heterogeneous catalyst is a catalytically active species that is in a different phase to the reagents within a reaction system, more often than not a solid in either liquid or vapor phase synthesis. There are a number of green chemistry related advantages of using heterogeneous catalysts as opposed to the homogeneous equivalents: Safety – An important consideration in the practice of green chemistry. Heterogeneous catalysts often tend to be environmentally benign and easy/safe to handle. This is due to the active species being adhered to a support, (often forming a powder) essentially reducing its reactivity with the surrounding environment; Reusability – Due to the difference in phases, the catalyst is simply filtered off (through centrifugation on industrial scales) and reactivated for reuse many times over; Activity – In many cases increased activity is observed when supporting an active homogeneous species on a support, due to the complex but unique surface characteristics found with a variety of different supports; and Selectivity – heterogeneous catalysts can give an increased degree of selectivity in reaction pathway. This can simply be a consequence of adsorption of substrates and the consequential restricted freedom of movement of the reacting molecules. However, there are a handful of disadvantages. The quantity of solid catalyst required is often higher than that of the homogeneous equivalent, due to the lower concentration on the support surface, (reusability makes this less of an issue though). Blocking of the pores/support channels can occur with narrow pores sizes and reduce efficiency over time in some liquid phase reactions; nevertheless, this is often twinned with high stereoselectivity.

Green Chemistry Metrics: Material Efficiency and Strategic Synthesis Design – Application of green metrics forces the precise itemization of what constitutes waste so that targeted reductions of these components can be made. The mass of waste of any chemical reaction is the sum of the masses of unreacted starting materials, byproducts produced as a mechanistic consequence of making the desired target product, side products produced from competing side reactions other than the intended reaction, reaction solvent, all work-up materials, and all purification materials used. Simple first generation waste reduction strategies target the last three items in the list since they contributed the bulk of the overall mass of waste. Waste reduction strategies targeting the first three items are based on synthesis design and are necessarily more challenging to implement. The connecting green metrics are atom economy (AE), environmental E-factor (E), and reaction mass efficiency (RME). These metrics and applications are described in detail in this entry.

Green Chemistry with Microwave Energy – An alternative heating technique using microwaves is useful for targeted energy introduction directly into polar reactants in chemical syntheses and transformations. This entry summarizes noteworthy greener methods that use microwaves that have resulted in the development of sustainable synthetic protocols for drugs and fine chemicals. Microwave assisted organic transformations are presented such as: solid-supported reagents based processes; greener reaction media including aqueous, ionic liquid, and solvent-free for the synthesis of various heterocycles; and oxidation-reduction and also coupling reactions.

Nanotoxicology in Green Nanoscience – The unique properties that make nanomaterials an attractive technology (surface chemistry, surface area, size, shape, core material functionalization, aggregation, etc.) may also contribute to novel biological effects as a result of nanomaterial exposure. Toxicology will play an important role in elucidating the mechanisms of those interactions. This entry contains an exploration of the role of toxicology in implementing green nanoscience, the methodology of incorporating nanotoxicology in order to directly and indirectly address the principles of green chemistry in green nanoscience, and the importance of utilizing robust models for nanotoxicity testing.

Organic Batteries – Development of sustainable processes for energy storage and supply is one of the most important worldwide concerns today. Primary batteries, such as alkaline manganese and silver oxide batteries produce electric current by a one-way chemical reaction and are not rechargeable and hence useless for reversible electricity storage. Portable electronic equipment, electric vehicles, and robots require rechargeable secondary batteries. Li-ion, lead acid, and nickel-metal hydride batteries are generally used at the present to power them. Solar cells and wind-power generators expect a parallel use of rechargeable batteries for leveling and preserving their generated electricity. Ubiquitous electronic devices such as integrated circuit smart cards and active radio-frequency identification tags need rechargeable batteries that are bendable or flexible and environmentally benign for durability in daily use. Designing of soft portable electronic equipment, such as rollup displays and wearable devices, also require the development of flexible batteries. It is essential to find new, low-cost, and environmentally benign electroactive materials based on less-limited resource for electric energy storage and supply.[1] Reversible storage materials of electric energy or charge that are currently under use in electrodes of rechargeable batteries are entirely inorganic materials, such as Li ion-containing cobalt oxide, lead acid, and nickel-metal hydride.

Oxidation Catalysts for Green Chemistry – Catalysis is at the heart of Green Chemistry as it is the means to increase efficiency and efficacy of chemical and energy resources while promoting environmental friendliness and intensifying time and cost savings in chemical synthesis. A catalyst's function simply is to provide a pathway for chemicals (reactants) to combine in a more effective manner than in its absence. In the absence of a catalyst, heat is usually the way to overcome the energy barrier but this increases energy consumption and often results in unwanted side reactions. A catalyst cannot make an energetically unfavorable reaction occur or change the chemical equilibrium of a reaction because the catalyzed rate of both the forward and the reverse reactions are equally affected. Oxidation processes are used for odor control, bleaching of pulp for paper production, wastewater treatment, disinfection, bulk and specialty chemical production, aquatics and pools, food and beverage processing, cooling towers, agriculture/farming, and many others. There are a great many oxidation reactions that are catalytically driven. Such reactions are presented in this entry with emphasis on implementation of green strategies.

New Polymers, Renewables as Raw Materials – Recent advances in genetic engineering, composite science, and natural fiber development offer significant opportunities for developing new, improved materials from renewable resources that can biodegrade or be recycled, enhancing global sustainability. A wide range of high-performance, low-cost materials can be made using plant oils, natural fibers, and lignin. These materials have economic and environmental advantages that make them attractive alternatives to petroleum-based materials.

Supercritical Carbon Dioxide (CO_2) as Green Solvent – A supercritical fluid (SCF) is created when the temperature and pressure are higher than its critical values. Therefore, CO_2 becomes supercritical when its temperature and pressure are higher than $31.1^\circ C$ (critical temperature, Tc) and pressure 7.38 MPa (critical pressure, P_c). SCFs have many unique properties, such as strong solvation power for different solutes, large diffusion coefficient comparing with liquids, zero surface tension, and their physical properties can be turned continuously by varying the pressure and temperature because the isothermal compressibility of SCFs is very large, especially in the critical region. These unique properties of SCFs lead to great potential for the development of innovative technologies. Besides these common advantages of SCFs, supercritical CO_2 (scCO_2) has some other advantages, such as nontoxic, nonflammable, chemically stable, readily available, cheap, and easily recyclable, and it has easily accessible critical parameters. Therefore, scCO_2 can be used as green solvent in different fields. The basic properties of scCO_2 and its applications in extraction and fractionation, chemical reactions, polymeric synthesis, material science, supercritical chromatography, painting, dyeing and cleaning, and emulsions related with CO_2, are discussed in this entry.

Chapter 2
Gas Expanded Liquids for Sustainable Catalysis

Bala Subramaniam

Glossary

Carbon selectivity	Refers to the fraction of the carbon in a hydrocarbon-based feed that is utilized in making the desired product.
Compressible gas	A gas in the vicinity of its critical temperature wherein it is highly compressible with pressure, either condensing or attaining liquid-like densities as the pressure approaches or exceeds its critical pressure. Below the critical temperature, a compressible gas will typically condense at sufficiently high pressures to produce a liquid phase.
Gas-expanded liquids	When a liquid such as an organic solvent is pressurized with a compressible gas, the liquid phase will volumetrically expand if the gas dissolves in it. The volumetrically expanded liquid phase is termed as a gas-expanded liquid. When the pressure is released, the dissolved gas will escape from the liquid phase causing the liquid phase to contract to its original volume.
Homogeneous catalysis	Refers to a process wherein the reactants, catalyst, and products are soluble in a single phase and the catalytic reaction occurs in that phase.
Multiphase catalysis	Refers to a process wherein the reactants, catalyst, and products are present in two or more different immiscible

This chapter was originally published as part of the Encyclopedia of Sustainability Science and Technology edited by Robert A. Meyers. DOI:10.1007/978-1-4419-0851-3.

B. Subramaniam (✉)
Department of Chemical and Petroleum Engineering, The Center for Environmentally Beneficial Catalysis, University of Kansas, Lawrence, KS 66045, USA
e-mail: bsubramaniam@ku.edu

P.T. Anastas and J.B. Zimmerman (eds.), *Innovations in Green Chemistry and Green Engineering*, DOI 10.1007/978-1-4614-5817-3_2,
© Springer Science+Business Media New York 2013

5

	phases separated by phase boundary(ies). The reaction typically occurs at a boundary between two phases.
Renewable feedstock	Refers to a feedstock from nature whose use has minimal adverse impact on the ecosystem and that has the ability to manifest itself again in nature in a matter of a few months to a few years rather than in hundreds to thousands of years.
Solvent engineering	Exploiting the synergies between catalysis and solvent media for enhancing rates, selectivity, and separations in a sustainable manner.
Supercritical fluid	A substance that is above its critical pressure (P_c) and critical temperature (T_c). For applications in catalysis and separations, the near-critical region [0.9–1.2 T_c (in K) and 0.9–2 P_c] wherein small changes in temperature and/or pressure yield relatively large changes (from gas-like to liquid-like values) in density and transport properties, is generally of interest.
Sustainability	Sustainability of a catalytic process refers to the long-term environmental, social, and economic viability of the process.
Turnover frequency (TOF)	Quantifies the intrinsic activity of a catalyst in converting the reactants to products. It is usually expressed in terms of the rate at which the reactants are converted to products [(moles of substrate converted)/(gram atoms of catalyst used)/(time)].

Definition of the Subject

The modern-day chemical industry relies mostly on fossil fuel (such as petroleum, natural gas, and coal)–based feedstock. There are several megaton industrial catalytic processes that produce essential commodities for everyday life but present challenges with respect to reducing environmental footprints and enhancing sustainability. Examples of such processes include the homogeneous hydroformylation of higher olefins, the selective oxidation of light olefins to their corresponding epoxides, and the oxidation of p-xylene to produce terephthalic acid. For a targeted product, there are several possible scenarios for developing sustainable alternatives to conventional technologies. These include (a) developing greener process technologies based on existing feedstock, (b) replacement of fossil fuel–based feedstock with renewable ones such as those derived from biomass (which will also entail the development of new chemistries and process technologies), or (c) replacement of the target product itself with alternate candidates from renewable feedstocks. This entry will discuss the potential of gas-expanded liquids (GXLs), a relatively new class of solvents, for developing alternative and more sustainable catalytic processes.

A gas-expanded liquid (GXL) phase is generated by dissolving a compressible gas such as CO_2 or a light olefin into the traditional liquid phase at mild pressures (tens of bar) [1]. When CO_2 is used as the expansion gas, the resulting liquid phase is termed a CO_2-expanded liquid or CXL. GXLs combine the advantages of compressed gases such as CO_2 and of traditional solvents in an optimal manner. GXLs retain the beneficial attributes of the conventional solvent (polarity, catalyst/reactant solubility) but with higher miscibility of permanent gases (O_2, H_2, CO, etc.) compared to organic solvents at ambient conditions and enhanced transport rates compared to liquid solvents [2–5]. The enhanced gas solubilities in GXLs have been exploited to alleviate gas starvation (often encountered in homogeneous catalysis with conventional solvents), resulting in a one to two orders of magnitude greater rates than in neat organic solvent or $scCO_2$. *Environmental advantages* include substantial replacement of organic solvents with environmentally benign CO_2. *Process advantages* include reduced flammability due to CO_2 presence in the vapor phase and milder process pressures (tens of bar) compared to $scCO_2$ (hundreds of bar). GXLs thus satisfy many of the attributes of an ideal alternative solvent.

Introduction

Solvent usage has often been linked to waste generation and associated environmental and economic burdens [6, 7]. Within the last two to three decades, many research groups have investigated benign alternate media for performing chemical reactions [8–12]. Examples of such media include supercritical CO_2 ($scCO_2$) [13–20], water [21–23], gas-expanded liquids (GXLs) [1, 24–27], ionic liquids (ILs) [28–30], and switchable solvents [31–35]. While $scCO_2$ is "generally regarded as safe," its nonpolar nature renders it unsuitable for most homogeneous catalysis involving polar transition metal complexes. Further, $scCO_2$ media require operating pressures in excess of 100 bar. The use of either supercritical or near-critical water (P_c = 220.6 bar; T_c = 373.9°C) requires rather harsh operating pressures and temperatures. The use of ionic liquids as tunable media for catalysis either alone or in combination with $scCO_2$ [36] shows much promise.

The ideal alternative solvent, in addition to being considered green, should typically satisfy the following criteria: (a) retain the beneficial aspects of the conventional solvent (polarity, catalyst/reactant solubility) being replaced, (b) facilitate facile product/catalyst separation, (c) enhance process safety, and (d) operate at mild pressures (tens of bar) for economic viability. The qualitative principles of green chemistry [37] and green engineering [38] provide valuable guidelines for developing greener process alternatives. Some of these principles include the use of renewable and abundant resources as feedstock, nonhazardous reagents as reaction and separation media, inherently safe process design, and process intensification at mild conditions. However, a reliable assessment of overall "greenness" and sustainability requires quantitative comparison with conventional processes using metrics such as atom economy, the E-factor (amount of waste produced/unit of desired product) [6], toxic emissions potential [39], and process economics. Such analyses also provide guidance in identifying potential process improvement opportunities and establishing performance metrics for sustainability.

For a targeted product from a given feedstock, the basic elements of a catalytic process or system include the catalyst, media, multiphase reactor, and separation. Effective integration of these elements into a sustainable technology requires a multiscale approach [40]. This entry highlights reported examples of catalytic process concepts with gas-expanded liquids and multiscale approaches to develop such processes for large-scale catalytic technologies. The examples include catalytic hydrogenations, hydroformylations, and selective oxidations. Specifically, it is shown how the tunable physicochemical properties of GXLs can be effectively exploited to promote sustainable catalysis. An example of quantitative economic and environmental impact analyses is also presented to show how such an assessment validates and facilitates the design and development of sustainable systems that are also practically viable.

Thermodynamic and Physical Properties of GXLs

A gas in the vicinity of its critical temperature [0.9–1.2 T_c (K)] is highly compressible and attains liquid-like densities when sufficiently compressed to near-critical or supercritical pressures. As an example, consider CO_2 ($P_c = 71.8$ bar; $T_c = 304.1$ K) as an expansion gas. At 40°C (313.15 K and $T/T_c = 1.03$), CO_2 is highly compressible attaining liquid-like densities as its critical pressure is approached. In such a compressed state, CO_2 dissolves in a liquid phase and volumetrically "expands" the liquid phase forming a GXL. When the system pressure is released, the dissolved CO_2 escapes from the liquid phase contracting the liquid phase. Gases whose critical temperatures are far removed from the reaction temperature do not exhibit such compressibility and are generally incapable of expanding solvents.

Liquids expand to different extents in the presence of CO_2 pressure, depending on the ability of the liquids to dissolve CO_2. The equipment and experimental procedures for precise measurements of volumetric expansion and phase equilibria are provided elsewhere [41, 42]. As such, liquids are divided into three general classes [43]. Class I liquids such as water have insufficient ability to dissolve CO_2, and therefore do not expand significantly. Glycerol and other polyols also fall into this class (Fig. 2.1a).

Class II liquids, such as ethyl acetate, acetonitrile, methanol, and hexane, dissolve large amounts of CO_2 and hence expand appreciably (Fig. 2.1) undergoing significant changes in physical properties. As shown in Fig. 1, conventional solvents such as acetonitrile and ethyl acetate are volumetrically expanded several fold by compressed CO_2 at 40°C, reaching as high as eight for ethyl acetate at around 70 bar. Regardless of the solvent, the volumetric expansion of class II solvents is strongly dependent on the mole fraction of CO_2 in the liquid phase [41]. Class III liquids, such

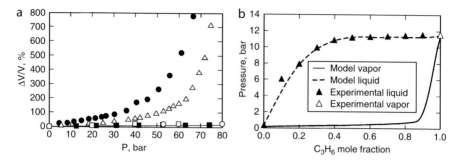

Fig. 2.1 (**a**) Expansion of solvents as a function of the pressure of CO_2 at 40°C, for ethyl acetate (•), MeCN (Δ), [1-butyl-3methylimidazolium]BF_4 (filled squares), polypropylene glycol (❑), and polyethylene glycol (O) [1]. (**b**) Vapor–liquid equilibrium for C_3H_6 + MeOH binary system at 21°C [46]

as ionic liquids and liquid polymers, dissolve relatively smaller amounts of CO_2 and therefore expand only moderately (Fig. 2.1a, [44]). By using a solvent such as acetonitrile or methanol that exhibits mutual solubility in CO_2 and water, it is possible to create CO_2-expanded ternary systems containing water [45]. Gases such as propylene can also expand organic solvents. Figure 2.1b shows that for the propylene + methanol binary system, the propylene mole fraction in the liquid phase approaches approximately 30 mol% at 10 bar, implying that compressed propylene at ambient temperatures can cause significant volumetric expansion of the methanol phase. Such behavior has been exploited in epoxidation of light olefins, as explained in a later section.

The expansions of Class II liquids have been successfully modeled by the Peng-Robinson Equation of State (PR-EoS) [47] and by molecular simulations [48, 49]. Gases such as ethane, fluoroform, and other similar compressible gases are also capable of expanding liquids. As discussed in later sections of this entry, gaseous substrates (such as propylene and ethylene) and gaseous oxidants (ozone) have also been exploited as expansion gases to overcome solubility limitations in the liquid phase where reaction occurs [50].

The presence of CO_2 in the liquid phase affects the physicochemical properties. For example, the viscosity of methanol decreases with CO_2 pressure at 40°C, by nearly 80% from pure methanol to CXL methanol at 77 bar CO_2 [51]. This viscosity reduction is especially striking for ionic liquids, which often have higher viscosity than organic solvents [52, 53]. The diffusivity of benzene in CXL methanol at 40°C and 150 bar increases by over 200% on replacing pure methanol with 75% CO_2 [54]. Similarly, the diffusion coefficients of benzonitrile in CO_2-expanded ethanol were observed to increase with increasing CO_2 mole fraction [55].

As regards polarity, Roškar et al. [56] have measured the dielectric constant of methanol with CO_2 at 35°C and found that from ambient conditions to approximately 40 bar of CO_2, the dielectric constant changes very little. However, the dielectric constant significantly decreases by about an order of magnitude at 76 bar of CO_2 pressure. It has been shown that the Kamlet–Taft parameters (for acidity, basicity, and polarizability) for mixed CO_2 + organic solvents are tunable by CO_2 addition [57–59]. This tunability has potential ramifications for the stabilization of charged and polar compounds in CXLs. The following sections provide examples of how GXLs have been systematically exploited to develop greener process concepts that display improved reaction performance with respect to rates, selectivity, and separations.

Catalytic Reactors for Investigating GXLs

Either batch or continuous-flow reactors may be employed for performing catalytic conversion, selectivity, and kinetic studies. In all cases, it is essential to know the phase behavior of the reaction mixture to rationally interpret the results. Equipment must be designed and pressure-tested according to standardized design and testing procedures such as those prescribed by the American Society for Testing Materials. All equipment must incorporate adequate pressure relief and inherent safety measures. For example, in the case of exothermic reactions, the amount of reactants fed should be such that the estimated adiabatic temperature rise at total conversion does not lead to "thermal runaway" conditions and unsafe pressures. Similarly, enough inerts should be added to the vapor phase to avoid the formation of explosive organic vapors in the presence of air. During continuous reactor operation, safety shutdown measures should be incorporated. Described below are examples of batch and continuous reactors reported in the literature.

Batch Reactors

Proof-of-concept batch studies are typically done in 5–10 mL view cells, as shown in Fig. 2.2, typically rated to operate at 150°C and 200 bar. A schematic drawing of an experimental reactor is shown in Fig. 2.2. The reactor is a low-volume (10 mL) hollow cylinder with sapphire windows at each end, sealed by o-rings and screw caps. The sapphire windows allow visual inspection of the cell contents and permit in situ spectroscopic studies. The cell body has as many as five ports. Two of the ports are used for introducing reactants such as O_2 or H_2 and the compressed gas medium such as CO_2. Oxygen or hydrogen is typically introduced via a mass flow controller. The third port is used for injecting liquid reactant into the reactor and is connected to a safety head containing a rupture disk. A pressure transducer that continuously monitors the reactor pressure and a thermocouple that monitors the reactor temperature are connected to the remaining two ports. A magnetic stirrer provides adequate mixing of the reactor contents. Fiber optics may be attached to the sapphire windows and connected to a UV-Vis spectrophotometer. These facilitate temporal in situ monitoring of chemical reactions over broad spectral ranges.

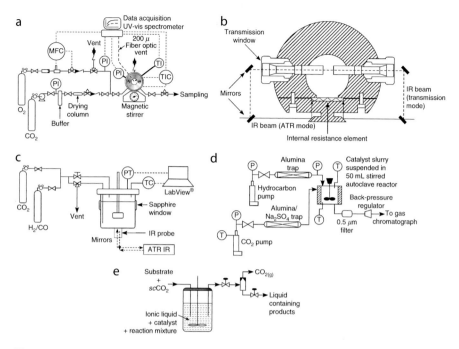

Fig. 2.2 Schematics of experimental reactors used for investigations of catalytic reactions in CXLs: (**a**) 5–10 mL view cell; (**b**) ATR-IR view cell; (**c**) 50 mL stirred reactor with ReactIR (*PT* pressure transducer, *TC* thermocouple); (**d**) continuous CXL reactor; (**e**) continuous flow reactor using a supercritical fluid–ionic liquid biphasic system

At the end of a batch run, the fluid phase cell contents may be depressurized into suitable sample traps for off-line analysis.

Baiker and coworkers [60] employed in situ high-pressure attenuated total reflectance infrared (ATR-IR) spectroscopy to elucidate the molecular interactions between dissolved CO_2 and ionic liquids. This cell, shown in Fig. 2.2c, consists of a horizontal stainless steel cylinder with sapphire windows on each end, one of which is attached to a piston to control the reactor volume and pressure [61]. It can operate at temperatures up to 100°C and pressures up to 20 bar. The IR beam can be directed by a combination of four mirrors either through the ZnSe internal resistance element for ATR-IR measurements of the dense phase or through cylindrical ZnSe windows for transmission spectroscopy of the upper phase. In addition, the internal resistance element can be coated with a layer of catalyst to measure the interactions between the reactants and the catalyst during the reaction. Thus, the cell can be used to simultaneously provide information about dissolved and adsorbed reacting species and by-products as well as the catalyst itself.

For larger-scale batch investigations, adequate mechanical mixing must be ensured to avoid mass transfer limitations. Mechanically stirred reactors (50–300 mL such as Autoclave and Parr) with stirrer speeds up to 1500 rpm are well suited for this purpose. For example, a 50-cm³ high-pressure autoclave reactor equipped with an in situ attenuated total reflectance (ATR) IR probe (Mettler Toledo Inc.) was employed to investigate the hydroformylation of 1-octene [62]. Figure 2.2c shows a schematic of the apparatus (entitled "ReactIR"). Mixing is provided by a magnetic stirrer with a maximum agitation rate of 1700 rpm. Pressure and temperature are monitored by a Labview® data acquisition system and controlled with a Parr 4840 controller. Syngas is introduced from a gas reservoir, which is equipped with a pressure regulator that is used to admit syngas and maintain a constant total pressure in the reactor. The pressure transducer monitors the total pressure in the reservoir. The IR probe, placed at the bottom of the reactor, monitors concentration profiles of various species in the CXL phase. The maximum working pressure of the probe is approximately 140 bar. The temporal IR data are then processed using ConcIRT software to extract the absorbance profiles for each species, based on their characteristic peaks (identified with standards).

Continuous Reactors

For performing heterogeneous catalytic reactions in GXLs, a continuous stirred tank reactor (CSTR) is better suited for isothermal pressure-tuning studies [63]. A schematic is shown is Fig. 2d wherein the experiments are conducted in a 50 mL reactor from Autoclave Engineers, rated to 344 bar and 350°C. Catalyst particles are suspended in the reaction mixture by an impeller operating at 1200 rpm. Reaction pressure is maintained with a dome-loaded back-pressure regulator.

Cole-Hamilton and coworkers [64] demonstrated a continuous reactor system for investigating homogeneous catalytic reactions in $IL/scCO_2$ systems (Fig. 2e). Here, the IL phase containing dissolved catalyst is retained in the reactor. The $scCO_2$ is used to transport the soluble substrate into and products from the IL phase continuously. The challenge is to minimize catalyst leaching from the IL phase by the $scCO_2$ + reaction mixture exiting the reactor.

Applications in Homogeneous Catalysis

Hydrogenations

Hydrogenations have been shown to benefit from CO_2 expansion of the solvent in several ways. For example, $CO_2 + H_2$ mixtures are just as effective chemically as pure H_2 at the same total pressure, and also desirable from a safety standpoint. Jessop's group [65] showed that the properties of the expanding gas can have a major effect on reaction rate. Investigating the effect of expansion gas on the rate of CO_2 hydrogenation in MeOH/NEt$_3$ mixture as solvent (Eq. 2.1), it was found that the turnover frequency was 770 h^{-1} with no expansion gas but 160 h^{-1} with ethane (40 bar) and 910 h^{-1} with CHF$_3$ (40 bar) as the expansion gas. The addition of liquid hexane produced a similar decrease in the rate. Hence, the low polarity of ethane was implicated for the decreased rate in the ethane-expanded solvent.

$$CO_2 + H_2 \xrightarrow[\text{NEt}_3/\text{MeOH}]{\text{RuCl(OAc)(PMe}_3)_4} HCO_2H$$

50°C

NEt$_3$/MeOH   (2.1)

40 bar H$_2$, 10 bar CO$_2$

40 bar expansion gas

Solvent expansion as a method to enhance the H$_2$ availability in the liquid phase of homogeneous catalytic reactions is of interest in asymmetric hydrogenations. Foster's group [66] investigated the asymmetric hydrogenation of 2-(6′-methoxy-2′-naphthyl) acrylic acid, an atropic acid, in CO_2-expanded methanol with [RuCl$_2$(BINAP) (cymene)]$_2$, finding the reaction faster but less selective than in neat methanol. A subsequent study using RuCl$_2$(BINAP) catalyst reported that the reaction in expanded methanol was slower than in normal methanol [67]. More recently, the asymmetric hydrogenation of methyl acetoacetate to methyl (R)-3-hydroxybutyrate by [(R)-RuCl(binap)(p-cymen)]Cl was investigated in methanol-dense CO_2 solvent systems [68]. Although the CO_2-expanded methanol system resulted in a reduction of both reaction rate and product selectivity, this changed in the presence of water. High selectivities were obtained with the optimized methanol–CO_2–water–halide system.

Contrasting effects are also demonstrated during asymmetric hydrogenations in ionic liquids (ILs). The enantioselectivity during hydrogenation of tiglic acid in [BMIm][PF$_6$] (where BMIm is L-n-butyl-3-methylimidazolium) is superior [H$_2$ = 5 bar; ee = 93%] to that in CO_2-expanded [BMIm][PF$_6$] (H$_2$ = 5 bar; CO$_2$ = 70 bar; ee = 79%) wherein the H$_2$ availability is improved with CO_2 [69]. In contrast, the hydrogenation of atropic acid is greatly improved in selectivity (ee increases from 32% to 57%) when the IL is expanded with 50 bar CO_2 [70]. Leitner and coworkers [69] demonstrated that hydrogenation of N-(1-phenylethylidene)aniline proceeded to only 3% conversion in [EMIm][Tf$_2$N] (1-ethyl-3-methylimidazolium bis(trifluoromethyl-sulfonyl)imide) and to >99% in CO_2-expanded [EMIm][Tf$_2$N], in which the H$_2$ solubility is significantly enhanced with CO_2 pressure (Fig. 2.3).

Facile hydrogenation of solid substrates such as vinylnaphthalene with a RhCl (PPh$_3$)$_3$ catalyst was demonstrated by melting the solid with compressed CO_2 and performing the reaction in the melt phase at 33°C, which is well below the normal melting point of the solid [71]. Scurto and Leitner [72] showed that the addition of compressed CO_2 induced a melting point depression of an ionic solid salt (tetrabutylammonium tetrafluoroborate) at greater than 100°C, well below its normal melting point of 156°C. They used this CO_2-induced melting technique to   demonstrate   hydrogenation,   hydroformylation,   and   hydroboration   of

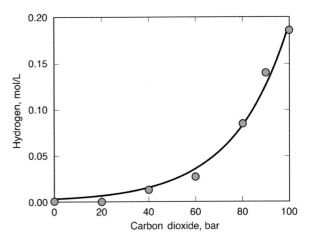

Fig. 2.3 Enhancement of H_2 solubility in [EMIm][Tf_2N] by CO_2 pressurization at a constant H_2 partial pressure (30 bar at $T = 297$ K). Taken from [69]

vinylnaphthalene using rhodium complexes. Employing homogeneous octene hydrogenation with rhodium complexes as a model system, Scurto and coworkers demonstrated the importance of understanding the phase behavior, mass transfer, and intrinsic kinetics effects for the reliable interpretation and design of hydrogenations in IL + CO_2 media [73].

Hydroformylations

Hydroformylations benefit from CXLs by virtue of the fact that syngas + CO_2 mixtures are more effective than syngas alone at a fixed pressure. The use of CO_2 helps generate CXLs, which enhance both the syngas solubility as well the tunability of the H_2/CO ratio at milder pressures. Jin and Subramaniam [74] demonstrated the advantages of homogeneous hydroformylation of 1-octene in CXLs using an unmodified rhodium catalyst $(Rh(acac)(CO)_2)$. The turnover numbers (TONs) in CO_2-expanded acetone were significantly higher than those obtained in either neat acetone or $scCO_2$. In subsequent reports, the performances of several rhodium catalysts, $Rh(acac)(CO)_2$, $Rh(acac)[P(OPh)_3]_2$, $Rh(acac)(CO)[P(OAr)_3]$, and two phosphorous ligands, PPh_3 and biphephos, were compared in neat organic solvents and in CXLs [75]. For all catalysts, enhanced turnover frequencies (TOFs) were observed in CXLs. For the most active catalyst, $Rh(acac)(CO)_2$ modified by biphephos ligand, the selectivity to aldehyde products was improved from approximately 70% in neat solvent to nearly 95% in CXL media. Such improved catalyst performance for hydroformylation in CO_2-based reaction mixtures were also reported recently for rhodium catalysts modified with triphenylphosphine, triphenyl phosphite, and tris(2,4-di-*tert*-butylphenyl) phosphite [76], with turnover numbers as high as $3(10^4)$ mol aldehyde/(mol Rh)/h. In contrast, the Rh complex-catalyzed hydroformylation of 1-hexene in CO_2-expanded toluene was more rapid than in

Table 2.1 Comparison with commercial processes and other reported work

Process parameters	BASF (Co)	Shell (Co/P)	SCF-IL (Rh/P)	SCF (Rh/P)	CXL (Rh/P)
Substrate	1-octene	1-octene	1-octene	1-octene	1-octene
P, bar	300	80	200	125	38
T, °C	150	200	100	100	60
TOF, h^{-1}	35	20	517	259	316
$S_{n\text{-aldehyde}}$, %	50	80	75	75	89

$scCO_2$ but slower than in normal toluene [77]. The high CO_2 pressure was believed to shift some of the 1-hexene out of the liquid phase into the CO_2 phase, thereby lowering the concentration of hexene available to the catalyst.

Ionic liquid + CO_2 media has also been used for hydroformylation [64, 78]. Constant activity for up to 3 days was demonstrated during continuous 1-octene hydroformylation with a Rh-based catalyst in [BMIm][PF$_6$]/$scCO_2$ medium. Compared to the industrial cobalt-catalyzed processes, higher TOFs were observed; the selectivity to linear aldehyde (70%) is comparable to those attained in the industrial processes (70–80%). However, the air/moisture sensitivity of the ILs and the ligands may lead to the deactivation and leaching of the rhodium catalyst. Cole-Hamilton and coworkers also reported a solventless homogeneous hydroformylation process using dense CO_2 to transport the reactants into and transfer the products out of the reactor leaving behind the insoluble Rh-based catalyst complex in the reactor solution [79].

Table 2.1 compares the TOFs, n/i ratio, and operating conditions (P&T) of some existing industrial processes and recently published work that employ CXLs, supercritical CO_2, and ionic liquids as reaction media [64]. Clearly, hydroformylation in CO_2-expanded octene appears to be promising in terms of both TOF ($\sim$300 h^{-1}) and selectivity ($\sim$90%, $n/i > 10$). In addition, the required operating conditions (60°C and 38 bar) are much milder compared to other processes. The further development of this process, including quantitative sustainability analysis, is provided in a following section.

Another example involves enantioselective hydroformylation of solid vinylnaphthalene with a Rh catalyst in a CO_2-based phase, created by melting the solid with compressed CO_2. This technique is similar to the melt hydrogenation described earlier.

O_2-Based Oxidations

CXLs provide both rate and safety advantages for homogeneous catalytic oxidations. For the homogeneous catalytic O_2-oxidation of 2,6-di-*tert*-butylphenol, DTBP, by Co(salen*) in $scCO_2$, in CO_2-expanded acetonitrile and in the neat organic solvent [80, 81], it was found that the TOF in the CO_2-expanded acetonitrile is between one and two orders of magnitude greater than in $scCO_2$. Additionally, the 2,6-di-*tert*-butyl quinine (DTBQ) selectivity is lower in neat acetonitrile, clearly

Table 2.2 Vapor–liquid equilibrium for CO_2 (1)–O_2 (2)–acetonitrile (3) [2]

T (°C)	P_{total} (bar)	Liquid-phase composition		Vapor-phase composition		f_{O2} (bar)	x_{O2}, pure gas[a] mole fraction	EF
		x_1	x_2	y_1	y_2			
40.0	13.7	0.03	0.004	0.27	0.68	9.2	0.0046	0.77
40.0	23.7	0.08	0.007	0.35	0.63	14.9	0.0074	0.92
40.0	26.8	0.08	0.009	0.30	0.68	17.9	0.0089	0.98
40.0	48.9	0.15	0.016	0.36	0.62	29.8	0.0148	1.05
40.0	69.4	0.20	0.022	0.37	0.61	41.6	0.0207	1.08
40.0	82.2	0.24	0.028	0.38	0.61	48.6	0.0242	1.18
40.0	6.2	0.02	0.001	0.40	0.52	3.2	0.0016	0.59
40.0	16.7	0.09	0.003	0.55	0.42	7.0	0.0035	0.86
39.8	29.6	0.16	0.005	0.60	0.38	11.1	0.0055	0.99
40.0	52.7	0.29	0.012	0.64	0.34	18.0	0.0089	1.31

[a]Horstmann et al. 2004 [82]

demonstrating that CXLs are optimal media for this reaction. In another example, cyclohexene oxidation by O_2 was investigated with a non-fluorinated iron porphyrin catalyst, (5,10,15,20-tetraphenyl-21H,23H-porphyrinato)iron(III) chloride, Fe (TPP)Cl, and a fluorinated catalyst (5,10,15,20-tetrakis(pentafluorophenyl)-21H,23H-porphyrinato)iron(III) chloride, Fe(PFTPP)Cl [80]. While Fe(TPP)Cl is insoluble and displays little activity in scCO$_2$, it displays high activity and selectivity in CO_2-expanded acetonitrile. The enhanced TOFs in the CXL media were attributed to their tunable polarity compared to scCO$_2$ and the approximately two orders of magnitude greater O_2 solubility in CO_2-expanded solvents compared to neat solvents at ambient conditions (1 atm and 25 °C). These results clearly show that CO_2-expanded solvents advantageously complement scCO$_2$ as reaction media by broadening the range of conventional catalyst/solvent combinations with which homogeneous oxidations by O_2 can be performed.

An important safety aspect when using CXLs is that the dominance of dense CO_2 in the vapor phase reduces the flammability envelope. Brennecke and coworkers showed that either pure O_2 or O_2 + CO_2 mixtures at the same total pressure showed more or less similar O_2 solubilities in pressurized acetonitrile or methanol [2]. Table 2.2 shows the vapor–liquid equilibrium data for the CO_2/O_2/acetonitrile system at approximately 40 °C and pressures between 6 and 83 bar. The CO_2 and O_2 mole fractions are shown in the table, with the balance being acetonitrile. To better understand the effect of CO_2 presence on O_2 solubility in acetonitrile, the authors use the Enhancement Factor (EF), defined as the ratio of the solubility of O_2 in the liquid phase of the ternary mixture to the solubility of pure O_2 in the solvent at the same O_2 fugacity and temperature.

$$EF = \frac{x_{gas}^{mixture}}{x_{gas}^{solvent}}$$

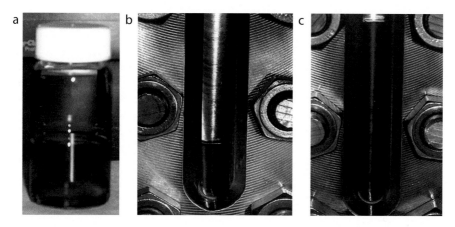

Fig. 2.4 Example of CO_2 expansion of a homogenous oxidation catalyst mixture in acetic acid. (**a**) room temperature, (**b**) 120°C with 20 bar N_2 and 18 bar CO_2, (**c**) 120°C with 20 bar N_2 and 159 bar CO_2; [Co] = 60 mM, [Mn] = 1.8 mM, [Br] = 60 mM prior to the expansion

Clearly, EF is >1 implies that the presence of the CO_2 increases the solubility of the O_2. As can be inferred from Table 2.2, the EF values increase with an increase in pressure and reached values greater than 1 at pressures above 50 bar. EF values >1 can be attributed to the high CO_2 solubility in the liquid phase, which increases free volume in the solution and enhances the solubility of the O_2. If the binary and ternary systems are compared at the same total pressure (rather than the same O_2 fugacity), then the oxygen solubility in the ternary mixture liquid phase is lower than the solubility in the binary system. This finding is significant because it confirms that replacing CO_2 with O_2 in the gas phase does not drastically change the O_2 solubility in the CXL phase but serves to significantly mitigate vapor-phase flammability and thereby to enhance process safety.

The Mid-Century (MC) process for the oxidation of methylbenzenes to carboxylic acids represents an important industrial process for the synthesis of polymer intermediates for producing fibers, resins, and films. For example, terephthalic acid (TPA) is manufactured by homogeneous liquid-phase oxidation of *p*-xylene around 200°C and 20 bar with Co/Mn/Br catalysts in acetic acid medium. The air (source of oxygen) is vigorously sparged into the liquid phase of a stirred tank reactor. The purity of the solid TPA product obtained from this reactor is typically 99.5% pure, the major impurity being 4-carboxybenzaldehyde. For obtaining polymer-grade TPA, further purification is necessary to eliminate the intermediate oxidation products suggesting that the vigorous mixing does not completely overcome the O_2 availability limitations. Recently, the Co/Mn/Br catalyzed oxidation of *p*-xylene to TPA in CO_2-expanded solvents was demonstrated at temperatures lower than those of the traditional MC process [82]. As inferred from Fig. 2.4, the acetic acid solution containing the Co/Mn/Br-based catalyst undergoes significant volumetric expansion by CO_2 addition.

As compared with the traditional air (N_2/O_2) oxidation system, the reaction with CO_2/O_2 at 160°C at a pressure of 100 bar increases both the yield of TPA and the purity of solid TPA via a more efficient conversion of the intermediates, 4-carboxybenzaldehyde and p-toluic acid. Further, the amount of yellow colored by-products in the solid TPA product is also lessened. Additionally, the burning of the solvent, acetic acid, monitored in terms of the yield of the gaseous products, CO and CO_2, is reduced by approximately 20% based on labeled CO_2 experiments. These findings show that the use of CXLs promotes sustainability by maximizing the utilization of feedstock carbon for desired products while simultaneously reducing carbon emissions.

H_2O_2-Based Oxidations

Many of the current commercial propylene oxide (PO) processes are energy intensive and yield coproducts [83]. They either consume vast amounts of chlorine and lime producing large quantities of wastewater containing HCl and salt, or use expensive reactants that lead to equimolar amounts of coproducts. A novel propylene epoxidation process has been disclosed wherein titanium-substituted silicalite (TS-1) catalysts [84–87] catalyze propylene epoxidation with reasonable efficiency using an O_2/H_2 mixture to generate H_2O_2 in situ. TS-1 catalysts have high catalytic activity and selectivity. However, the catalyst deactivates rapidly and requires high temperatures for regeneration.

Eckert/Liotta and coworkers showed that the reaction between H_2O_2 and dense CO_2 (benign reactants) yields a peroxycarbonic acid species, an oxidant that facilitates olefin epoxidation [88]. Using such a system with NaOH as a base, Beckman and coworkers [89] demonstrated propylene conversion to PO at high selectivity albeit at low conversion (3%). The low conversion is typical of interphase mass transfer limitations between the immiscible aqueous and CO_2 phases. By using a third solvent such as acetonitrile that shows mutual solubility with water and dense CO_2, CO_2-expanded $CH_3CN/H_2O_2/H_2O$ homogeneous mixtures were created. Subramaniam and coworkers [45] showed that olefin epoxidation reactions may be intensified in such a homogeneous phase containing the olefin, CO_2, and H_2O_2 (in aqueous solution). One to two orders of magnitude enhancement in epoxidation rates (compared to the biphasic system without acetonitrile) was achieved with >85% epoxidation selectivity. This is an example of a system where the CO_2 is used as a solvent (for the olefinic substrate) and as a reactant to generate the peroxy carbonic acid in situ. Recently, a similar concept was proposed for styrene epoxidation in CO_2-based emulsions [90].

Propylene Epoxidation in Propylene-Expanded Liquid Phase

Methyltrioxorhenium (CH_3ReO_3, abbreviated as MTO) is known to be an exceptional homogeneous catalyst for alkene epoxidation at relatively mild

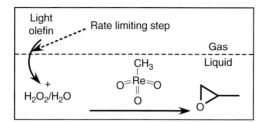

Scheme 2.1 Homogeneous catalytic epoxidation of propylene. The methyltrioxorhenium (MTO) catalyst and oxidant (H_2O_2) are dissolved in the liquid phase and the olefin is supplied from the gas phase

temperatures using H_2O_2 as the oxidant [91–94]. This catalytic system was recently shown to work elegantly for the industrially significant epoxidations of propylene [95] and ethlyene [96]. As shown in Scheme 2.1, the light olefin (either propylene or ethlyene) from the gas phase has to first dissolve into the aqueous liquid phase before undergoing reaction. However, the aqueous solubility of the olefins is typically low and limits the epoxidation rate. For example, propylene solubility in water is $1.36 \, (10^{-4})$ M at 1 bar, 21°C [97]. The key to enhancing the reaction rate was to exploit the compressibility of the light olefins at reaction temperatures to enhance their solubilities in the liquid phase. By dissolving a suitable amount of methanol to the liquid phase, the solubility of the light olefins can be substantially enhanced.

For propylene epoxidation, it was found that if a N_2/C_3H_6 mixture at 14 bar was used in the gas phase, remarkably high activity (92% propylene oxide yield in 1 h) was achieved [95]. When pressurizing with N_2 in a closed system at ambient temperatures, C_3H_6 ($P_c = 46.1$ bar; $T_c = 92.5$°C) condenses at 14 bar [98]. In the presence of a solvent such as methanol, pressurization beyond 10 bar increases the C_3H_6 mole fraction in the liquid phase to >30 mol% [46]. This increased concentration of propylene results in increased epoxidation rates. This is an example of how to utilize a light olefin as the expansion medium and exploit a *substrate-expanded liquid phase* for process intensification.

Table 2.3 compares the various GXL-based processes with the commercial chlorohydrin and hydroperoxide processes. The pressure-intensified propylene epoxidation process (last entry in Table 2.3) satisfies the sustainability principles of waste minimization, use of benign reagents, and process intensification at mild conditions. Little or no waste is produced compared to the commercial processes.

Ethylene Epoxidation in Ethylene-Expanded Liquid Phase

Conventional ethylene oxide manufacture emits CO_2 as by-product (roughly 18 MM tons/year) from the combustion of both the ethylene (feed) and ethylene oxide (EO) [100]. An alternate technology that is exclusively selective toward EO

Table 2.3 Comparison with other processes (substrate is propylene in all cases)

Process/reference	Reagents	Conditions	PO yield (Y) or selectivity (S)
Hancu et al. [89]	$CO_2 + H_2O_2 + H_2O$ (Biphasic system), NaOH	–	$Y < 3.0\%$
Danciu et al. [85]	$CO_2 + H_2 + O_2$, Pd/TS-1 catalyst	45°C; 131 bar	$Y \sim 7.3\%$; $S = 77\%$; 4.5 h
Biphasic CXL process, Lee et al. [99]	$CO_2 + CH_3CN + H_2O_2 + H_2O$ (monophasic), pyridine	40°C; 48 bar	$Y = 7.1\%$; 6 h
Chlorohydrin process, Trent [83]	Cl_2, caustic, lime	45–90°C; 1.1–1.9 bar	$S = 88–95\%$
Hydroperoxide[a] process, Trent [83]	W, V, Mo	100–130°C; 15–35.2 bar	$Y = 95\%$; $S = 98\%$; 2 h
Pressure-intensified CXL process [95]	$CH_3OH–H_2O_2–H_2O$, MTO, pyNO	30°C; 17 bar N_2	$Y = +98\%$; $S = 99\%$; 2 h

[a]Isobutane to TBHP: $95 \sim 150$°C; 20.7–55.2 bar; $Y = 20–30\%$; $S = 60–80\%$

Table 2.4 Comparison of commercial EO process with new GXL-based homogeneous catalytic process

Process	Catalyst, oxidant	Pressure, temperature	EO selectivity	CO_2 selectivity	EO productivity (g/gcat/h)
Shell	Ag/Al_2O_3, air or O_2	10–20 bar, 200–300°C	85–90%	10–15%	3–6
GXL based	MTO, H_2O_2	10–50 bar, 20–40°C	99+%	No CO_2 detected	~4.5

(eliminating the formation of CO_2) would dramatically reduce the carbon footprint of this large-scale industrial process. It was recently shown that the MTO-based liquid-phase epoxidation process works remarkably well for ethlyene as well. At ambient temperatures, the gaseous ethylene ($P_c = 50.6$ bar; $T_c = 9.5$°C) is just above the critical temperature. Hence, by compressing the ethylene gas beyond the critical pressure (>50 bar), it is possible to significantly increase its solubility (by one to two orders of magnitude) in a liquid reaction phase containing methanol or another suitable alcohol [101]. By employing MTO, H_2O_2, and a methanol/water mixture as solvent, a homogeneous catalytic system was demonstrated that eliminates CO_2 formation while producing ethylene oxide at >99% EO selectivity at near-ambient temperatures and EO productivities [~4 g EO/(g Re)/(h)] that are comparable to the conventional EO process (Table 2.4). *No* CO_2 was detectable in either the liquid or the vapor phases. Furthermore, since H_2O_2 does not decompose at typical reaction temperatures (<40°C), the vapor phase is void of O_2 and the formation of explosive vapors is impeded. In addition to mitigating the carbon footprint of a large-scale industrial process, the demonstrated technology concept is another example of how the synergy afforded by the facile compressibility of a substrate such as ethylene and the accompanying enhanced solubility in low molecular weight alcohols can be exploited to enhance selectivity and productivity.

Acid Catalysis

The replacement of mineral acids with benign and less hazardous alternatives has long been a grand challenge in chemicals synthesis. In situ generation of acids by the reaction of CO_2 with alcohols or water is desirable because depressurization leads to self-neutralization [102–104]. The formation of the dimethyl acetal of cyclohexanone is up to 130 times faster in CO_2-expanded methanol than in normal methanol without any added acid. Such in situ acids also catalyze the hydrolysis of β-pinene to terpineol and other alcohols with good selectivity for alcohols rather than hydrocarbons [103].

Addition of CO_2 to pressurized hot water accelerates reactions that can proceed by acid catalysis. CO_2 dissolved in water at 250°C promotes the decarboxylation of benzoic acid [105], the dehydration of cyclohexanol to cyclohexene, and the alkylation of p-cresol to 2-*tert*-butyl-4-methylphenol [106]. The hydration of cyclohexene to cyclohexanol at 300°C showed a fivefold rate increase as the CO_2 pressure was increased from 0 to 55 bar.

Miscellaneous

Ozonolysis

Ozone has high oxidation potential (E^o = 2.075 V in acid and 1.246 V in base) and has been extensively investigated as a potent oxidant. Though considered toxic, ozone does not persist in the environment eventually decomposing to molecular oxygen. Ozone is effective for the cleavage of carbon–carbon double bonds. This oxidation reaction is believed to proceed via metastable intermediates that upon further catalytic oxidation or reduction yield products that are suitable as building blocks for chemical synthesis. For example, the ozonolysis of unsaturated fatty acids can yield a range of both monoacids and diacids [107–110]. Ozone attacks most common organic solvents used as reaction media, creating undesirable waste products and consuming the ozone away from the desired reaction. It has been recently shown that ozone can be used effectively in liquid CO_2, in which it is not only stable but also remarkably soluble [111]. At typical ozonolysis temperatures (0–20°C), ozone is sufficiently close to its critical temperature (−12.1°C) such that the ozone density can be increased to liquid-like values by compression beyond its critical pressure (55.6 bar). Conveniently, in the 0–20°C range and beyond 50 bar, the CO_2 (P_c = 73.8 bar; T_c = 31.1°C) exists as liquid and the O_3 content in this liquid phase is easily tuned with pressure. An order of magnitude increased dissolution of O_3 in liquid CO_2 creates an O_3-*expanded liquid phase*. The O_3 half-life in liquid CO_2 was found to be approximately 6 h at −1.2°C [111]. Further, substrates such as methyl oleate and *trans*-stilbene when dispersed in liquid CO_2 (containing dissolved O_3) undergo complete conversion in minutes to the corresponding aldehyde and acid products (nonanal, nonanoic acid in the case of methyl oleate; and benzaldehyde, benzoic acid in the case of *trans*-stilbene). The foregoing results clearly show that ozonolysis in liquid CO_2 is a facile, clean, and inherently safe oxidation route.

Carbonylations

Exploiting the enhanced and tunable CO solubility in CXLs, it was recently shown that selective mono or double carbonylations could be achieved by using CO_2-expanded liquids during $[2 + 2 + 1]$ carbonylative reactions of alkenes or acetylenes with allyl bromides catalyzed by Ni(I) [112].

Polymerizations

Catalytic chain transfer polymerizations are free radical polymerizations that use a homogeneous catalyst to terminate one chain and initiate a new one. During the polymerization of methyl methacrylate, the chain transfer step is believed to be a diffusion-controlled reaction in which the Co(II) catalyst abstracts a hydrogen atom from the polymer radical (R•) and transfers it to a monomer to start the growth of a new chain. Zwolak et al. [113] reported that the rate of chain transfer was fourfold greater in CO_2-expanded methyl methacrylate (60 bar, 50°C) than in neat monomer. The improvement was attributed to the lower viscosity of the CO_2-expanded solution.

Biomass Conversions

Plant-based biomass has the potential to completely displace fossil fuel–based feedstocks for chemicals production in a sustainable manner. However, new catalytic technologies are needed for the fledgling biorefinery to convert plant-based biomass feedstocks to chemicals and chemical building blocks. Recently, the acid-catalyzed transesterification of soybean flakes in CO_2-expanded methanol containing sulfuric acid was demonstrated for producing fatty acid methyl esters (FAME) [114]. The authors report that the introduction of CO_2 into the system increases the rate of reaction by as much as 2.5 fold in comparison to control reactions without CO_2.

1,2-glycerol carbonate was formed from glycerol and carbon dioxide in methanol using (Bu$_2$SnO)-Bu-*n* (dibutyltin(IV)oxide, 1) as a catalyst [115]. The yield of 1,2-glycerol carbonate was as high as 35%. The reaction proceeds upon activation of the catalyst by methanol forming dibutyltindimethoxide followed by dibutyltinglycerate, which undergoes CO_2 insertion to ultimately yield glycerol carbonate.

Multiphase Catalysis

The attractive properties of GXLs are also applicable in heterogeneous catalysis. For example, adding CO_2 to an organic liquid phase in a fluid–solid catalytic system should enhance gas solubilities and improve the mass transfer properties of the expanded liquid phase. Reviews of supercritical phase heterogeneous catalysis may be found elsewhere [116–118].

Hydrogenations

Employing dense CO_2 as the solvent medium in a stirred reactor, it was reported that the Pd/Al_2O_3 catalyzed hydrogenation of an unsaturated ketone proceeded faster in CO_2-expanded ketone than in the unswollen ketone [119]. A similar observation was reported during the Pd/C catalyzed hydrogenation of pinene wherein the reaction rate was higher at lower pressures in a condensed CXL phase compared to single-phase operation at supercritical conditions [120]. Roberts and coworkers reported that the rate constant for the hydrogenation of the aromatic rings in polystyrene (PS) was found to be higher in CO_2-expanded decahydro-naphthalene (DHN) than in neat DHN [121]. For the $Pt/\gamma\text{-}Al_2O_3$ catalyzed hydrogenation of tetralin to decalin, Chan and Tan [122] reported enhanced rates in the presence of a CO_2-expanded toluene phase in a trickle bed reactor. For the Pd/C-catalyzed hydrogenation of CO_2-expanded alpha-methylstyrene, it was shown that the rate-enhancing effect of CO_2 is influenced by two competing factors: solvent strength and reactant concentration [123]. The presence of CO_2 modifies the solvent strength of the liquid phase, resulting in more favorable adsorption equilibrium for the surface reaction. However, the diluting effect of CO_2 leads to reduced reaction rates. Thus, there exists an optimum CO_2 level in the liquid phase that is tunable by reactor pressure.

During the $NiCl_2$-catalyzed reduction of benzonitrile to benzylamine by $NaBH_4$, Xie et al. [124] showed that CO_2 expansion of the reaction mixture converts the primary amine to a carbamate salt, thereby preventing its further reaction to secondary amines. The carbamic species release CO_2 upon gentle heating. Thus, the benzylamine yield was 98% in CO_2-expanded ethanol but <0.01% in normal ethanol.

Heterogeneous catalysis without solvent relies on both the substrate and product being liquids or gases. If the product is a solid at the reaction temperature, then the reaction will not proceed to completion because the reaction mixture will solidify before full conversion is obtained. Normally, this problem is solved by adding a solvent or using an elevated temperature, but a third option is to lower the melting point of the product by expansion with CO_2 [71]. For example, the Pt-catalyzed hydrogenation of oleic acid at 35°C stalls at 90% conversion even with extended reaction times (25 h). However, in the presence of 55 bar CO_2, the reaction proceeds to 97% conversion after only 1 h.

Selective Oxidations

During the Pd/Al$_2$O$_3$ catalyzed partial oxidation of octanol, Baiker and coworkers report significantly enhanced oxidation rates at intermediate pressures where a condensed CXL phase probably exists compared to higher pressure where a single supercritical phase exists [125]. The reaction in the condensed phase benefits from increased concentrations of the substrate relative to the single supercritical phase while also enjoying adequate O$_2$ availability in the liquid phase.

During the O$_2$-based oxidation of cyclohexene on a MCM-41 encapsulated iron porphyrin chloride complex, conversion and selectivity in CO$_2$-expanded acetonitrile ($\sim$30 mol% CO$_2$) almost doubled compared to the neat organic solvent [126]. For the oxidation of 2,6-di-*tert*-butylphenol (DTBP) to 2,6-di-*tert*-butyl-1,4-benzoquinone   (DTBQ)   and   3,5.3',5'-tetra-*tert*-butyl-4,4'-diphenoquinone (TTBDQ), a series of porous materials with immobilized Co(II) complexes were screened as catalysts in neat acetonitrile, supercritical carbon dioxide (*sc*CO$_2$), and CO$_2$-expanded acetonitrile [127]. In this case, the highest conversions were found in *sc*CO$_2$ suggesting that scCO$_2$, rather than CXL or liquid reaction media, provides the best mass transfer of O$_2$ and of substrates through the porous catalysts.

Hydroformylation and Carbonylation

Abraham and coworkers [77] investigated 1-hexene hydroformylation over a rhodium–phosphine catalyst immobilized on a silica support and found that the rates in CO$_2$-expanded toluene and in scCO$_2$ were comparable but faster than in normal toluene. However, the activity declined with time due to possible catalyst leaching.

Leitner and coworkers have demonstrated that compressed CO$_2$ can be used to effectively overcome mass transfer limitations encountered during solid-phase organic synthesis with pressurized gaseous reagents [128]. Depending on the relative importance of mass-transfer limitations and catalyst/substrate concentration, CXLs may provide the optimum conditions for both hydroformylation and carbonylation reactions. For example, the catalytic carbonylation of norbornene (Pauson–Khand reaction) supported on a polymer support proceeds nearly quantitatively in CXL media. Here, the CXL medium provides the optimum combination of catalyst concentration and CO availability to maximize the reaction rate.

Solid Acid Catalysis

The acylation of anisole with acetic anhydride was investigated in a continuous slurry reactor over mesoporous-supported solid acid catalysts such as Nafion® (SAC-13) and heteropolyacids [129]. The CXL media gave lower conversion and, surprisingly, faster deactivation compared to a liquid-phase reaction despite the use of polar cosolvents. The deactivation is possibly due to retention of heavy molecules (possibly di- and tri-acylated products) formed by the interaction of acetic anhydride with para-methoxyacetophenone (p-MOAP) in the catalyst micropores. The addition of CF_3CO_2H has been shown to catalyze a Friedel–Crafts alkylation of anisole in CO_2-expanded anisole (95°C, 42 bar), but the reaction was no faster than that in normal anisole [130].

An Example of CXL-Based Process Development Including Sustainability (Economic and Environmental Impact) Analyses

For developing sustainable catalytic processes, a multiscale approach involving concurrent catalyst design, solvent engineering, and reactor engineering is essential [40]. An example of such an approach, involving chemists and engineers, is presented in this section for an industrially significant reaction.

For the hydroformylation of higher olefins, cobalt-based catalysts are employed. The cobalt catalysts require rather harsh operating conditions (140–200°C, 50–300 bar) and the catalyst recovery steps involve much solvents, acids, and bases [131]. The challenges for a sustainable technology alternative are to develop a process that operates at milder temperatures (<100°C) and pressures (<100 bar), and requires a simple yet environmentally friendly catalyst recovery method. The use of a Rh catalyst for 1-octene hydroformylation in CXL media provides exceptional TOF ($\sim$316 h^{-1}) and regioselectivity ($n/i \sim 9$) at very mild pressure ($\sim$40 bar) and temperature (30–60°C) compared to conventional Co-based processes [75]. This markedly enhanced regioselectivity in CXLs during homogeneous 1-octene hydroformylation is partly attributed to the beneficial tunability of the H_2/CO ratio in the CXL phase.

It is well known that the concentrations of the syngas components (CO and H_2) in the liquid phase are major determinants of the reaction pathways and therefore the product selectivity. In general, higher H_2 concentrations are needed for catalyst activation and lower CO concentrations are required to achieve higher rates and avoid inhibition effect due to formation of inactive carbonyl species [132]. Because CO is generally more soluble than hydrogen in most conventional solvents [133], the resulting H_2/CO ratio in the liquid phase is less than that in the feed syngas. However, when CO_2 is added to either 1-octene or nonanal (to create a CXL), it was observed that the H_2 is more soluble than CO in the CXLs [5]. This means that the H_2/CO ratio in the liquid phase should be greater in CXLs (based on the organic solvent and extent of CO_2 addition) compared to the ratio in the feed.

Gas solubility measurements showed the presence of CO_2 at hydroformylation conditions ($T = 40–80$°C and pressures up to 90 bar) enhances the solubilities of both CO and H_2 in the liquid phase. The enhancement factor, defined as the ratio of the gas (CO or H_2) mole fraction in the neat solvent relative to that in the CXL at identical

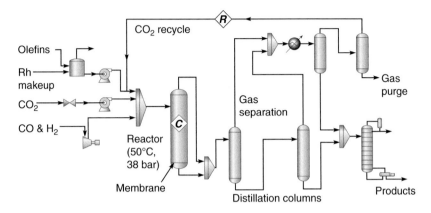

Fig. 2.5 Process flow diagram for CXL-based hydroformylation concept

temperature and gas (CO or H_2) fugacities in the vapor phase, is greater for hydrogen (around 1.8) compared to carbon monoxide (around 1.5). This unique tunability of the H_2/CO ratio in CXL media is believed to enhance both the TOF and n/i ratio.

A detailed engineering model that takes into account kinetics, phase equilibrium, and mass transfer effects provided a better understanding of the mass transfer and kinetic effects in the CXL-based media [134]. A plant-scale simulation of the CXL process concept (Fig. 2.5) was constructed to facilitate economic and environmental impact analyses. Economic analysis revealed that >99.8% rhodium has to be recovered per pass for the CEBC hydroformylation process to be competitive with a simulated Co-based commercial process [135]. Environmental impact analysis revealed that the CEBC process produces half as much waste with lower overall toxicity compared to the simulated conventional process [135].

To develop Rh catalysts that meet the quantitative criterion for economic viability, a soluble polymer-attached, recyclable rhodium(I) catalyst with chelate-capable phosphite functionality, that was used to produce a polymer ligand, was synthesized (Scheme 2.2). By controlling the molecular weight, the polymer is designed such that it is completely soluble in the hydroformylation reaction medium yet bulky enough to diffuse through a nanofiltration membrane [136, 137]. The polymer support was designed to bind Rh in a bidentate fashion to provide better site isolation for the rhodium catalysts as well as to inhibit decomplexation and subsequent leaching of rhodium from the polymer.

Using the soluble polymer-attached ligand to bind the Rh precursor, a continuous hydroformylation process concept that uses Rh-based homogeneous catalysts and operates at mild pressures (tens of bar) and temperatures less than 100°C has been demonstrated. The reactor schematic is shown in Fig. 2.6. Syngas and compressed CO_2 (to generate CXLs) are added to the 1-octene + nonanal reaction mixture in which the polymer-attached rhodium complex is dissolved. The product stream is continuously withdrawn while maintaining the reactor pressure constant.

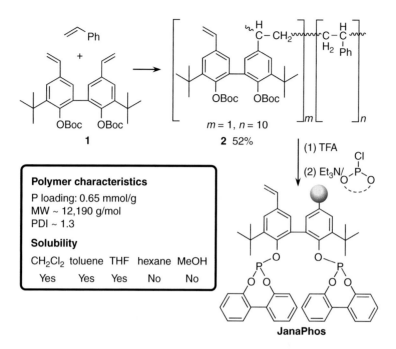

Scheme 2.2 Synthesis of polymer-attached ligand [136]

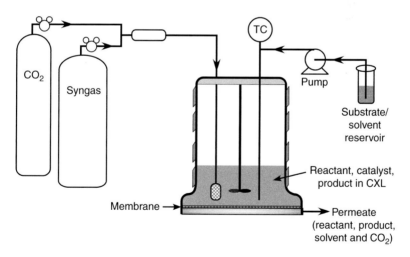

Fig. 2.6 Schematic of continuous homogeneous hydroformylation in CXL with catalyst retention by membrane nanofiltration

Continuous operation is characterized by steady membrane flux, constant conversion, and constant product selectivity (Fig. 2.7a). At steady state, the permeate stream contains the unreacted 1-octene and CO_2 (which is separated by depressurization and may be recycled) products. As seen from Fig. 2.7b, the Rh and

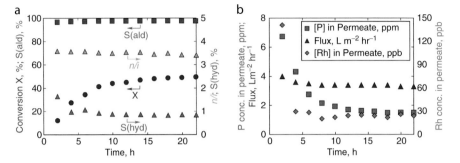

Fig. 2.7 Demonstration of a continuous CXL-based homogeneous hydroformylation reactor with effective catalyst retention by membrane nanofiltration. $T = 50°C$, $P_{syngas} = 30$ bar, $CO/H_2 = 1:1$, P/Rh = .6, Oct/Rh = 2200, toluene/1-octene = 70:30 by volume, stirrer speed: 1000 rpm [138]

P concentrations in the permeate phase, quantified using ICP analysis, were on the order of a few tens of ppb [138]. The cost of the makeup Rh is 0.4 cent/lb product, which exceeds the economic viability criterion.

The foregoing example represents a systems-based, multiscale research approach developing novel, environmentally beneficial, and economically viable process concepts. A patent pertaining to the CXL-based hydroformylation concept [139] has been licensed to a company for a defined field of use. The demonstrated technology concept, when fully optimized, should find applications in a variety of other applications in homogeneous catalysis, including hydrogenation and carbonylation of conventional and biomass-based substrates.

Summary and Future Directions

The novel science and technology advances discussed in this entry demonstrate the various unique ways in which GXL media may be exploited to develop greener process concepts for oxidations, hydroformylations, hydrogenations, ozonolysis, and other chemistries. The demonstrated advantages include *process intensification* at mild conditions by increasing dissolution of the limiting reagent in the GXL reaction phase; the *efficient utilization* of feedstock and reactive gases such as O_3 due to the inertness of CO_2, an often used expansion medium; enhancing *inherent safety* of the process by suppression of flammable vapors; and *waste minimization* by suppression of side reactions that generate undesired products such as CO_2 and reduced usage of volatile organic solvents.

The alternative concepts presented herein address chemistries underlying large-scale catalytic technologies, including the homogeneous hydroformylation of higher olefins and the epoxidations of light olefins such as ethylene and propylene. The global capacity of these processes is growing at 4–6% annually. Hence, the deployment of viable alternative technologies for future expansion of these processes (and also as replacement of existing units when needed) would have a significant impact in reducing environmental footprints. For these promising concepts to be developed further and considered for commercialization, continued fundamental investigations that integrate catalysis, phase behavior involving GXL media (for reactions and separations), kinetic and reactor modeling are essential. These investigations must be complemented by ongoing quantitative economic and environmental impact analyses to provide research and process engineering guidance for developing practically viable process concepts.

The use of renewable feedstocks has the potential to completely displace petroleum crude and coal for producing industrial chemicals. Further, reducing the carbon footprint of consumer products will be an important expectation of a majority of next-generation consumers. The deployment of green technologies is also essential for renewable feedstocks in order to fulfill their promise for producing sustainable fuels and chemicals. Renewable feedstocks cannot yet be processed with existing technologies because their chemical structures and compositions (their relative C, H, and O contents) are varied and diverse compared to conventional crude oil. The fledgling biorefinery industry is uniquely positioned to accept new technology concepts. Chemical catalysis is a promising avenue to develop atom-economical and energy-efficient biomass conversion technologies. To address this challenge, multidisciplinary expertise is needed for the design and synthesis of novel metal-based catalysts, high-pressure chemistry, solvent engineering, multiphase reaction engineering, and multiscale modeling. Specifically, the various system elements (i.e., catalysts, solvents, reactors, and separators) should be designed such that when integrated, the resulting system displays enhanced rates and selectivity with reduced environmental footprint.

Acknowledgments Much of the author's work described in this entry was made possible by NSF ERC Grant EEC-0310689, the Kansas Technology Enterprise Corporation, and the support of the University of Kansas through the Dan F. Servey Distinguished Professorship.

Bibliography

Primary Literature

1. Jessop PG, Subramaniam B (2007) Gas-expanded liquids. Chem Rev 107:2666–2694
2. Lopez-Castillo ZK, Aki SNVK, Stadtherr MA, Brennecke JF (2006) Enhanced solubility of oxygen and carbon monoxide in CO_2-expanded liquids. Ind Eng Chem Res 45:5351–5360
3. Lopez-Castillo ZK, Aki SNVK, Stadtherr MA, Brennecke JF (2008) Enhanced solubility of hydrogen in CO_2-expanded liquids. Ind Eng Chem Res 47:570–576
4. Zevnik L, Levec J (2007) Hydrogen solubility in CO_2-expanded 2-propanol and in propane-expanded 2-propanol determined by an acoustic sensor. J Supercrit Fluids 41:335–342
5. Xie Z, Snavely WK, Scurto AM, Subramaniam B (2009) Solubilities of CO and H_2 in neat and CO_2-expanded hydroformylation reaction mixtures containing 1-octene and nonanal up to 353 K and 9 Mpa. J Chem Eng Data 54:1633–1642
6. Sheldon RA (1994) Consider the environmental quotient. Chem Tech 24:38–47

7. Sheldon RA, Arends IWCE, Hanefeld U (2007) Green chemistry and catalysis. Wiley, Weinheim
8. Tundo AP, Black DS, Breen J, Collins T, Memoli S, Miyamoto J, Poliakoff M, Tumas W (2000) Synthetic pathways and processes in green chemistry: introductory overview. Pure Appl Chem 72:1207–1228
9. DeSimone JM (2002) Practical approaches to green solvents. Science 297:799–803
10. Adams DJ, Dyson PJ, Tavener SJ (2004) Chemistry in alternative reaction media. Wiley, Chichester
11. Eckert CA, Liotta CL, Bush B, Brown JS, Hallett JP (2004) Sustainable reactions in tunable solvents. J Phys Chem B 108:18108–18118
12. Seki T, Baiker A (2009) Catalytic oxidations in dense carbon dioxide. Chem Rev 109:2409–2454
13. Morgenstern DA, LeLacheur RM, Morita DK, Borkowsky SL, Feng S, Brown GH, Luan L, Gross MF, Burk MJ, Tumas W (1996) Supercritical carbon dioxide as a substitute solvent for chemical synthesis and catalysis. In: Anastas PT, Williamson TC (eds) Green chemistry: designing chemistry for the environment, ACS symposium series vol 626. American Chemical Society, Washington, DC, pp 132–151
14. Jessop PG, Leitner W (1999) Chemical synthesis using supercritical fluids. Wiley, Weinheim
15. Amandi R, Hyde J, Poliakoff M (2003) Heterogeneous reactions in supercritical carbon dioxide. In: Aresta M (ed) Carbon dioxide recovery and utilization. Kluwer, Dordrecht, pp 169–180
16. DeSimone JM, Tumas W (2003) Green chemistry using liquid and supercritical carbon dioxide. Oxford University Press, New York
17. Gordon CM, Leitner W (2004) Supercritical fluids as replacements for conventional organic solvents. Chim Oggi 22:39–41
18. Beckman EJ (2002) Using CO_2 to produce chemical products sustainably. Environ Sci Technol 36:347A–353A
19. Licence P, Poliakoff M (2005) Economics and scale-up. In: Cornils B, Hermann WA, Horváth IT, Leitner W, Mecking S, Olivier-Bourbigou H, Vogt D (eds) Multiphase homogeneous catalysis, vol 2. Wiley, Weinheim, pp 734–746
20. Arai M, Fujita SI, Shirai M (2009) Multiphase catalytic reaction in/under dense phase CO_2. J Supercrit Fluids 47:351–356
21. Li CJ, Chan TH (1997) Organic reactions in aqueous media. Wiley, New York
22. Cornils B, Herrmann WA (1998) Aqueous-phase organometallic catalysis. Wiley, Weinheim
23. Savage PE (2009) A perspective on catalysis in sub- and supercritical water. J Supercrit Fluids 47:407–414
24. Musie G, Wei M, Subramaniam B, Busch DH (2001) Catalytic oxidations in carbon dioxide-based reaction media, including novel CO_2-expanded phases. Coord Chem Rev 219–221:789–820
25. Hutchenson KW, Scurto AM, Subramaniam B (2009) Gas-expanded liquids and near-critical media: green chemistry and engineering, vol 1006, ACS symposium series. American Chemical Society, Washington, DC
26. Akien GR, Poliakoff M (2009) A critical look at reactions in class I and II gas-expanded liquids using CO_2 and other gases. Green Chem 11:1083–1100
27. Scurto AM, Hutchenson KW, Subramaniam B (2009) Gas-expanded liquids (GXLs): fundamentals and applications. In: Hutchenson KW, Scurto AM, Subramaniam B (eds) Gas-expanded liquids and near-critical media: green chemistry and engineering, vol 1006, ACS symposium series. American Chemical Society, Washington, DC, pp 3–37
28. Wasserscheid P, Welton T (2002) Ionic liquids in synthesis. Wiley, Weinheim
29. Rogers RD, Seddon KR, Volkov S (2003) Green industrial applications of ionic liquids. Kluwer, Dordrecht
30. Pârvulescu VI, Hardacre C (2007) Catalysis in ionic liquids. Chem Rev 107:2615–2665

31. Jessop PG, Heldebrant DJ, Xiaowang L, Eckert CA, Liotta CL (2005) Reversible nonpolar-to-polar solvent. Nature 436:1102
32. Liu Y, Jessop PG, Cunningham M, Eckert CA (2006) Liotta CL, switchable sufactants. Science 313:958–960
33. Phan CD, Heldebrant DJ, Huttenhower H, John E, Li X, Pollet P, Wang R, Eckert CA, Liotta CL, Jessop PG (2008) Switchable solvents consisting of amidine/alcohol or guanidine/alcohol mixtures. Ind Eng Chem Res 47:539–545
34. Phan L, Brown H, White J, Hodgson A, Jessop PG (2009) Soybean oil extraction and separation using switchable or expanded solvents. Green Chem 11:53–59
35. Phan L, Jessop PG (2009) Switching the hydrophilicity of a solute. Green Chem 11:307–308
36. Ahosseini A, Ren W, Scurto AM (2009) Understanding biphasic ionic liquid/CO_2 systems for homogeneous catalysis: hydroformylation. Ind Eng Chem Res 48:4254–4265
37. Anastas P, Warner JC (1998) Green chemistry: theory and practice. Oxford University Press, New York
38. Anastas PT, Zimmerman JB (2003) Design through the 12 principles of green engineering. J Environ Sci Technol 37:95A–101A
39. Allen DT, Shonnard DR (2001) Green engineering: environmentally conscious design of chemical processes. Prentice Hall, New York
40. Dudukovic MP (2009) Frontiers in reactor engineering. Science 325:698–701
41. Kordikowski A, Schenk AP, Van Nielen RM, Peters CJ (1995) Volume expansions and vapor-liquid equilibria of binary mixtures of a variety of polar solvents and certain near-critical solvents. J Supercrit Fluids 8:205–216
42. Ren W, Scurto AM (2007) High-Pressure phase equilibria with compressed gases. Rev Sci Instrum 78:125104–125107
43. Heldebrant DJ, Witt H, Walsh S, Ellis T, Rauscher J, Jessop PG (2006) Liquid polymers as solvents for catalytic reductions. Green Chem 8:807–815
44. Ren W, Sensenich B, Scurto AM (2010) High-pressure phase equilibria of carbon dioxide (CO_2) + n-alkyl-imidazolium bis(trifluoromethylsulfonyl)amide ionic liquids. J Chem Thermodyn 42:305–311
45. Rajagopalan B, Wie M, Musie GT, Subramaniam B, Busch DH (2003) Homogeneous catalytic epoxidation of organic substrates in CO_2-expanded solvents in the presence of water soluble oxidants and catalysts. Ind Eng Chem Res 42:6505–6510
46. Ohgaki K, Takata H, Washida T, Katayama T (1988) Phase equilibria of four binary systems containing propylene. Fluid Phase Equilibr 43:105–113
47. Peng DB, Robinson DT (1976) A new two-constant equation of state. Ind Eng Chem Fund 15:59–64
48. Houndonougbo Y, Jin H, Rajagopalan B, Wong K, Kuczera K, Subramaniam B, Laird BB (2006) Phase equilibria in carbon dioxide-expanded solvents: experiment and molecular simulations. J Phys Chem B 110:13195–13202
49. Swalina C, Arzhantsev S, Li HP, Maroncelli M (2008) Solvation and solvatochromism in CO_2-expanded liquids. 3. the dynamics of nonspecific preferential solvation. J Phys Chem B 112:14959–14970
50. Subramaniam B (2010) Gas-expanded liquids for sustainable catalysis and novel materials. Coord Chem Rev 254:1843–1853
51. Sih R, Dehghani F, Foster NR (2007) Viscosity measurements on gas expanded liquid systems-methanol and carbon dioxide. J Supercrit Fluids 41:148–157
52. Kelkar MS, Maginn EJ (2007) Effect of temperature and water content on the shear viscosity of the ionic liquid 1-ethyl-3-methylimidazolium bis(trifluoromethanesulfonyl)imide as studied by atomistic simulations. J Phys Chem B 111:4867–4876
53. Ahosseini A, Ortega E, Sensenich B, Scurto AM (2009) Viscosity of n-alkyl-3-methyl-imidazolium bis(trifluoromethylsulfonyl)amide ionic liquids saturated with compressed CO_2. Fluid Phase Equilibr 286:62–68

54. Maxey NB (2006) Transport and phase-transfer catalysis in gas-expanded liquids. PhD Dissertation, Georgia Institute of Technology, Atlanta
55. Lin IH, Tan CS (2008) Diffusion of benzonitrile in CO_2-expanded ethanol. J Chem Eng Data 53:1886–1891
56. Roškar V, Dombro RA, Prentice GA, Westgate CR, McHugh MA (1992) Comparison of the dielectric behavior of mixtures of methanol with carbon dioxide and ethane in the mixture-critical and liquid regions. Fluid Phase Equilibr 77:241–259
57. Wyatt VT, Bush D, Lu J, Hallett JP, Liotta CL, Eckert CA (2005) Determination of solvatochromic solvent parameters for the characterization of gas-expanded liquids. J Supercrit Fluids 36:16–22
58. Abbott AP, Hope EG, Mistry R, Stuart AM (2009) Probing the structure of gas expanded liquids using relative permittivity, density and polarity measurements. Green Chem 11:1530–1535
59. Ford JW, Janakat ME, Liu J, Liotta CL, Eckert CA (2008) Local polarity in CO_2-expanded acetonitrile: A nucleophilic substitution reaction and solvatochromic probes. J Org Chem 73:3364–3368
60. Seki TJ, Grunwaldt JD, Baiker A (2009) In situ attenuated total reflection infrared spectroscopy of imidazolium-based room-temperature ionic liquids under "supercritical" CO_2. J Phys Chem B 113:114–122
61. Burgi T, Baiker A (2006) Attenuated total reflection infrared spectroscopy of solid catalysts functioning in the presence of liquid-phase reactants. Adv Catal 50:227–283
62. Guha D, Jin H, Dudukovic MP, Ramachandran PA, Subramaniam B (2007) Mass transfer effects during homogeneous 1-octene hydroformylation in CO_2-expanded solvent: modeling and experiments. Chem Eng Sci 62:4967–4975
63. Lyon CJ, Subramaniam B, Pereira CJ (2001) Enhanced isooctane yields for 1-butene/Isobutane alkylation on SiO2-supported Nafion® in supercritical carbon dioxide. In: Spivey JJ, Roberts GW, Davis BH (eds) Catalyst deactivation 2001. Studies in surface science and catalysis, vol 139. Elsevier, Amsterdam, pp 221–228
64. (a) Webb PB, Kunene TE, Cole-Hamilton DJ (2005) Continuous flow homogeneous hydroformylation of alkenes using supercritical fluids. Green Chem 7:373–379; (b) Sellin MF, Webb PB, Cole-Hamilton DJ (2001) Chem Commun 781
65. Thomas CA, Bonilla RJ, Huang Y, Jessop PG (2001) Hydrogenation of carbon dioxide catalysed by ruthenium trimethylphosphine complexes: effect of gas pressure and additives on rate in the liquid phase. Can J Chem 79:719–724
66. Combes GB, Dehghani F, Lucien FP, Dillow AK, Foster NR (2000) Asymmetric catalytic hydrogenation in CO_2 expanded methanol-an application of gas anti-solvent reaction (GASR). In: Abraham MA, Hesketh RP (eds) Reaction engineering for pollution prevention. Elsevier, Amsterdam, pp 173–181
67. Combes G, Coen E, Dehghani F, Foster NR (2005) Dense CO_2 expanded methanol solvent system for synthesis of naproxen via enantioselective hydrogenation. J Supercrit Fluids 36:127–136
68. Floris T, Kluson P, Muldoon MJ, Pelantova H (2010) Notes on the asymmetric hydrogenation of methyl acetoacetate in neoteric solvents. Cat Lett 134:279–287
69. Solinas M, Pfaltz A, Cozzi P, Leitner W (2004) Enantioselective hydrogenation of imines in ionic liquid/carbon dioxide media. J Am Chem Soc 126:16142–16147
70. Jessop PG, Stanley R, Brown RA, Eckert CA, Liotta CL, Ngo TT, Pollet P (2003) Neoteric solvents for asymmetric hydrogenation: supercritical fluids, ionic liquids, and expanded ionic liquids. Green Chem 5:123–128
71. Jessop PG, DeHaai S, Wynne DC, Nakawatase D (2000) Carbon dioxide gas accelerates solventless synthesis. Chem Commun 8:693–694
72. Scurto AM, Leitner W (2006) Melting point depression of organic ionic solids/liquids with carbon dioxide for enhanced catalytic processes. Chem Commun 3681–3683

73. Ahosseini A, Ren W, Scurto AM (2009) Hydrogenation in biphasic ionic liquid/CO_2 systems. In: Hutchenson KW, Scurto AM, Subramaniam B (eds) Gas expanded liquids and near-critical media: green chemistry and engineering, vol 1006, ACS symposium series. American Chemical Society, Washington, DC, pp 218–234

74. Jin H, Subramaniam B (2004) Catalytic hydroformylation of 1-octene in CO_2-expanded solvent media. Chem Eng Sci 59:4887–4893

75. Jin H, Subramaniam B, Ghosh A, Tunge J (2006) Intensification of catalytic olefin hydroformylation in CO_2-expanded media. AIChE J 52:2575–2591

76. Koeken ACJ, Benes NE, van den Broeke LJP, Keurentjes JTF (2009) Efficient hydroformylation in dense carbon dioxide using phosphorus ligands without perfluoroalkyl substituents. Adv Syn Catal 351:1142–1450

77. Hemminger O, Marteel A, Mason MR, Davies JA, Tadd AR, Abraham MA (2002) Hydroformylation of 1-hexene in supercritical carbon dioxide using a heterogeneous rhodium catalyst. 3. Evaluation of solvent effects. Green Chem 4:507–512

78. Webb PB, Sellin MF, Kunene TE, Williamson S, Slawin AMZ, Cole-Hamilton DJ (2003) Continuous flow hydroformylation of alkenes in supercritical fluid-ionic liquid biphasic systems. J Am Chem Soc 125:15577–15588

79. Frisch AC, Webb PB, Zhao G, Muldoon MJ, Pogorzelec PJ, Cole-Hamilton DJ (2007) Solventless continuous flow homogeneous hydroformylation of 1-octene. Dalton Trans 47:5531–5538

80. Wei M, Musie GT, Busch DH, Subramaniam B (2002) CO_2-expanded solvents: unique and versatile media for performing homogeneous catalytic oxidations. J Am Chem Soc 124:2513–2517

81. Wei M, Musie GT, Busch DH, Subramaniam B (2004) Autoxidation of 2,6-di-tertbutylphenol with cobalt Schiff base catalysts by oxygen in CO_2-expanded liquids. Green Chem 6:387–393

82. Zuo X, Niu F, Snavely WK, Subramaniam B, Busch DH (2010) Liquid phase oxidation of p-xylene to terephthalic acid at medium-high temperatures: multiple benefits of CO_2-expanded liquids. Green Chem 12:260–267

83. Trent DT (1996) Propylene oxide Kirk-Othmer encyclopedia of chemical technology, vol 20, 4th edn. Wiley, New York, pp 271–302

84. Thiele GF, Roland E (1997) Propylene epoxidation with hydrogen peroxide and titanium silicalite catalyst: activity, deactivation and regeneration of the catalyst. J Mol Catal A Chem 117:351–356

85. Danciu T, Beckman EJ, Hancu D, Cochran RN, Grey R, Hajnik DM, Jewson J (2002) Direct synthesis of propylene oxide with CO_2 as the solvent. Angew Chem Int Ed 42:1140–1142

86. Laufer W, Meiers R, Holderich W (1999) Propylene epoxidation with hydrogen peroxide over palladium containing titanium silicalite. J Mol Catal A Chem 141:215–221

87. Jenzer G, Mallat T, Maciejewski M, Eigenmann F, Baiker A (2001) Continuous epoxidation of propylene with oxygen and hydrogen on a Pd-Pt/TS-1 catalyst. Appl Catal A 208:125–133

88. Nolen SA, Lu J, Brown JS, Pollet P, Eason BC, Griffith KN, Glaser R, Bush D, Lamb DR, Liotta CL, Eckert CA, Thiele GF, Bartels KA (2002) Olefin epoxidations using supercritical carbon dioxide and hydrogen peroxide without added metallic catalysts or peroxy acids. Ind Eng Chem Res 41:316–323

89. Hancu D, Green H, Beckman EJ (2002) H_2O_2 in CO_2/H_2O biphasic systems: green synthesis and epoxidation reactions. Ind Eng Chem Res 41:4466–4474

90. Zha YJ, Zhang JL, Han BX, Hu SQ, Li W (2010) CO_2-controlled reactors: epoxidation in emulsions with droplet size from micron to nanometre scale. Green Chem 12:452–457

91. Herrmann WA, Fischer RW, Marz DW (1991) Methyltrioxorhenium as catalyst for olefin metathesis. Angew Chem Int Ed Engl 30:1636–1638

92. Rudolph J, Reddy KL, Chiang JP, Sharpless KB (1997) Highly efficient epoxidation of olefins using aqueous H_2O_2 and catalytic methyltrioxorhenium/pyridine: pyridine-mediated ligand acceleration. J Am Chem Soc 119:6189–6190

93. Wang WD, Espenson JH (1998) Effects of pyridine and its derivatives on the equilibria and kinetics pertaining to epoxidation reactions catalyzed by methyltrioxorhenium. J Am Chem Soc 120:11335–11341

94. Yin G, Busch DH (2009) Mechanistic details to facilitate applications of an exceptional catalyst, methyltrioxorhenium: encouraging results from oxygen-18 isotopic probes. Catal Lett 130:52–55

95. Lee HJ, Shi TP, Busch DH, Subramaniam B (2007) A greener, pressure intensified propylene epoxidation process with facile product separation. Chem Eng Sci 62:7282–7289

96. Lee HJ, Ghanta M, Busch DH, Subramaniam B (2010) Towards a CO_2-free ethylene oxide process: homogeneous ethylene epoxidation in gas-expanded liquids. Chem Eng Sci 65: 128–134

97. Azarnoosh A, Mcketta JJ (1959) Solubility of propylene in water. J Chem Eng Data 4:211–212

98. Yorizane M, Sadamoto S, Yoshimura S (1968) Low-temperature vapor-liquid equilibriums. Nitrogen-propylene and carbon dioxide-methane systems. Kagaku Kogaku Ronbun 32:257–264

99. Lee HJ, Shi TP, Subramaniam B, Busch DH (2006) Selective oxidation of propylene to propylene oxide in CO_2 expanded liquid system. In: Schmidt SR (ed) Catalysis of organic reactions. CRC Press, Boca Raton, pp 447–451

100. Weissermel K (2003) Industrial organic chemistry, 4th edn. Wiley, Weinheim, pp 145–153

101. Haneda A, Seki T, Kodama D, Kato M (2006) High-pressure phase equilibrium for ethylene + methanol at 278.15 K and 283.65 K. J Chem Eng Data 51:268–271

102. West KN, Wheeler C, McCarney JP, Griffith KN, Bush D, Liotta CL, Eckert CA (2001) In situ formation of alkylcarbonic acids with CO_2. J Phys Chem A 105:3947–3948

103. Chamblee TS, Weikel RR, Nolen SA, Liotta CL, Eckert CA (2004) Reversible in situ acid formation for -pinene hydrolysis using CO_2 expanded liquid and hot water. Green Chem 6:382–386

104. Gohres JL, Marin AT, Lu J, Liotta CL, Eckert CA (2009) Spectroscopic investigation of alkylcarbonic acid formation and dissociation in CO_2-expanded alcohols. Ind Eng Chem Res 48:1302–1306

105. Alemán PA, Boix C, Poliakoff M (1999) Hydrolysis and saponification of methyl benzoates. Green Chem 1:65–68

106. Hunter SE, Savage PE (2004) Recent advances in acid- and base-catalyzed organic synthesis in high-temperature liquid water. Chem Eng Sci 59:4903–4909

107. Throckmorton PE, Hansen LI, Christensen RC, Pryde EH (1968) Laboratory optimization of process variables in reductive ozonolysis of methyl soyate. J Am Oil Chem Soc 45:59–62

108. Nickell EC, Albi M, Privett OS (1976) Ozonization products of unsaturated fatty acid methyl esters. Chem Phys Lipids 17:378–388

109. Nishikawa N, Yamada K, Matsutani S, Higo M, Kigawa H, Inagaki T (1995) Structires of ozonolysis products of methyl oleate obtained in a carboxylic acid medium. J Am Oil Chem Soc 72:735–740

110. O'Brien M, Baxendale IR, Ley SV (2010) Flow ozonolysis using a semipermeable Teflon AF-2400 membrane to effect gas-liquid contact. Org Lett 12:1596–1598

111. Subramaniam B, Busch DH, Danby A, Binder TP (2008) Ozonoysis reactions in liquid CO_2 and CO_2-expanded solvents. U. S, Patent Application, 20090118498

112. Del Moral D, Osuna AMB, Cordoba A, Moreto JM, Veciana J, Ricart S, Ventosa N (2009) Versatile chemoselectivity in Ni-catalyzed multiple bond carbonylations and cyclocarbo-nylations in CO_2-expanded liquids. Chem Commun 31:4723–4725

113. Zwolak G, Jayasinghe NS, Lucien FP (2006) Catalytic chain transfer polymerisation of CO_2-expanded methyl methacrylate. J Supercrit Fluids 38:420–426

114. Wyatt VT, Haas MJ (2009) Production of fatty acid methyl esters via the in situ transesterification of soybean oil in carbon dioxide-expanded methanol. J Am Oil Chem Soc 86:1009–1016
115. George J, Patel Y, Pillai SM, Munshi P (2009) Methanol assisted selective formation of 1, 2-glycerol carbonate from glycerol and carbon dioxide using (Bu2SnO)-Bu-n as a catalyst. J Mol Catal A Chem 304:1–7
116. Baiker A (1999) Supercritical fluids in heterogeneous catalysis. Chem Rev 99:453–474
117. Grunwaldt JD, Wandeler R, Baiker A (2003) Supercritical fluids in catalysis: opportunities of in situ spectroscopic studies and monitoring phase behavior. Catal Rev Sci Eng 45:1–96
118. Beckman EJ (2004) Supercritical and near-critical CO_2 in green chemical synthesis and processing. J Supercrit Fluids 28:121–191
119. Devetta L, Giovanzana A, Canu P, Bertucco A, Minder B (1999) Kinetic experiments and modeling of a three-phase catalytic hydrogenation reaction in supercritical CO_2. Catal Today 48:337–345
120. Chouchi D, Gourgouillon D, Courel M, Vital J, Nunes da Ponte M (2001) The influence of phase behavior on reactions at supercritical conditions: the hydrogenation of α-pinene. Ind Eng Chem Res 40:2551–2554
121. Xu D, Carbonell RG, Kiserow DJ, Roberts GW (2005) Hydrogenation of polystyrene in CO_2-expanded solvents: catalyst poisoning. Ind Eng Chem Res 44:6164–6170
122. Chan JC, Tan CS (2006) Hydrogenation of tetralin over Pt/γ-Al_2O_3 in trickle-bed reactor in the presence of compressed CO_2. Energy Fuels 20:771–777
123. Phiong H-S, Cooper CG, Adesina AA, Lucien FP (2008) Kinetic modelling of the catalytic hydrogenation of CO_2-expanded alpha-methylstyrene. J Supercrit Fluids 46:40–46
124. Xie X, Liotta CL, Eckert CA (2004) CO_2-protected amine formation from nitrile and imine hydrogenation in gas-expanded liquids. Ind Eng Chem Res 43:7907–7911
125. Jenzer G, Schneider MS, Wandeler R, Mallat T, Baiker A (2001) Palladium-catalyzed oxidation of octyl alcohols in "supercritical" carbon dioxide. J Catal 199:141–148
126. Kerler B, Robinson RE, Borovik AS, Subramaniam B (2004) Application of CO_2-expanded solvents in heterogeneous catalysis: a case study. Appl Catal B 49:91–98
127. Sharma S, Kerler B, Subramaniam B, Borovik AS (2006) Immobilized metal complexes in porous hosts: catalytic oxidation of substituted phenols in CO_2 media. Green Chem 8:972–977
128. Stobrawe A, Makarczyk P, Maillet C, Muller JL, Leitner W (2008) Solid-phase organic synthesis in the presence of compressed carbon dioxide. Angew Chem Int Ed 47:6674–6677
129. Sarsani VSR, Lyon CJ, Hutchenson KW, Harmer MA, Subramaniam B (2007) Continuous acylation of anisole by acetic anhydride in mesoporous solid acid catalysts: reaction media effects on catalyst deactivation. J Catal 245:184–190
130. Chateauneuf JE, Nie K (2000) An investigation of a Friedel-Crafts alkylation reaction in homogeneous supercritical CO_2 and under subcritical and splitphase reaction conditions. Adv Environ Res 4:307–312
131. Garton RD, Ritchie JT, Caers RE (2003) Oxo process. PCT International Application, WO 2003/082789 A2
132. Bhanage BM, Divekar SS, Deshpande RM, Chaudhari RV (1997) Kinetics of hydroformylation of 1-dodecene using homogeneous HRh(CO)(PPh$_3$)$_3$ catalyst. J Mol Catal A Chem 115:247–257
133. Purwanto P, Deshpande RM, Delmas H, Chaudhari RV (1996) Solubility of hydrogen, carbon monoxide, and 1-octene in various solvents and solvent mixtures. J Chem Eng Data 41:1414–1417
134. Guha D, Jin H, Dudukovic MP, Ramachandran PA, Subramaniam B (2007) Mass transfer effects during homogeneous 1-octene hydroformylation in CO_2-expanded solvent: modeling and experiments. Chem Eng Sci 62:4967–4975

135. Fang J, Jin H, Ruddy T, Pennybaker K, Fahey D, Subramaniam B (2007) Economic and environmental impact analyses of catalytic olefin hydroformylation in CO_2-expanded liquid (CXL) media. Ind Eng Chem Res 46:8687–8692
136. Jana R, Tunge JA (2009) A homogeneous, recyclable rhodium(I) catalyst for the hydroarylation of Michael acceptors. Org Lett 11:971–974
137. Wang R, Cai F, Jin H, Xie Z, Subramaniam B, Tunge JA (2009) Hydroformylation in CO_2-expanded media. In: Hutchenson KW, Scurto AM, Subramaniam B (eds) Gas-expanded liquids and near-critical media: green chemistry and engineering, vol 1006, ACS symposium series. American Chemical Society, Washington, DC, pp 202–217
138. Fang J, Jana R, Tunge JA, Subramaniam B (2011) Continuous homogeneous hydroformylation with bulky rhodium catalyst complexes retained by nano-filtration membranes. Appl Catal A: Gen 393:294–301
139. Subramaniam B, Tunge JA, Jin H, Ghosh A (2008) Tuning product selectivity in catalytic hydroformylation reactions with CO_2-expanded liquids. US Patent 7.365,234, 29 Apr 2008
140. Horstmann S, Grybat A, Kato R (2004) Experimental determination and prediction of gas solubility data for oxygen in acetonitrile. J Chem Thermodyn 36:1015–1018

Books and Reviews

Anastas P, Eghbali N (2010) Green chemistry: principles and practice. Chem Soc Rev 39:301–312
McHugh MA, Krukonis VJ (1994) Supercritical fluid extraction: principles & practice. Butterworth-Heinemann, Boston
Muldoon MJ (2010) Modern multiphase catalysis: new developments in the separation of homogeneous catalysts. Dalton Trans 39:337–348
Olivier-Bourbigou H, Magna L, Morvan D (2010) Ionic liquids and catalysis: recent progress from knowledge to applications. Appl Cat A 373:1–56

Chapter 3
Green Catalytic Transformations

James H. Clark, James W. Comerford, and D.J. Macquarrie

Glossary

Atom economy	Calculates the percentage of atoms in the reagents used in the final product.
Chemisorption	Where a species is chemically bound to a surface.
Clay	A pure material, either synthesized or natural, used as a heterogeneous catalyst or support.
Dehydroxylation	Where two isolated silanols condense to form a siloxane bridge, with loss of water.
E-factor	A measure of waste produced per quantity of product.
Envirocats[TM]	A series of commercially available catalysts developed by the University of York and Contract Chemicals Ltd.
Heterogeneous	Where two species in a reaction, catalyst and reagents, are in a different phase, i.e., solid and vapor phase.
Homogeneous	Where species in a reaction are in the same phase.
Ion exchange	A process where ions, typically metal cations, are exchanged with other ions on a surface using a suitable solvent.
Physisorption	Where a species interacts strongly with a surface through electrostatic interaction, hydrogen bonding.
Support/supported	Referring to a species interacting with or bound to a surface.
Zeolite	A diverse aluminosilicate structure commonly used in heterogeneous catalysis.

This chapter was originally published as part of the Encyclopedia of Sustainability Science and Technology edited by Robert A. Meyers. DOI:10.1007/978-1-4419-0851-3.

J.H. Clark (✉) • J.W. Comerford • D.J. Macquarrie
Department of Chemistry, University of York, Heslington, York, UK
e-mail: james.clark@york.ac.uk; jwc@uniofyorkspace.net; duncan.macquarrie@york.ac.uk

P.T. Anastas and J.B. Zimmerman (eds.), *Innovations in Green Chemistry and Green Engineering*, DOI 10.1007/978-1-4614-5817-3_3,
© Springer Science+Business Media New York 2013

Definition of the Subject and Its Importance

With ever-increasing demand of chemical products on a global scale, as well as poor public image in recent years, there has been increasing pressure for chemistry industry to become more efficient and sustainable. Processes and catalytic cycles on large scales are being scrutinized by companies to enhance efficiency, reduce environmental impact and associated costs using state-of-the-art research and technology. An overwhelming number of new and improved catalytic transformations are reported in many different fields on a daily basis. However, a catalytic transformation must fulfill a number of criteria to be deemed as "green." The process/catalytic cycle must exhibit a notable improvement on existing syntheses, not only in terms of activity, but as an overall process by assessing waste produced throughout (cradle to grave concept including even the synthesis of the catalyst itself), potential reusability of catalyst, as well as ease of product isolation and potential user risk, where the safety/toxicity of chemicals involved in the process are assessed.

Introduction

The necessitated development of green synthetic procedures has grown from a number of different socioeconomic factors over recent decades. Increasing demand of chemical products worldwide has meant that the environmental impact of industry (particularly within the EU), is being assessed more harshly than ever. This has had a distinct effect on the relationship between chemical industry and the environment, and as such green chemistry has received huge attention. The key factor of concern is the quantity and nature of the waste being produced, a number of different strategies have been implemented by the EU and ECHA to reduce waste by restricting the use of hazardous/potentially toxic chemicals on an industrial scale, through either banning the substance completely or imposing substantial tax deterrents on certain chemical wastes (REACH legislation). This provides a financial incentive for companies to employ waste-minimizing techniques for their processes. An excellent overview of waste minimization is given in "Chemistry of Waste Minimization" which gives a detailed insight into how environmental factors are affected by the production and restriction of chemical waste [1]. Focus has been increasingly diverted toward the prevention of waste rather than the treatment, through a combination of replacing homogeneous catalysts with heterogeneous catalysis, utilizing renewable raw materials, and investing in new technologies such as continuous flow.

Homogeneous Versus Heterogeneous

By definition, a heterogeneous catalyst is a catalytically active species that is in a different phase to the reagents within a reaction system, more often than not a solid in either liquid or vapor phase synthesis. There are a number of advantages of using heterogeneous catalysts as opposed to the homogeneous equivalents, detailed below:

- Safety – An important consideration in the practice of green chemistry. Heterogeneous catalysts often tend to be environmentally benign and easy/safe to handle. This is due to the active species being adhered to a support (often forming a powder), essentially reducing its reactivity with the surrounding environment.
- Reusability – Due to the difference in phases, the catalyst is simply filtered off (through centrifugation on industrial scales) and reactivated for reuse many times over. For instance, zeolites used for petroleum refining can be reactivated and reused for many years – sometimes up to a decade – before disposal.
- Activity – In many cases increased activity is observed when supporting an active homogeneous species on a support, due to the complex but unique surface characteristics found with a variety of different supports.
- Selectivity – Heterogeneous catalysts can give an increased degree of selectivity in reaction pathway. This can simply be a consequence of adsorption of substrates and the consequential restricted freedom of movement of the reacting molecules. More famously, the pores of a solid support can cause size and shape restrictions with regard to the substrates, intermediates, and products. For example, the substitution of aromatic rings leads to products with different geometries and bulkier intermediates may be less likely to form and/or bulkier products may not be able to leave the pores leading to orientational selectivity. Shape selectivity can also affect stereoselectivity through control over reaction pathways.

However, there are a handful of disadvantages. The quantity of solid catalyst required is often higher than that of the homogeneous equivalent, due to the lower concentration on the support surface (reusability makes this less of an issue though). Blocking of the pores/support channels can occur with narrow pores sizes and reduce efficiency over time in some liquid phase reactions; nevertheless, this is often twinned with high stereoselectivity. Despite this, the advantages found with heterogeneous catalysis far outweigh the minor drawbacks.

Green Chemistry Metrics

Green chemistry metrics are important tools in assessing the effectiveness and efficiency of reactions. Increasingly, they are becoming common place in the designing of synthetic routes/chemical syntheses, the main metrics are defined below:

- E-Factor – Developed by Sheldon, the environmental factor determines the quantity of waste (in Kg) produced per Kg of product. This takes into account all waste throughout the synthesis, not only solvents used in reactions, etc., but solvent used for recrystallisation. An E-Factor of 1 is considered ideal.

$$Ef = \frac{Total\,Quantity\,of\,waste\,(Kg)}{Total\,quantity\,of\,product\,(Kg)}$$

- Atom economy – Formulated by Trost, calculates the percentage of atoms in the starting material present in the final product, clean syntheses aim for 100%.

$$Atom\,economy\,(\%) = \frac{Molecular\,weight\,of\,product}{Molecular\,weight\,of\,reagents} \times 100$$

- Reaction mass efficiency – Developed by GSK, similar to atom economy but assesses difference in mass.

$$Mass\,efficiency\,(\%) = \frac{Mass\,product}{Total\,mass\,reagents} \times 100$$

- Carbon efficiency – Developed by GSK, again similar in format, determines difference in carbon mass between reagents and products.

$$Carbon\,efficiency\,(\%) = \frac{Quantity\,of\,carbon\,in\,product}{Quantity\,of\,carbon\,in\,reagents} \times 100$$

Overall, recent research has aimed to reduce the quantity and toxicity of waste produced by industrial processes and substantial focus has been centered on the type of catalysts used. Below is a detailed discussion of heterogeneous catalysts covering structural and mechanistic aspects as well as highlighting applications, interesting developments, and increased green credentials of many "classic" syntheses.

Zeolites

Background

The term "zeolite" was first employed by the Swedish mineralogist Cronstedt in 1756, literally meaning boiling stone in Greek [2]. Natural zeolites have had applications in chemistry over past centuries, such as odor control, gas separation, desiccants, water treatment, agriculture, etc., but the development of synthetic zeolites has been hallmarked as one of the greatest discoveries in science, allowing microporous structures to be "tailored to fit" a huge number of various applications such as extraction, purification, and heterogeneous catalysis. The replication of a natural zeolite was first achieved by Barrer in the early 1940s and further research of entirely synthetic zeolites was continued by Milton at the Union Carbide Linde Division's Research Laboratory in Buffalo, New York, in 1949. The developed

Zeolite A and B(P) were used for the separation of gases based on size, in particular air, to produce high purity oxygen for steel mills and other large scale applications [3], although this was not commercialized until the 1970s. Following research by Mobil focused on zeolites for large scale petroleum cracking and dozens of zeolites were produced that were able to act as acid catalysts.

Composition and Structure

Zeolites are crystalline three dimensional microporous structures with etrahedral building blocks comprised of $[SiO_4]^{-4}$ and $[AlO_4]^{-5}$. The chemical composition is often represented by $M_{2/n}$ $O-Al_2O_3-ySiO_2-wH_2O$, where, due to the trivalent aluminum species, the zeolite has a negative charge proportional to each Al and requires either a pentavalent element, such as a neighboring phosphorus (P^{5+}) or a cation to stabilize this. Non-framework cations are found in the porous network and can vary; alkaline metals from group IA and IIA such as Na^+ are often used. The tetrahedral building blocks form structures known as secondary building units, (SBU), of which there is a large variety (around 16). It is these units that form the larger unit cells (cages) which make up the zeolitic framework, commonly used to classify zeolites, although they can be classified through the type of SBU, framework density (the number of T atoms per unit cell), or pore size. For instance, sodalite (SOD), faujasite (FAU), and zeolite A (LTA) share the same sodalite unit cell, however the SOD cages are of a cubic arrangement, LTA is similar in arrangement but linked together through the 4–4 SBU and FAU comprises of a diamond like structure of sodalite cages, linked together through the hexagonal face [4], Fig. 3.1. The specific naming of the frameworks is complex and will not be discussed.

Another common classification mentioned above was that of aperature size, referring to the diameter of the rings forming the zeolite channel. Table 3.1 shows a range of zeolite spanning the typical pore sizes from small to ultralarge, found with common zeolites [1].

Pore sizes up to ~14Å are categorized as microporous in accordance with the IUPAC classification of materials [5]. The size and shape of the zeolite for catalytic applications is crutial, potentially giving exceptional shape size selectivity. Aside from pairing framework with application, it is possible to tune the pore size by changing the Si:Al ratio; a reduction will produce a smaller unit cell, yet require fewer stabilizing cations, freeing up the zeolite channel. In addition, modification post-synthesis is possible simply through ion exchange, the choice of ions can increase the effective size of the pore opening. Despite the selectivity obtained with smaller pore size zeolites, there can be diffusional and blocking problems in liquid phase reactions. As a result there has been much research into mesoporous zeolites, allowing increased diffusional rates and reduced deactivation.

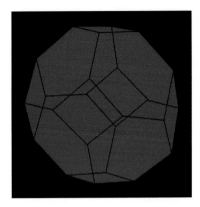

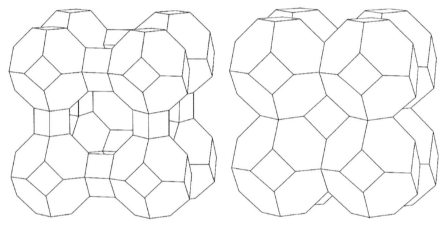

Fig. 3.1 *Top* – SOD unit cell; *Left* – LTA framework; *Right* – SOD framework

Table 3.1 Pore sizes of common zeolites

Zeolite	Number of tetrahedral in ring	Diameter of main channels Å
Sodalite	4	very small – <4
Zeolite A	8	Small – 4
ZSM-5	10	Medium – 5.6
Faujasite	12	Large – 7.4
Cloverlite	20	Ultra large – 13.2
MCM-41	>20	Macroporous – 100

Synthesis

The synthesis of zeolites typically involves mixing a silica and alumina source under basic conditions. Also incorporated into the mixture is a suitable cation of choice, such as either an alkaline metal in an oxide, hydroxide, or salt, with dual function either as counterions in the final material or acting as structure

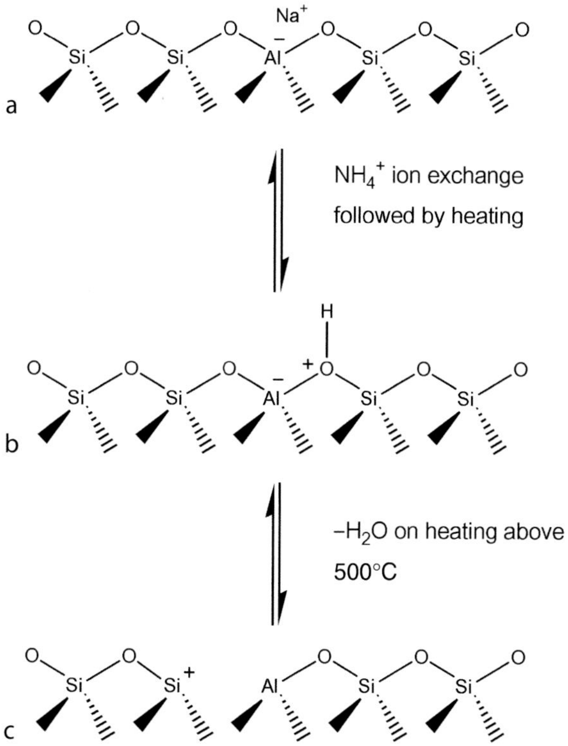

Fig. 3.2 Ion exchanged zeolite and subsequent heating to give Lewis basicity. (**a**) Zeolite as synthesized; (**b**) Bronsted acid zeolite; (**c**) Lewis acid zeolite

directing templates. Aside from the properties of the mixture, temperature, pressure, and crystallization time are also important variables and vary considerably. Typical syntheses crystallize between room temperature to 200°C under autoclave pressure, the time required ranging from hours to days.

Acidity

Counterions within the porous network (Fig. 3.2a) can be readily exchanged using salt solutions, a process referred to as *ion exchange*. To give the zeolite Bronsted acidity, ammonium salt solution (such as ammonium hydroxide) can be used to exchange the metal ions with ammonium. On heating the material deprotonates giving hydrogen as the stabilizing cation (Fig. 3.2b). The Bronsted acid strength is dependent on the Si:Al ratio, where a high ratio of silicon to aluminum results in stronger acidity, as well as increased hydrophobicity, i.e., zeolite B150 is a stronger solid acid and more hydrophobic than B25 (with a ratio of Si 25: Al 1). Further heating above 500°C results in the reversible dehydroxylation of the silanol adjacent to the Al site, giving an aluminum Lewis acid site and a charge silicon species (Fig. 3.2c).

Table 3.2 Showing applications and corresponding zeolites

Application	Zeolite	Notes
Catalytic Cracking	HY-high silica zeolite REY Medium pore zeolites	Selectivity and high conversion rate
Hydrocracking	H-ZSM-5, Faujasite (X, Y), metals on Mordenite (Co,Mo, W, Ni) USY, CAMgY	Conversion rate high
Dewaxing	Pt-mordenite, ZSM-5, silicalite, high silica zeolites (ferrierite)	Low pour and cloud points
Hydroisomerisation	Pt-Mordenite	Low octane pentanes and hexanes converted to high octane yields
Aromatization	Pt, K, Ba silicalite, Pt, Ga, Zn-ZSM-5	Aromatization of C3–C8 cuts, aromatization of LPG
Benzene alkylation	ZSM-5	Production of ethylbenzene and styrene with low quantity of by-products
Xylene isomerisation	ZSM-5	High selectivity to para yield
Toluene disproportionation	ZSM-5, Mordenite	Production of xylenes and benzene
Methanol to petroleum	ZSM-5, erionite	High olefin yields (high octane rating) and high petroleum yield
Fischer-Tropsch	Metal-ZSM-5 (Co, Fe)	Natural gas to petroleum

The basicity of zeolite tends to be less well documented. Lewis basicity stems from a negative charge on the oxygen where Bronsted is due to an extraframework cation on the surface [3].

Catalysis

There are an overwhelming number of examples of solid zeolite acid, base, and oxidation catalysis throughout the past 60 years. As such, the following section will discuss the most popular applications throughout the decades as well as recently published examples. A common observation with much of the research is the impressive shape selectivity and structural diversity found with zeolitic catalysis.

Catalytic Cracking

It has been around 50 years since Mobil first developed the fluid catalytic cracking (FCC) process using zeolite technology and this still remains the dominant catalytic application for zeolites today. Nevertheless, there are many other important transformations using zeolites, of which, a selection are shown in Table 3.2 [3].

Catalytic cracking typically employs fluidized catalyst bed technology and is simple in concept, but a detailed insight of the process and chemistry is outside the scope of this encyclopedia. As an example take the basics of petroleum refining; the initial feedstock comprises a range of carbons chains/aromatics (paraffins, naphthenic and aromatic hydrocarbons), light being C_{10}–C_{20} and heavy being C_{15}–C_{25}. This is pre-heated to ~370–400°C and mixed with powdered solid catalyst at a ratio of 2:1 oil:catalyst, to give a slurry which is injected into a reaction vessel at ~480–550°C [6]. At this temperature the oil partially evaporates and reacts with the catalyst, after 1–4 s the mixture is separated into catalyst, product for distillation and unreacted product (~20–25% of the starting material). The deactivated catalyst is simply reactivated in a stream of air ~700°C and re introduced back into the process. Impressively, the catalyst can be reused for many years in this highly efficient continuous process. The most utilized catalysts currently used in production consist of type Y faujasite zeolite with a Si:Al ratio of ~2.5 and ZSM-5 with a Si:Al ratio of ~15 to >100, allowing control of activity and selectivity. Typical stabilizing cations consist of H^+, Na^+, Ca^{2+}, Ba^{2+}, NH_4^+, NR_4^+ and various other rare earth metals. Despite some of their advantageous properties to the zeolite structure, Na^+ and particularly K^+ act as a catalyst poisons.

The ZSM-5 catalyst has since in developed in the field of FCC and cationic exchange with phosphorus has been of recent interest [7–13]. Enhanced activity was reported [14] in the cracking of propylene in the presence of phosphorus exchanged ZSM-5 prepared through hydrothermal dispersion. Classic modification techniques such as impregnation and ion exchange result in a material that can easily lose phosphorus, leading to contamination and loss of activity/selectivity. It is thought that the phosphorus compounds interact with bridged OH groups on the surface reducing the zeolites acidity, consequently increasing activity and modifying shape selectivity.

Isomerization

Skeletal isomerisation again plays an important role in the refining industry. A well researched transformation is that of n-butene into isobutene, an intermediate to methyl-tert butyl ether (MTBE), one of the most important oxygenated additives used in lead free petroleum, Fig. 3.3. Initially the reaction is thought to be acid catalyzed consisting of a double bond shift to 2-butene, where the second isomerisation step to isobutene requires a stronger acid. Conversely, an increase in acid strength, as found with classic methods employing homogeneous acids such as HCl, also increases quantity of by-products, such as dimerization to octenes, cracking to C_2, C_3, C_5 and C_6 olefins and formation of coke [17]. Potential catalysts, therefore, were of medium acidity to increase selectivity to isobutene and reduced deactivation. As such there has been much research into ferrierite (FER), which shown extremely high selectivity under mild conditions [18–22].

Direct comparison between FER and zeolites of similar acidity (i.e., ZSM-5 – MFI framework) has shown the former to be consistently more selective. This has

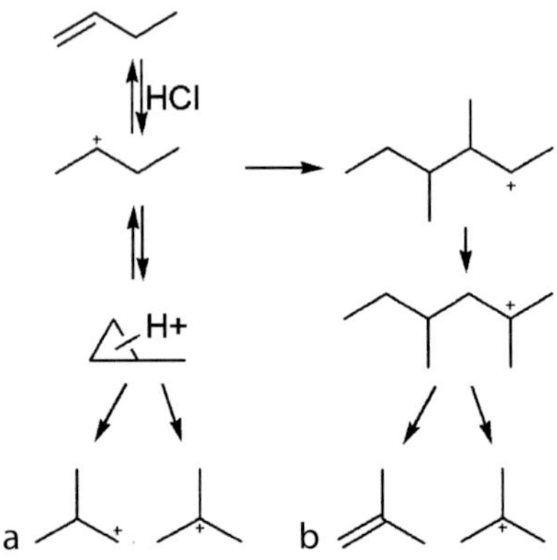

Fig. 3.3 Isomerisation of butene to isobutene. (**a**) Monomolecular mechanism with cyclopropanic intermediate proposed by Brouwer and Hogeveen [15]; (**b**) Bimolecular mechanism proposed by Guisnet et al. [16]

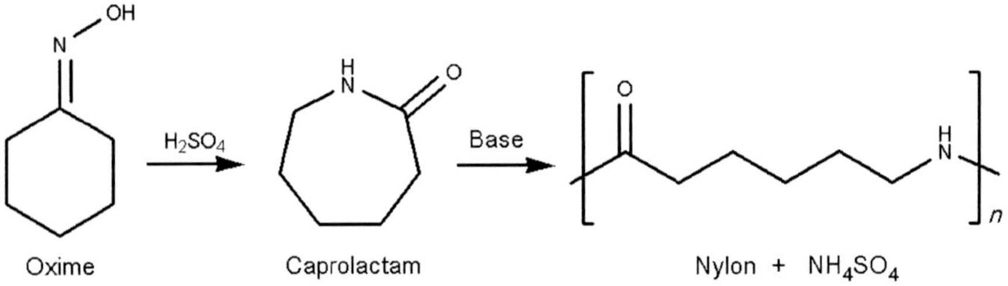

Fig. 3.4 Rearrangement of oxime to give ε-caprolactam and nylon

been ascribed to two main factors; firstly the low number of acid sites (high Si/Al ratio) is thought to prevent undesired bimolecular processes [6] such as aromatization, cyclization, hydrogen transfer and oligomerization, reducing by-products and coke formation (therefore increasing selectivity) and secondly the unique structure of ferrierite. However, despite the high performance of this catalyst, the role of coke formation as well as the precise mechanism in the isomerisation of butane to isobutene is still in debate [19].

Rearrangement

The synthesis of ε-caprolactam through the Beckmann rearrangement of cyclohexanone oxime is industrially important in the synthesis of nylon-6 and resins, Fig. 3.4. Previously, the synthesis would involve the conversion of cyclohexanone with hydroxylamine sulfate or phosphate, to give cyclohexanone oxime. Subsequently, strong acids such as fuming sulphuric were used as an acid catalyst, followed by treatment with ammonia yielding the final ε-caprolactam product. However, large quantities of

Table 3.3 Examples of catalysts reported for the synthesis of ε-caprolactam [23–27]

Catalyst	TOS[a]	Yield (%)	Selectivity (%)
FSM-16	0.5 h	98	49
ZnO/FSM-16	″	99	69
Al2O3/FSM-16	″	98	63
Si-MCM-41	3 h	>99	32
Al-MCM-41(A)[b]	″	>99	48
Al-MCM-41(B)	″	>99	65
Al-MCM-41(C)	″	>99	87
β-MFI	4 h	95	95
H3BO3/Hβ	″	>99	80
H-LTL	6 h	>99	97
H-OFF-ERI	″	>99	96
Silicalite-1 MFI	″	77	95
High Si MFI	″	>99	95
H-USY	″	>99	82
H-MOR	″	92	90

[a]Time on stream
[b]Increasing Si:Al ratio

$(NH_4)_2SO_4$ salt by-product are formed, along with a high reactor corrosion factor and high risk factor (handling large volumes of concentrated acid), making the process environmentally unsound.

As such this has been studied intensely, with a number of developed catalysts being used to varying degrees of success, Table 3.3. Out these, titanium silicate (TS-1) has shown exceptional activity and selectivity, its development over the past decade gives an excellent example of green synthesis. The old two step process of ammoximation followed by rearrangement has been telescoped into a single continuous flow process [28] and has been used industrially by Sumitomo Chemical Co., Ltd since 2003, (operating on 60,000 tons per year scale!). TS-1 is used to convert cyclohexanone [29] to the oxime giving a conversion of >99% with a selectivity of 98%. This is followed by vapor phase rearrangement of the oxime to the lactam using high silica MFI zeolite, giving a yield on >99% with a selectivity of 95%. The improvement of this reaction can be shown by the increase in atom economy for ammoximation, from 36% using $H_2SO_4 + 1.5NH_3$ to 100% using TS-1 [11].

Schüth recently reported [30] that by crosslinking the TS-1 structure, both improved catalytic activity and reduced deactivation over time was observed. The preparation of the zeolite uses siloxane linkers, resulting in an increased in mesoporous surface character similar to that of silicalite. Interestingly activity correlates with increased concentration of linkers, as well as Ti concentration. Deactivation of that catalyst was found to be substantially improved compared with standard TS-1, up to 32% increased productivity. The observed effects are thought to be due to both the presence of titanium and siloxane crosslinkage, cooperatively reducing coke formation. The benefits of these heterogeneous catalysts aside from high activity and selectivity include simple separation of product and any by-products though filtration, recovery, reuse, and low toxicity.

Fig. 3.5 Restrictive ZSM-5
channels

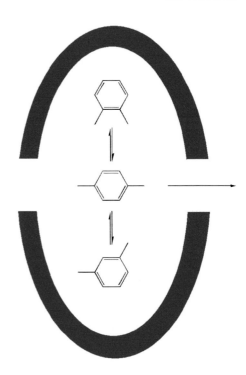

Transformations of Aromatics over Zeolites

The production of ethylbenzene from benzene and ethylene is an important step in
the synthesis of styrene and subsequently polystyrene [3]. Previously the process
used AlCl$_3$ in a Friedel-Crafts alkylation and as with many strong homogeneous
Lewis acids, there are problems with corrosion, waste separation, and contamination. A cleaner, continuous process using fixed beds of ZSM-5 gives high yields. In
particular, ZSM-5 is an excellent example of shape selective properties and
tuneability available with zeolite catalysis. The alkylation of toluene and methanol,
for example, gives p-xylene, industrially important for its use in the manufacture of
terephthalic acid and dimethyl terephthalate (starting material for polyester fibers,
vitamins and other pharmaceuticals) [1]. The exceptional selectivity observed with
ZSM-5 is due to the different diffusional rates of the ortho, meta, and para isomers
through the porous zeolite channels, Fig. 3.5. It has been demonstrated through
variable temperature reactions that the diffusion of p-xylene is around 1,000 times
faster than the ortho and meta isomers [31].

Despite this high selectivity, further research into ion exchange of ZSM-5 has
produced even more active catalysts. By exchanging ions of different sizes into the
porous network the size of the channels can vary, becoming increasingly restrictive
with large metal ions. Sotelo [32] discusses ion exchange with magnesium in H-
ZSM-5, where para selectivity approaches 100%, with isomer by-products being
due to reaction on the external surface of the catalyst.

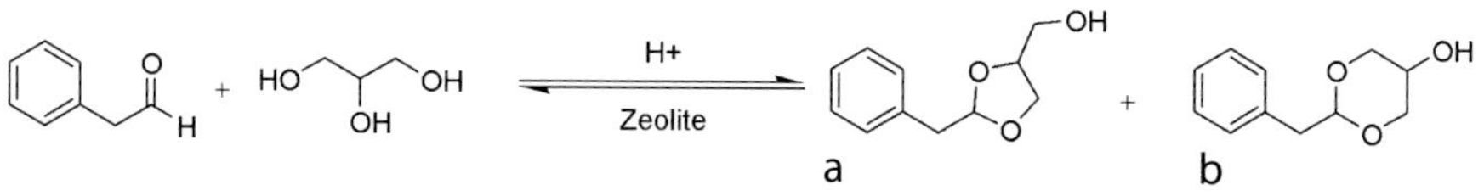

Fig. 3.6 Synthesis of 2-benzyl-4-hydroxymethyl-1,3-dioxolane and 2-benzyl-5-hydroxy-1,3-dioxane

Zeolites in Fine Synthesis of Speciality Chemicals

As well as their application to a number cracking, fuel and polymer reactions, zeolites are also able to catalyze a number of syntheses in fine chemical industry, such as in the synthesis of perfumes, fragrances, flavors, and pharmaceuticals, a few examples are given below.

Fragrances

Vanillin propylene glycol acetal and phenylacetaldehyde glycerol acetals are important flavoring compounds with a vanilla and hyacinth fragrance. The propylene glycol acetal of vanillin is often used to imitate vanilla flavors as it causes flavor attenuation, this can easily be reversed through hydration back to the aldehyde, restoring full flavor [33]. Typical homogeneous acid catalysts are often used in the common transformation of the aldehyde to the acetal and suffer similar drawbacks with waste, separation, necessitated neutralization, and expense. A number of zeolites have therefore been applied to many fragrant syntheses, such as Jasminaldehyde (n-amylcinnamaldehyde), a violet scent obtained through the condensation of heptanal and benzaldehyde [34] and Fructone (ethyl 3,3-ethylendioxybutyrate), a strong fruity scent of apple/pineapple/wood synthesized thorough the acetalization of ethyl acetoacetate with ethylene glycol [35]. Climent et al. reported of synthesizing 2-benzyl-4-hydroxymethyl-1,3-dioxolane (Fig. 3.6a) and 2-benzyl-5-hydroxy-1,3-dioxane (Fig. 6b) from the acetalization of glycerol (soon to be a by-product on a million ton/pa scale) and benzaldehyde [33], both give a hyacinth scent.

Out of a number of zeolites investigated (Table 3.4), USY-2 and beta were particularly active and selective, comparable to p-toluene sulphonic acid. The secondary product (B) is thought to be converted from the acid catalyzed rearrangement of the less stable (A). Interestingly, the ratio of product to by-product is lower for Mordenite and ZSM-5, though overall conversion is low, again demonstrating shape selectivity and narrow pore size of the zeolites. Selectivity is not of huge importance in this case as both are active fragrants and are industrially acceptable, despite A having slightly increased potency than B.

Table 3.4 Conversion and selectivity to acetal product using various zeolites

Catalysts	Conversion (%)[a]	Selectivity (%)	
		A	B
USY-2	93	58	35
Beta-2	92	61	31
Mordenite	33	28	5
MCM-41	36	26	10
ZSM-5	54	46	8
PTSA	97	66	31

[a]After 1 h in refluxing toluene

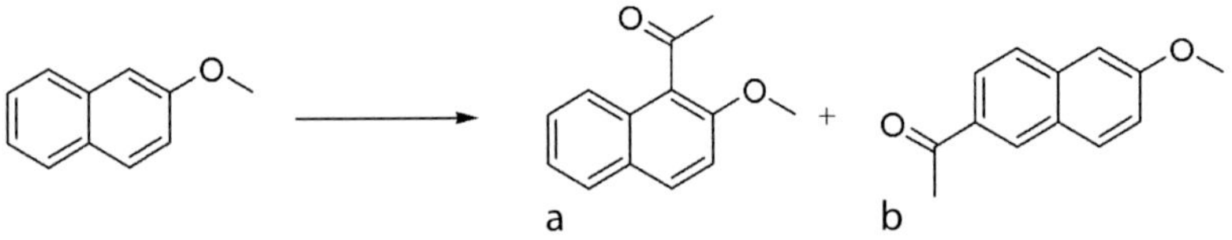

Fig. 3.7 1-acetyl-2-methoxynaphthalene and the 2-acetyl-6-methoxynaphthalene precursors

Pharmaceuticals

Ibuprofen, [(±)-2-(4-isobutylphenyl)] propionic acid, is a readily available and commonly used nonsteroidal anti-inflammatory/pain killer, able to treat many ailments such as muscular injury, headaches, and cold and flu symptoms (the list goes on). Its synthesis comprises of a six step process first patented by Boots in the 1960s. Typically isobutylbenzene is acylated with acetic anhydride using $AlCl_3$ as a Lewis acid catalyst; this causes problems with waste and to a certain extent selectivity, only isobutyl acting as a directing group to the para position (the ortho and meta are observed less so due to steric and electron effects, respectively). The resulting 2-(4-isobutyl phenyl)-ethanone is then reacted with Propionic acid (1-chloro-ethyl ester) and sodium ethoxide resulting in an epoxide, which, under acid catalysis, eliminates methyl ethanoate to give 2-(4-Isobutyl-phenyl)-propionaldehyde. There has been research into the replacement of $AlCl_3$ with zeolites for synthesis of the intermediate 2-(4-isobutyl phenyl)-ethanone. Beta zeolite is active as an acylation catalyst and interestingly increased surface area through grinding or from formation of small particle sizes gives increased yields to around 20–30% [36, 37]. Higher yields, however, are observed in the synthesis of 2-acetyl-6-methoxynaphthalene ethanone, a precursor to (2-(6-methoxy-2-naphthyl)propionic acid) otherwise known as Naproxen, a similar nonsteroidal anti-inflammatory [38, 39].

Despite the fact that the 1-acetyl-2-methoxynaphthalene (A) Fig. 3.7 is kinetically favored [40], beta zeolite channels show selectivity toward the 2-acetyl-6-methoxynaphthalene (B). The increased selectivity is generated by the narrow pores, subsequently giving lower yields, potentially due to blocking, Table 3.5 [36].

Table 3.5 Acylation of isobutylbenzene with acetic anhydride

Catalyst	Yield (%)	Product distribution	
		1-Ac-2-Mn (A)	6-Ac-2-Mn (B)
Beta zeolite	68	30	70
Beta zeolite[a]	75	8	92
Beta zeolite[b]	75	24	76
BZ-1[c]	81	35	65
BZ-1	86	25	75
BZ-2 [d]	82	35	65
Ce^{3+}-BZ-2[e]	88	22	78

[a]Using propionic anhydride
[b]Pre-activation at 750°C
[c]Microcrystalline beta zeolite of particle size 1–10 μm, obtained through mechanical disintegration
[d]Microcrystalline beta zeolite of particle size 10–50 μm obtained through shortening the crystallization time to 48 h instead of one week
[e]Ion exchanged zeolite using 5%wt metal chloride solution

After modification (grinding) an increased yield is observed thought to be due to an increase in surface area and accessibility to the acidic sites previously deep within the porous network. However, without the restrictive nature of the pores, selectivity to the 2-acetyl-6-methoxynaphthalene decreases, suggesting that the synthesis of the 1-acetyl-2-methoxynaphthalene largely occurs on the external surface of the zeolite. The incorporation of Ce^{3+} gives higher selectivity thought to be due to increased Lewis acidity.

Limitations of Zeolites as Catalysts

Despite the numerous advantages found with zeolite catalysis there are still a handful of downsides. Most zeolites tend to have an aperture size between 4 and 8Å which limits application to less bulky substituents, mesoporous zeolites alleviate this for the most part, yet they too have their own disadvantages. The formation of coke (known as coking) as well as the buildup of products/reagents can cause rapid deactivation of the zeolite. The cost of the zeolite is often higher than that of an equivalent homogeneous acid, but as the cost of disposal of toxic chemical waste increases this will become less of an issue.

Clays

Background

Clays are a class of soil <2 μm in diameter primarily composed of fine-grained phyllosilicates (Greek for leaf, "phyllon"and Latin for flint, "silic"), which when wet is generally plastic and sticky in appearance and can be dried out to give a hard, cohesive material [41]. The large majority of applications in the catalytic field are focused on acid catalysis. A particularly helpful book is "The Origin of Clay Minerals in Soils and Weathered Rocks" with a detailed discussion of clay structure and type [42].

Structure of Clays

The subject of clay structure is a complex one [2]; as this chapter is designed to give an overview of clean synthesis, the following discussion will focus on the basics. Clay structure consists of sheets, tetrahedral and octahedral, which together can form layers, 1:1 and 2:1. The tetrahedral sheet is made up of $[SiO_4]^{-4}$ groups linked together through three *basal* oxygens, the fourth however is termed apical and points away from the layer toward the adjacent octahedral layer. The basal "structural" oxygens form a hexagonal lattice, characteristic of the tetrahedral layer. It should be noted that through ionic substitution, the Si^{4+} center can be replaced with a cation of similar size, such as Fe^{3+} or Al^{3+}. The octahedral layer comprises of $AlO_4(OH)_2$ units in sixfold coordination (although the cation center can be Al^{3+}, Fe^{3+}, Fe^{2+}, or Mg^{2+} depending on the clay). Depending on the oxidation state of the metal center (2+ or 3+), the structure can fall into a further two categories, dioctahedral or trioctahedral. The former type refers to the state where only two out of a possible three cation sites are filled, as a result of a divalent cation, this is often referred to as a gibbsite sheet (based on the naturally occurring material). Trioctahedral clays have trivalent cations; subsequently all potential cationic sites are filled.

The sheets can form layers in two ways (generally as there are exceptions), 1:1 and 2:1. The former refers to a single tetrahedral sheet bonding with an octahedral, whereas 2:1 refers to an octahedral sheet "sandwiched" between two tetrahedral sheets (Fig. 3.8).

The sheets bond through the apical oxygen via the interlayer cations, hydrogen bonding, or van der Waals/electrostatic forces. Furthermore, the spatial arrangement of the tetrahedral and octahedral sheets is not directly compatible, creating distortions in either one or both structures (classed as the polytype); however, this level of detail is outside the scope of this encyclopedia. The region between the layers is known as the *interlayer*, (X). This provides access to sites within the clay structure and contains equivalents of stabilizing ions to counteract charge created by isomorphic substitution within the tetrahedral layer, key to the clays activity. Lastly, layers found in various clays can be mixed, in that they contain different octahedral layers, i.e., one layer of chlorite and one of smectite, in regular or irregular quantities. Table 3.6 shows the structural makeup of some 1:1 and 2:1 clays.

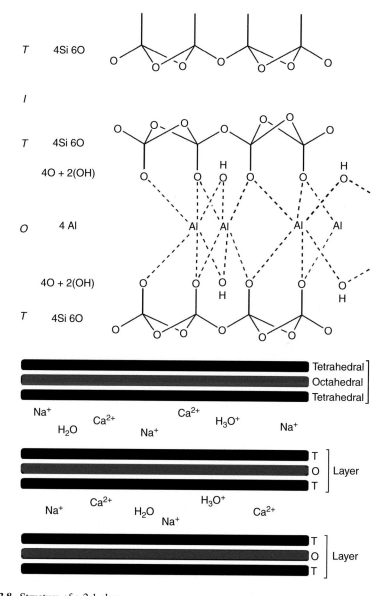

Fig. 3.8 Structure of a 2:1 clay

Montmorillonite

Out of the diverse range of clay types available, montmorillonite is the most utilized when applied to organic catalysis. It is comprised of a 2:1 structure of tetrahedral coordinated silicate $[SiO_4]^{-4}$ and octahedrally coordinated gibbsite $[Al_2(OH)_6]$. The material is particularly susceptible to isomorphic exchange, where Al^{3+}

Table 3.6 Composition of 1:2 and 2:1 clays [43, 44]

Group	Interlayer	Dioctahedral	Trioctahedral
1:1	None or only H_2O	Kaolinite $Al_2Si_2(O_5)(OH)_4$	Serpentine $Mg_3Si_2(O_5)(OH)_4$
2:1	None	Pyrophyllite $Al_2Si_4(O_{10})(OH)_2$	Talc $Mg_3Si_4(O_{10})(OH)_2$
Smectite $0.25 < x < 0.6$	Hydrated exchange-able cations	Montmorillonite $M_x[Al_{2-x}Mg_x](Si_4)$ $O_{10}(OH)_2$	Hectorite $M_x[Mg_{3-x}Li_x](Si_4)$ $O_{10}(OH)_2$
		Beidellite $M_x[Al_2](Si_{4-x}Al_xO(OH)_2$	Saponite $M_x[Mg_3](Si_{4-x}Al_x)$ $O_{10}(OH)_2$
Vermiculite $0.6 < x < 0.9$		Vermiculite (DIO) $M_x[Al_2](Si_{4-x}Al_x)$ $O_{10}(OH)_2$	Vermiculite (TIO) $M_x[Mg_3](Si_{4-x}Al_x)$ $O_{10}(OH)_2$
Mica	Non-hydrated cations	Muscovite $K[Al_2](Si_3Al)O_{10}(OH)_2$	Phlogopite $K[Mg_3](Si_3Al)O_{10}(OH)_2$

replaces Si^{4+} in the tetrahedral sheet and Mg^{2+} replaces Al^{3+} in the octahedral sheet. This creates a negatively charged layer that is compensated by interchangeable cations, such as Na^+ and Ca^{2+}, which are situated between the layers. In fact, large quantities of cations can be "held" or "stored" in the clay depending on the extent of isomorphic exchange. On hydration the clay swells as the layers separate away from each other, allowing easy exchange of stabilizing cations in and out of the clay.

Acidic Properties of Clays

Clays can exhibit both Bronsted and Lewis acidity, the latter being due either to structural cations on the surface of the sheets or ions exchanged in the interlayer region. The Bronsted acidity stems from strong dissociation of intercalated water molecules coordinated to the Lewis acid center, generating mobile H-bonded protons in a highly polarizing environment, given as,

$$\left[M(H_2O)_6\right]n+ \rightarrow \left[MOH(H_2O)_5\right](n-1)^+ + H^+$$

It follows that the weaker the Lewis acid, i.e., the more electron withdrawing the cation, the more acidic the Bronsted acid [45]. Despite being able to reach acidities close to that of 98% H_2SO_4, increased acidity as well as surface area and hydroxyl group concentration have been researched resulting in a common process, the acid treatment of clays. Typical acid treatment involves using a strong inorganic acid such as HCl, sulphuric, or phosphoric in various quantities. This not only replaces exchangeable cations with hydrogen, but also leaches Al out of the central octahedral layer, producing enhanced surface area and increased acidity [46, 47].

Pillared Clays

Despite high acid strength and good surface areas, basic and acid treated clays are still susceptible to thermal degradation; structural collapse of the clay is typically observed upon heating above $\sim$200°C. By *pillaring* the clays increased surface area, concentration of acid site, and thermal stability is achieved, due to a fixed metal oxide pillar. A common pillaring agent is polyoxocation Al_{13}, prepared by mixing a base with aqueous chlorohydrate, $AlCl_3$, or $Al(NO_3)_3$ to give a solution with OH/Al^{3+} ratios of up to 2.5. The polyoxocation complex has been reported to have a tridecamer structure of $[AlO_4Al_{12}(OH)_{24}(H_2O)_{12}]^{7+}$, also known as the Keggin ion [48]. The complex is then mixed with the clay, allowing diffusion of the structure between the layers and ion exchange. Once the clay is then filtered, washed, and calcined between 300°C and 500°C, the cations become bound supporting pillars and are able to release protons, Fig. 3.9.

Catalysis

As with zeolite, clays have found many applications, due to their high acidity, surface areas, high temperature tolerance (pillar only), and unique swelling properties. Discussed below are a few classic examples as well as more recent developments.

Esterification

Organic esters are important in a plethora of applications, from intermediates, pharmaceuticals to fragrances and perfumes to plastisizers (the list goes on). However, there are a number of downsides to conventional esterification process. If using a non-activated carboxylic acid, for instance, the reaction has to be catalyzed with a strong acid in the presence of a large quantity of alcohol, due to the reversible nature of the reaction. It is often the case where the acid is first activated, where electron density is withdrawn from the acid center, usually using thionyl chloride to produce an acid chloride. On reacting the alcohol with the acid chloride a large quantity of HCl waste is produced and a base is often added to the reaction mixture to "mop up" the inorganic acid. Overall the synthesis produces either a large quantity of alcohol or acid waste, or has more than a single step, activation of the acid with a toxic and potent lachrymator, thionyl chloride (PCl_5 has also been used). Avoid the use of these agents not only has

Fig. 3.9 Pillared clays

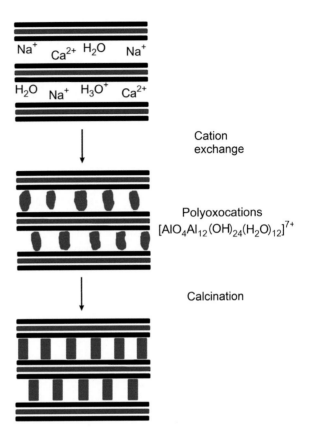

Cation exchange

Polyoxocations
$[AlO_4Al_{12}(OH)_{24}(H_2O)_{12}]^{7+}$

Calcination

Table 3.7 Synthesis of p-cresyl phenylacetate using various montmorillonite catalysts

Catalyst[a]	Calcination temperature (°C)	Reaction time (hours)	Yield (%)
Na^+	100	16	Nil
H^+	100	12	52
Al^{3+}	100	6	67
Al^{3+}	200	12	36
Al^{3+}	400	12	Nil

[a]Ion exchange on montmorillonite clay

a positive environmental impact, but reduces worker risk, and substantially reduces cost. Consequently there has been much research into alternative solid acids and Al^{3+} exchanged montmorillnite clay in particular has shown promising developments. Recent investigation [49, 50] into the synthesis of p-cresyl phenylacetates (industrially used for perfumes, soaps, etc.), using Al^{3+} exchange montmorillonites, has been reported, Table 3.7. The use of the p-cresyl, and aromatic alcohols in general, is of interest as they are weakly nucleophilic due to lone pair electron delocalisation around the ring, therefore requiring a strong delta positive charge on the acid center.

Fig. 3.10 Mechanism of acid catalyzed esterification

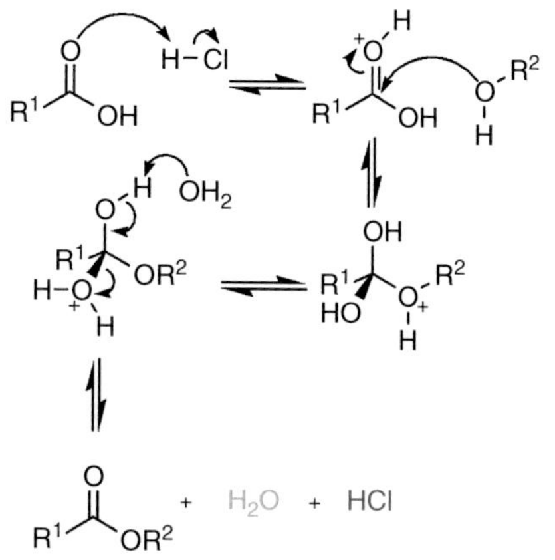

Interestingly, the Na$^+$ exchanged (raw) clay is not active whatsoever and is thought to be attributed to the low electronegativity of the Na+ ions being less able to polarize the interlamellar water and produce strong bronsted acidity. When exchanged with hydrogen, however, increased yields are observed. It is only with the highly polarizing Al^{3+} cation exchange that the highest yields are seen, due to the high acid strength required to protonate and give the oxonium ion intermediate [10], Fig. 3.10. Increased calcination temperatures cause loss of water from the catalyst surface, converting Bronsted into Lewis acidity, resulting in a loss of activity.

This is an excellent example of a clean synthesis, due to its low molar quantity of alcohol, one part acid to two parts alcohol, the nontoxic by-product (H$_2$O), as well as the reusability of the catalyst by simply washing with water and heating to 100°C.

Etherification

Common ether synthesis typically involves treating an alkoxide anion with an alkyl halide, where a strong homogeneous base such as NaH deprotonates an alcohol giving an alkoxide, which attacks an electrophile such as methyl iodide. A NaOH base will suffice when using alcohols such as phenol and dialkyl sulfates are often used as electrophiles, leaving the sulfate ion in the waste stream. An interesting application recently published looks at transforming the potentially large quantity of glycerol by-product from the biodiesel industry (predicted to reach millions of tons) to useful glycerols ethers [51]. Amberlyst-35, Beta zeolite, Montmorillonite K10, Niobic acid, and p-toluene sulphonic acid (PTSA) were directly compared as acid catalysts, Fig. 3.11.

The stronger acid catalysts, beta zeolite and amberlyst-35, gave higher yields selectively to the mono benzyl-glycerol ether (around 55% and 38%, respectively, after only 2 h) due to their narrow porous channels and shape selectivity.

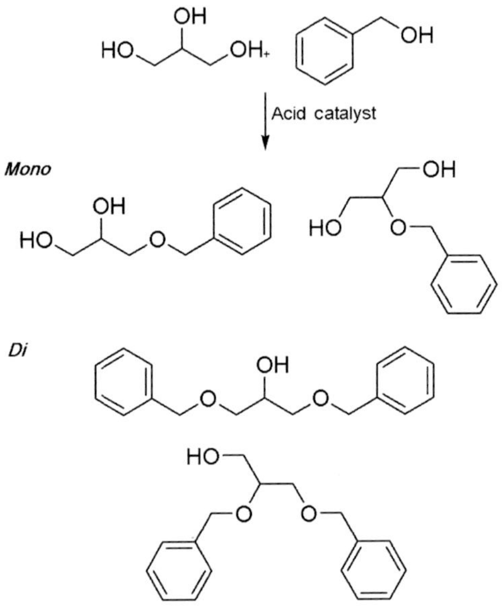

Fig. 3.11 Synthesis of
Glycerol ethers

Interestingly, montmorillonite K10, gave the overall highest conversion of the di benzyl-glycerol ether, ~56%, and also gave a high conversion of the mono benzyl-glycerol, of around 35%. Overall, the β-zeolite and montmorillonite K10 clay gave similar conversions of opposite selectivity, demonstrating that the observed acidity on the clay surface is only a part of is high activity. With a low stochimetric ratio of benzyl alcohol to glycerol, 3:1, this reaction gives high conversions in short reaction times, allowing easy separation and reuse due to it heterogeneous nature and creates a valuable material from a potentially large future waste stream.

Alkylation

The Friedel-Crafts alkylation is problematic when considering the frequency of its use in industry. The alkylation of aromatic rings typically involves reaction of pre-activated alkyl chain, such as 2-chloropropane, with powdered $AlCl_3$ catalyst or liquefied HF, which helps the alkyl halide to polarize, allowing π electrons from the aromatic benzene to attack the carbocation. Many clays have been investigated over past decades [52–54], and early work was focused on increasing the activity thorough ion exchange, (for the most part the base clay does not show appreciable activity). Laszlo and Mathy [55] reported increased activity with exchanged clays when applied to various reactions such as the alkylation of aromatics with alkyl halides, alcohols, and alkenes. The test reaction was that of benzyl chloride and benzene, activity of the metal exchanged clays followed the following reactivity series [56], $Fe^{3+} > Zn^{2+} > Cu^{2+} > Zr^{4+}, Ti^{4+} > Ta^{5+} > Al^{3+} > Co^{2+} > K10 > Nb^{5+}$. This not

Table 3.8 Application of solid acids to alkylations

Reaction type	Catalyst	Reactant	Reagent
Alkylation of aromatic hydrocarbons with	Wyoming montmorillonite expanded with AlCl$_3$ and/or silanized with (EtO)$_4$Si	Benzene	Ethane
i) Olefins	H-ZSM-5	Toluene	Ethane
ii) Saturated hydrocarbons	FeCl$_3$ doped montmorillonite K10	Benzene	Admantane
iii) Alcohols	ZnCl$_2$ clay, Al$_2$O$_3$ expanded smectite clay	Benzene, toluene naphthalene	Aliphatic alcohols cyclopentanol
	H-ZSM-5	Benzene	Ethanol
	Pillared montmorillonite, piller saponite, H-ZSM-5, HY	Toluene	Methanol
	H-ZSM-5	Naphthalene, methyl naphthalene	Methanol
iv) Halide	Transition metal salt deposited on montmorillonite K10	Benzene	Benzyl chloride
Alkylation of phenols	Activated clay	Phenol	Isobutene
	Bentonite		Oleic acid
	Supported/mixed oxides		Methanol
Alkylation of aromatic amines	Bentonite and Triton B	Diphenylamine	Styrene

only showed an excellent alternative for alkylation catalysis, but more importantly demonstrated how support surfaces, whether clay, zeolite, or amorphous oxide, provide a very different environments for reacting species when compared to that of homogeneous solutions. For instance, the activity displayed by Zn is much higher than that of Al, where the opposite is true for the homogeneous chloride salt equivalents [19]. The range in activity is thought to be related to water coordinating to the metal center in a highly polarizing environment, exhibiting varying degrees of Bronsted acidity. A number of examples of solid acid catalysts and subsequent reactants/reagents are shown in Table 3.8 [57]. By far one of the greatest breakthroughs was that of the clayzic catalyst for benzylations, this will be discussed later in the supported reagents section.

Sulfonylations

Use of SO$_3$/R$_2$SO with H$_2$SO$_4$ is commonly in the synthesis of sulfones. These are often used in pharmaceuticals, polymers, and agrochemicals. Aside from the typical drawbacks associated with homogeneous acid catalysis, there can be problems with selectivity between para/ortho isomer in the sulfonylated of aromatic species. Both bronsted acidic and ion exchanged zeolites [58, 59] and clays have been researched

Table 3.9 Sulfonylation of various species using zeolite and montmorillonite

Catalyst	Sulfonylating agent	Yield (%)	Selectivity		
			O	M	P
Fe^{3+} montmorillonite	Ms_2O	81	47	19	34
Fe^{3+} montmorillonite	$PhSO_2Cl$	84	5	0	95
Fe^{3+} montmorillonite	TsCl	86	1	0	99
Fe^{3+} montmorillonite	TsOH	42	13	7	80
Fe^{3+} montmorillonite	Ts_2OH	84	11	6	83
Beta Zeolite	Ms_2O	78	42	21	37
Beta Zeolite	$PhSO_2Cl$	72	11	3	86
Beta Zeolite	TsCl	53	1	0	99
Beta Zeolite	TsOH[a]	50	12	5	83
Beta Zeolite	Ts_2OH	88	14	6	80

[a]Standard reaction time 6 h, TsOH reaction time 24 h in refluxing toluene

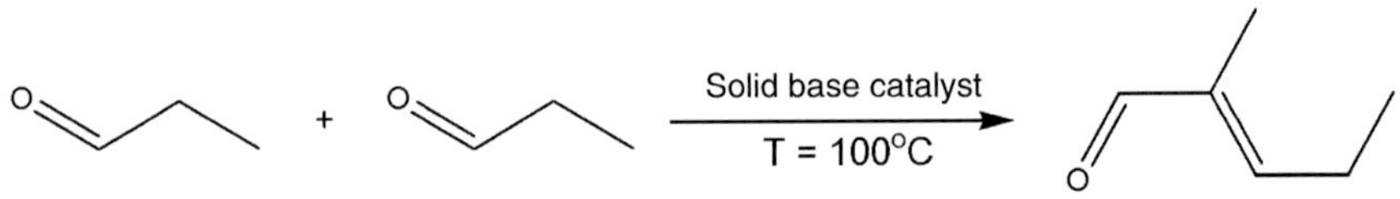

Fig. 3.12 Synthesis of 2-methylpentanal

as heterogeneous replacements, of which Fe^{+3} exchanged montmorillonite clay has shown high activity toward this type of transformation [60]. The activity of exchanged metals on montmorillonite follows $Fe^{3+} > Zn^{2+} > Cu^{2+} > Al^{3+} >$ K10 in the sulfonylation of m-xylene and toluene-p-sulfonic anhydride [61]. A screen of various reagents using beta zeolite and Fe^{3+} montmorillonite can be seen in Table 3.9. The selectivity in the methanesulfonation of toluene parallels that of $AlCl_3$, achieving 33% selectivity toward the para position, 99% para selectivity is observed in the arenesulfonylation (TsCl). The activity of the catalysts is thought to be to the mixed of Lewis and Bronsted acid catalysis. Reusability of the catalysts is found to be greater in systems that do not use Cl as an activating agent.

Aldol Condensation

The use of clays for acid catalyzed reactions is well researched; however, little investigation (in comparison) is focused on their applications as a base. This largely stems from the components looking to be replaced, in that, typically homogeneous acids, $AlCl_3$, HCl, H_2SO_4, tend to be more of a problem than, say, NaOH or KOH. Nevertheless similar problems still remain with stoichiomeric quantities of waste twinned with disposal cost, as well as difficulty of product/reagent separation. A good example of utilizing basic properties of zeolites is given in the solvent free aldol condensation of propanol to 2-methylpentanal [62], Fig. 3.12. Propanol itself tends to be limited to solvent applications; alternatively, 2-methylpentenal has

Table 3.10 Conversion of 2-methylpentanal

Catalyst	Conversion (%)	Selectivity (%)		
		2-methylpentanal	3-Hydroxy-2-methylpentanal	3-Pentanone
HT (1.5)	44	68	30	2
HT (2.0)	47	83	17	-
HT (2.5)	80	92	8	-
HT (3.0)	86	96	4	-
HT (3.5)	97	99	-	1

commercial importance in pharmaceuticals, fragrances, flavors, and cosmetics. With optimum reaction conditions using a homogeneous base such as NaOH, a high yield of 99% is obtained yet only a selectivity of 86% is achieved. As such, a number of solid bases have been investigated with hydrotalcite showing enhanced selectivity.

Unlike other materials, hydrotalcite (HT) has a typical composition of $Mg_6Al_2(OH)_{16}(CO_3)4(H_2O)$, its name based on its resemblance to talc and high water content. The activity of the HT is found to vary with Mg:Al ratio, from 1.5–3.5, directly proportional to the basicity of the material, Table 3.10. It is thought that two types of basic site exist in hydrotalcite, firstly a weak OH- Bronsted and secondly a stronger O^- Lewis base [63].

Limitations

There are few limitations of modern clay catalysis, most acid-treated clays show enhanced acidity in comparison to untreated and thermal decomposition of the clay structure can be avoided by pillaring the sheets. Increased selectivity is often seen and the ability of clay to ion exchange a variety of metals and make clay catalysis an efficient and green alternative to homogeneous activating agents.

Supported Reagents

Introduction

A supported reagent can be defined as a species that is supported, through either chemisorption or physisorption, onto an organic (such as ion exchange resin/carbon) or inorganic material. Typically supporting materials range from clays, to amorphous and structured silicas, alumina, and zeolites. Often the supported species is inorganic, such as precious metals, although there are a number of organic examples. These offer many advantages over their homogeneous equivalents:

- Reduced or nonexistent toxicity, leading to easier and safer handling
- Reusability
- Enhanced activity due to surface characteristic and active site distribution

Table 3.11 Typical surface areas of commonly used supports

Support	Surface area m^2/g	Notes
Amorphous Silica Gel	300–600	Weakly acid surface due to isolated silanols, although is available in neutral and basic forms. Pores sizes typically range from 4 to 10 nm
Amorphous Alumina	100–300	Weakly basic, but through treatment and variation of surface groups can be acidic or neutral. Pore sizes similar to silica
Acid treated montmorillonites	50–300	K10 commonly used, give enhanced acid character compared with untreated montmorillonite clay. Able to swell and ion exchange with suitable solvent
Pillared montmorillonites	200–500	Stable at higher temperatures and can give enhanced shape selectivity
Structured Silica (HMS, SBA)	300–800	A wide choice of pore shapes and sizes available, allowing the supported to be tailored to fit certain syntheses. High temperature tolerance, easy to functionalise surface though SiOH silanols
Zeolites	300–600	Highly selective crystalline microporous materials. Able to exchanges ions to tune acidity type and strength

- Increased selectivity though structural characteristic of supporting material
- Enantioselective potential

It should be noted, however, that complications related to poor diffusion, blocked pores due to substrates, potential poisoning of active species, and physical use on a large scale (fine silt and powders) sometimes create a few drawbacks.

Types of Support

Different supports have very different properties; however, for the majority of applications high surface areas of >100 m^2 g^{-1} are considered advantageous, as this typically parallels concentration of active species. Depending on the reaction, in particular, the properties of supports can be tailored to suit, such as the acidity/basicity of the surface (the majority of applications utilized an acidic surface), pore size, and channel regularity. This can range from around 4 Å for zeolitic supports to >150 Å for structured silicas, temperature stability from $\sim$200°C for standard acid-treated clays to $\sim$800°C for amorphous silica (high temperature catalysts are often preferred for obvious reasons). Shape selectivity is often acquired with small/restrictive pores found in microporous silicas, zeolites, and pillared clays; however, diffusional problems and blockages can cause loss of activity over time. To overcome this problem many different species are supported on mesoporous supports, such as large pore SBAs, where pore size can reach around 200 Å (some macroporous carbons can be above 500 Å), although shape selectivity is nearly always lost at this size. Table 3.11 shows commonly used supports and corresponding surface areas [64].

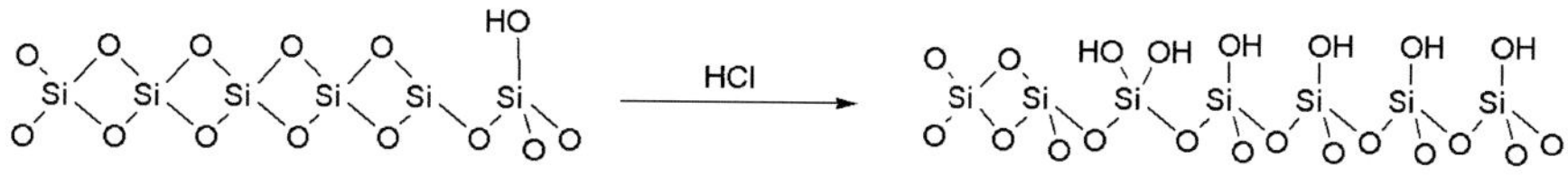

Fig. 3.13 Hydrolysis of siloxane bridges to give increased silanol quantity

Choice of support is an important consideration when designing clean synthetic routes. For instance, the acid properties of one support may destroy reagents whereas another may stabilize important intermediate species on the surface.

Preparation of Supported Species

Prior to loading active species on a support, certain pretreatments are commonly performed. Dehydration of the surface at around 120°C upward under vacuum, avoids hydrolysis complications and competitive adsorptions onto binding sites. In the synthesis of $AlCl_3$-Al_2O_3, for example, $AlCl_3$ is readily hydrolysed unless the alumina is dried at least 550°C before impregnation. Another common technique used increases the number of available surface hydroxyls able to react on silica supports (often using HCl), effectively increasing the concentration of active species able to bind to the surface, Fig. 3.13. A number of preparation methods exist, depending on the end use of the catalysts, different methods are more suited than others.

Wet impregnation/Evaporation is a commonly used technique for loading species onto a support, due to its suitability for a wide range of reagents and supports. The method is based on dissolving the species of choice in a solvent of reasonable volatility, i.e., DCM, acetone, ethanol and stirring the solution with the support (typically 1–2 h). Once even wetting and diffusion into the pores has occurred the solvent can be evaporated *slowly*, ensuring even distribution of the species on the surface, deep in the porous network of the material. Post-treatment often involves calcining the resulting material, (anything from 200°C to 600°C), allowing chemisorption of the species onto the surface, referred to as thermal activation. *Precipitation* is generally used when there are issues with solubility, often with of metal salts. Deposition of the active species on the surface is achieved through either reaction with a second species to form an insoluble salt, cooling a hot solution containing the solute, or addition of cosolvent in which the active species is insoluble. *Adsorption* methodology simply involves stirring the support and reagent together (sometimes under reflux), yet requires a strong interaction between the surface and supporting species, aiming to achieve chemisorption if possible. After adsorption, suitable post treatment ensures the species is bound to the surface. Other methods have also been used such as *mixing/grinding* support and reagent together, in situ supporting (where the two solids are introduced directly into the reaction and the supported reagent is formed and reacts as the reaction proceeds) and ultrasound, these tend to be used less often.

Aside from post-modification, it is possible to synthesize materials with a species of choice available and evenly dispersed on the surface, known as sol-gel synthesis. The technique allows organic functionality to be incorporated into the structure and gives high surface area materials with high ratios of active species per gram. This methodology uses soluble (generally monomeric) precursors to the "support" and the supported component, polymerising them together, often round templates to give highly structured materials. In addition, the organic species can often be reacted with many other substituents to tailor the functionality, amino functionalized silicas are particularly good at this.

Loading the Support

The dispersion of the active species becomes an important variable when physisorbing/chemisorbing a reactive species, such as a sulphonic acid or metal halide onto a surface. It should be noted that in other cases such as ion exchange of various cations with clays and zeolites, there are a defined number of sites that cannot be exceeded. Correct ratios of support to reagent can be estimated from the surface area and porosity of the material, optimal loadings tend to be between 0.5 and 2 mg per gram of support, aiming to achieve monolayer coverage. Porosimetry using N_2 adsorption onto the surface gives useful information when predicting potential concentration/loadings; however, problems can arise when using micro-porous supports, often is the case that N_2 can diffuse into pores that reagents cannot. As such, the exact requirement for various supports, active species and reactions, needs detailed investigation to achieve high efficiency. Overloading and underloading a support equally create a number of problems. Overloading can cause agglomerisation of reagents leading to blocked pores, reducing the overall surface area available for catalysis and hindering diffusion. After monolayer dispersion is achieved, a second (bi-layer) can form, diluting the unique surface-reagent properties, where second and third layer species tend toward homogeneous characteristics. Contrastingly, underloading the support can cause lower catalytic activity, therefore requires increase quantities of the catalyst. With a lower concentration of active species, an increase in exposed surface is observed, which can be reactive enough to catalyze by-product formation.

Catalysis

There has been extensive research into supporting Bronsted and Lewis acids on different supports, resulting in many different catalyst properties and applications. Below is a sample out of a number of examples, showing the effect of different supports, supported species, and reactions, giving a much generalized view of the

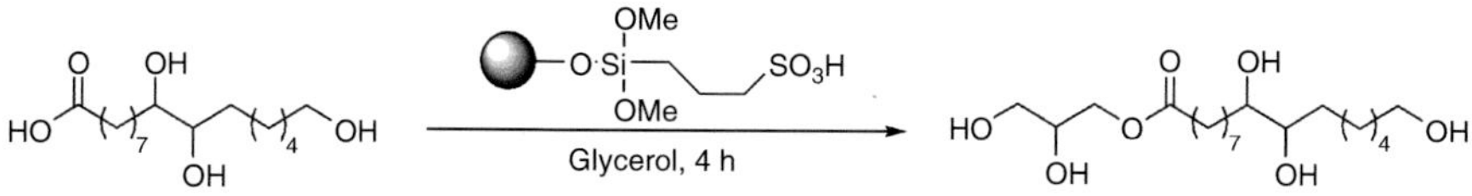

Fig. 3.14 Esterification using sulphonic acid on HMS

area. As with reactions previously discussed, through supporting active homogeneous species, the production of waste, risk factor, and cost is substantially reduced, giving high efficiency reactions with minimal impact on the environment and resources.

Bronsted Acid Catalysis

The acidity of supports, i.e., isolated silanols and alumina hydroxyls, tend to parallel organic acids, achieving a pKa of around 4–6. This is of little use for the majority of acid catalyzed reactions and as such often requires a species of stronger acidity to be bound to the surface. Typically there are several types of Bronsted acidity observed in relation to supports, the original acidity from mixed oxides (silica and alumina hydroxyls), zeolitic acidity, activated/dissociated H_2O molecules usually coordinated with a Lewis acid metal on the surface and homogeneous Bronsted acids tethered or chemisorbed onto the support. The latter allows very high acid strength species to become reusable and safe to handle, below are a few examples.

Sulphonic Acid on Silica

The hazards of using fuming acids on a large scale (as well as the subsequent tax deterrents on chemical waste) are well known and as a result there has been much investigation into supporting them on a variety of different media. One of the more "straight forward" supported acids is that of sulphuric acid/sulphonic acids supported on silica, demonstrating high activity and selectivity in a number of transformations [65–73]. Jérôme et al. have shown that sulphonic acid supported on hexagonal mesoporous silica (HMS) gives selective esterification of multifunctional carboxylic acids with glycerol (a by-product of the biodiesel process) [74]. The reaction choice relates to its poor selectivity and rapid polymerisation to many undesired products, for instance the use of homogeneous p-toluene sulfonic acid in the esterification of glycerol with 16-hydroxyhexadecanoic acid gives a low yield of the amphiphilic monomer, only 35%. By anchoring the sulphonic species to HMS and SBA-15, an impressive increase in selectivity to 98% was observed, due to the restrictive nature of the pores. On application to other, more complex acids, high activity and selectivity remain consistent, Fig. 3.14.

Table 3.12 Synthesis of 2-(4-Chloro-phenyl)-5-propyl-[1,3,4]oxadiazole [1]

Catalyst	Yield (%)
p-TsOH	42
NaHSO$_4$	33
NaHSO$_3$	28
H$_2$SO$_4$	25
Montmorillonite K10	48
Silica sulphuric acid	94

Other examples of sulphonic acids on silica gel include the synthesis of mono and disubstituted oxadiazoles from acyl hydrazides and orthoesters at room temperature [75]. Heterocycles are an important class of biologically active compounds often used as pharmaceutical intermediates. Out of a number of different solid acids investigated, sulphuric acid on silica gave high yields in as little as 10 min, Table 3.12. Interestingly, enhance activity was seen when compared to homogeneous H$_2$SO$_4$, showing how supported materials are often much more than the sum of their parts. Reuse of the catalyst showed that activity remained high even after five uses.

Shaabani [76] has also recently reported high yields using silica-H$_2$SO$_4$ in the synthesis of trisubstituted imidazoles by the one pot condensation of benzyl, benzoin, or benzylmonoxime with a substituted benzaldehyde and ammonium acetate. The synthesis used water as a solvent, whereas previously solvents such as methanol, ethanol, acetic acid, and DMSO were typically used. Excessive water in acid catalyzed reactions is known to deactivate acid centers; nevertheless yields of ~70% are achieved after 4–6 h.

Sulphated Zirconia

Zirconia as a support has received attention in recent years due to its high activity toward hydrocarbon conversions. In particular, sulphated zirconia (SZ, SO$_4^{2-}$ – ZrO$_2$) has shown catalytic potential toward light alkane isomerisation, as well as acylation, alkylation, cracking, and ring opening reactions [77–81]. As well as the many applications of mesoporous SZ to liquid phase reactions, standard microporous SZ is well suited to a number of vapor phase synthesis. It had previously been thought that the high activity of sulphated zirconia had been due to its super acidic properties, with an acid strength 1,000 times greater than homogeneous sulphuric acid [82]; however, more recent literature has shown that this is not the case [83].

FTIR spectroscopic titration with pyridine shows that surfaces activated between 500 and 600°C comprise of Bronsted acidity only [84], thought to be due to rapid adsorption of water, converting any Lewis acidity to Bronsted. Ratnam et al. recently reported of using SZ to catalyze the acylation of alcohols/phenols and amines, typically achieving yields of ~90% in 10 min [85].

Silica Gel as a Catalyst

The use of silica as a support has been widely published, yet there has been little report of using silica gel alone as an acid catalyst. Amide synthesis has been a consistent problem area in green chemistry for many years. The majority of the problems stem from the used of stoichiometric quantities of activating agents used to create a δ positive center on carboxylic acids, allowing nucleophilic attack from an amine. It should be noted that there are other type of reagents used for the synthesis of amides, such as nitriles, but the direct condensation of acid and amine is advantageous, due to their relatively low toxicity and by-product formation of water. The use of activating agent creates an equivalent of waste, for instance if an acid is activated to an anhydride, the second "low cost" acid is lost in the waste stream, excluding the proportion that has reacted with the amine, resulting in the formation of side products. There have been attempts at developing homogeneous catalysts, such as Yamamoto's functionalised 3,4,5-trifluorophenyl boronic acid catalysts, though these are often limited to simple reactions, cannot be recovered and their synthesis is not straightforward. However, Clark et al. recently demonstrated the activity of calcined silica gel in the direct synthesis from carboxylic acids and amines [86]. Initially a number of supported metals, $FeCl_3$ and $ZnCl_2$ on silica and montmorillonite clay were investigated, but it was found that the silica gel alone gave the best activity. Unactivated, silica gel is a mild desiccant and does not typically display any catalytic activity towards organic reactions. On heating, a number of surface changes occur, around 120–200°C physisorbed water is lost, above this temperature the silanols start to dehydroxylated to give siloxane bridges up to around 1,000°C (when all of the surface silanols are lost to give quartz). It was found that activation at 700°C gave a hydrophobic surface with mild Bronsted acidity and high activity toward amidation. A number of different reactions are shown below, Table 3.13.

Despite the use of high loadings of silica with particularly difficult syntheses, such as benzoic acid and aniline, the initial low cost, low toxicity and simple preparation and use makes the synthesis a substantial improvement on previous catalysts. Multiple reuse with reactivation (burning out organic buildup) is possible, without loss of activity. With a high atom economy of 93% and an E-factor of less than 1, the synthesis allows high conversions without the use of toxic and highly corrosive activating agents.

Table 3.13 Synthesis of amides using thermally activated silica gel

R^{1} [c]	R^2	Uncatalysed yield	Isolated yield (%)[a]
C_6H_5	C_6H_5	0	47[b]
$C_6H_5CH_2$	C_6H_5	10	81
$CH(CH_3)C_6H_5$	C_6H_5	4	72[b]
$CHClCH_3$	C_6H_5	4	70
$OCH_2C_6H_4Cl$	C_6H_5	0	73
$CH_2C_6H_5$	C_6H_4Cl	0	48
$CH_2C_6H_5$	$C_6H_4(CH_3)_2$	0	38[b]
$CH_2C_6H_5$	C_4H_8	2	89
CH_2CH_3	$CH_2CH_2CH_2CH_3$	89[c]	98[d]

[a]After 24 h reflux in toluene, 20%wt silica
[b]Using 50% wt silica
[c]After 12 h
[d]After 24 h

Lewis Acid Catalysis

Lewis acidity is often based on integrated surface cations acting as acid centers, where the strength largely depends on the surrounding environment. However, as previously discussed with zeolites and clays, water can coordinate to these acid centers and give Bronsted acidity, effectively "diluting" the Lewis acid effect. By tethering homogeneous species such as $ZnCl_2$ and $FeCl_3$ to various supports, increased Lewis acidity is observed. An excellent example of this is in the development and use of Envirocats[TM].

Envirocats[TM]

Designed as a series of commercially available clay and alumina-based catalysts, Envirocats[TM] were developed by collaboration between the University of York and Contract Chemicals Ltd. (CCL). They comprise of a metal salt supported on a high surface area inorganic support and have a range of applications, Table 3.14. Although they are often term "simple" in that the species being supported is not a complicated structure, the mechanism by which they catalyze many different reactions is complex [87].

A good example of their enhanced activity is of the benzylation of benzene and benzylchloride with EPZ10, also known as clayzic, Fig. 3.15. The diphenylmethane product is often used as a precursor to many pharmaceuticals, making this particular reaction of commercial interest. As with many Lewis acid catalyzed reactions,

Table 3.14 Envirocats[TM] and common uses

Catalyst	Supported species	Type of reactions	Acidity
EPZG	Iron/Clay[a]	Friedel-Crafts Benzoylations, some acylations such as etherification	Contains both Bronsted and Lewis acid sites
EPZ10	ZnCl$_2$/K10 Clay	Friedel-Crafts alkylation of aromatics, Benzylations	Very strong Lewis acid, few weakly Bronsted acidic sites
EPZE	ZnCl$_2$/Clay	Friedel-Crafts Sulfonylations and some Benzoylations	Prodominatly strong Lewis acid with some strong Bronsted acids. Different preparation method to EPZ10
EPIC	Phosphoric acid/ Clay	Esterifications, general Bronsted acid catalyzed reactions	Exclusively strong Bronsted acidity
EPAD	Cr(VI)/ alumina	Oxidations	Cr(VI) oxidant center, not suitable for used with peroxide – causes leaching of the metal

[a]Acid treated montmorillonite clay

Fig. 3.15 Benzylation of benzene and benzylchloride

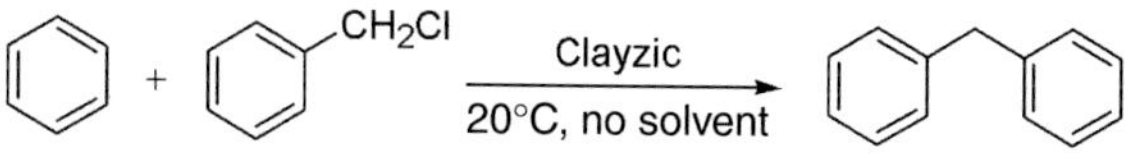

AlCl$_3$ was formerly the homogeneous catalyst of choice, exhibiting little selectivity and giving high ratios of polybenzylated product, unless very large excesses of benzene are used. Clayzic, on the other hand, can be used in similar conditions to AlCl$_3$, yet gives extraordinarily high yields of ~75% after 4 h at room temperature, even with a minimal excess of benzene. In addition, the catalyst can be simply filtered off, washed and reused many times over, showing an excellent example of green chemistry being integrated into industry.

AlCl$_3$ on Silica

Aluminum chloride is a widely used strong, in expensive homogeneous Lewis acid. There are, however, many problems associated with its use, including the necessitated destruction of the acid with water to aid separation of products, resulting high volumes of corrosive and toxic waste and formation of by-products due to its inherent activity. There have been a number of reports detailing the immobilization of aluminum chloride on a variety of supports to overcome these problems. Typical procedures involve dissolving the AlCl$_3$ in an organic solvent and slowly reacting this with the silica surface, theoretically creating a SiOAlCl$_2$ species capable of both Bronsted and Lewis acidity [86]. The resulting material has been applied to a range of reactions with success, early syntheses focus on the Fridel-Crafts acylation. Li et al. discusses the use of AlCl$_2$-silica in a one pot, three component Mannich type reaction, Table 3.15 [88]. Activating agents such as HCl,

Table 3.15 Mannich type synthesis using aniline, acetophenone, and benzaldehyde

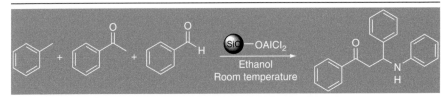

Catalyst	Time (hours)	Yield (%)
$FeCl_3 - SiO_2$	20	0
$ZnCl_2 - SiO_2$	20	0
$AlCl_3$	20	0
SiO_2	20	0
$AlCl_3/SiO_2$ dry mix	5	74
$AlCl_2 - SiO_2$	5	93

$InCl_3$, $Y(OTf)_3$, $Zn(BF_4)_2$, $Bi(OTf)_3$, PS-SO_3H, phosphorodiamidic acid, dodecylbenzene sulfonic acid have been reported, suffering from the usual homogeneous drawbacks. Interestingly, the reaction of aniline, benzaldehyde, and acetophenone to give a β-aminocarbonyl (used in numerous pharmaceutical syntheses and natural products) showed that homogeneous $AlCl_3$ gave lower yields when compared with the supported species. The optimum reaction conditions also add to the green credentials of this synthesis, the best solvent out of a range including, MeCN, CH_2Cl_2, THF, and toluene has been found to be ethanol, with a reaction temperature of $\sim 25°C$.

Supported Lewis Acids in Acylation

Freidel-Crafts reactions are a key target for green chemistry due to their frequent use in chemical industry and large volumes of waste produced. Two main problems areas of the synthesis are the initial activation to an acyl chloride, with subsequent generation of HCl and the use of a strong Lewis acid to cleave the Cl from the acyl chloride to produce an acylium ion, enabling nucleophilic attack from π electrons on the aromatic ring. Early developments focused on the direct replacement of the Lewis acid with a heterogeneous equivalent, examples include Ga_2O_3 and In_2O_3 supported on Si-MCM-41 [89, 90] and $InCl_2$, $GaCl_3$ and $ZnCl_2$ supported on both montmorillonite clay and Si-MCM-41 [91, 92]. Despite high yields, the catalyst required pre-activation to the acyl chloride, creating stoichiometric quantities of HCl waste. Further improvements have developed allowing anhydrides to be used as starting materials, still creating equivalents of waste, yet not as corrosive as previous mineral acid waste [36, 93–96]. However, recent publications discuss the ideal scenario, use of carboxylic acid staring material producing only water as a by-product [97, 98], P_2O_5/SiO_2 in particular has shown potential. Application to

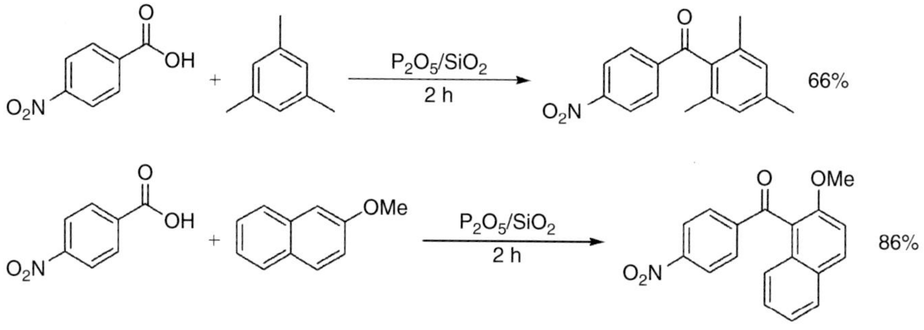

Fig. 3.16 Acylation using P$_2$O$_5$/SiO$_2$. (*Where solvent is 1,2-dichloroethane under reflux, % given is conversion)

Table 3.16 Supported basic species [64]

Solid base	Notes
Alumina – Na	Superbasic
Alumina – KOH	Widely used for many years
Alumina – KF	Widely used with variable basicity
Alumina – CsF	Little advantage over cheaper alumina-KF
Alumina – AlPO$_4$	Used in solventless reactions
Silica – Na	Superbasic
Silica – K	Superbasic
Silica – NaOH	Simplest form of basic silica
Zeolite-M$^+$	Can be very basic depending on alkali metal
Xonotlite-KOtBu	Unusual supports – similar to alumina-KF
MgO-K/Na	Lower surface area than alumina and silica

a range of aromatic acylations has given high yields (Fig. 3.16), on average ~20% higher when compared to the homogeneous P$_2$O$_5$ alone [99].

The inexpensive nature of the catalyst, simple preparation, and ease of handling, as well as the use of unactivated carboxylic acids, producing only water by-product, make this synthesis improved over previous homogeneous species, as well as a number of heterogeneous catalysts.

Solid Base Catalysis

As mentioned before, the push for research into "clean" and reusable heterogeneous base catalysts has not been as pronounced as that for solid acids; however there is still a wide range of basic catalysts available. Many supporting materials can be thermally treated to exhibit basic properties (alumina), though they are weak and often species with increased basicity are supported. The basicity is usually the result of basic hydroxyl groups, oxygen ions bound to the surface or basic ions bound to the surface. Table 3.16 gives an overview of commonly used solid bases.

Basic Species Supported on Alumina (KF/Alumina)

First discussed by Clark [100], potassium fluoride supported on alumina (KF/
Alumina) has been acknowledged as an extremely useful catalyst able to promote
a variety of base catalyzed reactions [101]. The strength and nature of the basic sites
on the surface has been a subject of much discussion over the years, but it has been
widely accepted that the basic strength can vary from weak-moderate to superbasic
depending on the preparation (drying condition of alumina post impregnation) and
subsequent exposure to CO_2, for example. The active species is thought to relate to
the "hard" fluoride anion as opposed to oxygen anions. It has been observed that
when K_2CO_3 and KOH are supported on alumina, the resulting activity is much
lower than that of KF and K_3AlF_6 species (formed at high loadings of KF, itself
exhibiting no activity) on the alumina surface, suggesting that the presence of
fluorine is essential to its activity [102]. Interestingly, unlike other fluoride ions
sources, there has never been any report of KF-alumina exhibiting any nucleophi-
licity, even in simple substitutions. Another peculiarity of the material is that high
activity is increasingly observed when KF loadings are higher than a monolayer
equivalent, at low loadings no crystallinity can be observed through X-ray diffrac-
tion. Further identification of the F-species has been researched using [21] F NMR,
yet the results have proved somewhat controversial [103]. KF-alumina remains one
of the most important base catalysts due to its wide range of applications to many
syntheses such as, isomerisations [103–105], condensations [102], Knoevenagel
[106], cyclization [107], as well as C-N [108], C-O [109–111], S-C [112–114] and
Si-O [115, 116] bond forming reactions. In particular, its application to Michael
additions has received much attention [117–127], a typical example of this is in the
addition of nitroalkanes to electron deficient alkenes [128]. The use of homogeneous catalysts (such as DBU) is often problematic due to the nitro acting as
a leaving group, yielding nitric acid and unsaturated alkene. In addition,
withdrawing groups on the α, β positions of the alkene make this a poor nucleo-
phile. However, the use of KF-alumina as a catalyst gives impressive results
compared to a range of alternative basic catalysts, Table 3.17.

Temperature control allows high selectivity between potential products, where
room temperature conversions give the Michael addition adduct. A high E:Z
stereoselective ratio of 95:5, is consistently found for the formation of the unsaturated 1,4 dicarbonyl derivatives from primary nitroalkanes. Overall the high activity, reusability, selectivity, and solvent-less nature of reactions utilizing KF/
Alumina make it an excellent alternative to classic homogeneous bases.

Basic Amine Species Supported on Silica

Aside from basic metals, there are a range of supported organic species able to
provide basicity, of which amine functionality has attracted much interest [129].
The amines in particular vary widely, from imines, phenolates [130] to

Table 3.17 Michael addition of methyl 4-nitrobutanoate with dimethyl maleate [102]

Catalyst	Yield of route A[c] (%)
KF/Basic Alumina	80
KF/Basic Alumina	79[d]
KF/Neutral Alumina	71
Basic Alumina	4
Hydrotalcite MG 70	7
Hydrotalcite MG 50	6
Amberlyst A-21	25
Amberlyst A-27	21
Silica supported-1,5,7-triazabicyclo-[4.4.0]dec-5-ene (TBD)	36
N,N-Diethylpropylamine supported on amorphous silica (KG-60-NEt$_2$)	19

[a]Catalysed with KF/Alumina at 55°C
[b]Catalysed using KF/Alumina at room temperature
[c]After 7 h, solventless reaction conditions
[d]Yield of route B

dialkylaminopyridine type species [131]. Hagiwara et al. discusses the use of amino functionalised silica gel in the Knoevenagel reaction [132]. Out of a range of different silicas of varying morphology, (i.e., pellets/powders) as well as different amines, 3-Aminopropyl supported on powdered silica has been found to be the most active. Typical yields of a range of different functionalized benzaldehydes reacted with ethyl cyanoacetate are in the region 90–99% in between 2–5 h. In addition, water is able to be used as a solvent, thought to allow both substrates to achieve close proximity to each other on the reverse phase silica gel, increasing reaction efficiency. This demonstrates how organic functionalized solid bases have many advantages over their inorganic equivalents. Often they are mild in reactivity, reducing potential side product formation, have high reproducibility, with flexibility in design allowing the catalyst to be "tailored to fit" a wide range of applications.

Future Directions

The value of catalysis in chemistry has been well known for many years with application in larger scale, petro-chemistry especially important and well established in industry. This was emphasized by the emergence in the second half of the twentieth century of more selective, typically zeolitic catalysts in continuous vapor phase processes that helped establish the petrochemical and commodity chemical industries as the engines of the great manufacturing industries of the world. An increasing awareness of the need to reduce the environmental impact of chemical manufacturing has caused a much wider range of industrial sectors to seek process improvements in chemical, pharmaceutical and polymer manufacturing. More efficient and selective processes that give less waste and greater resource efficiency and the avoidance of large quantities of often hazardous traditional reagents including acids, bases, and stoichiometric oxidants are important goals. Here, the transfer of long lifetime, selective, and reusable catalytic technologies from large-scale petrochemical processes to smaller scale, fine, and speciality (including pharmaceutical) chemical manufacturing has become and continues to be very important. It is needed to quickly move away from old, admittedly reliable but often dangerous and almost always very wasteful chemistry in all sectors of manufacturing. In research, it is required to work ever harder to find effective replacements for the great reagents of the twentieth century – aluminum chloride, chromic acid, hydrogen fluoride, caustic soda, and many others – all cheap, versatile in numerous reactions, and readily available yet also all hazardous, unselective, and leading to more waste than product especially after work-up; progress to date has been limited and most processes continue little changed from decades in the past. The use of heterogeneous catalysts is required to extend the use of continuous production and intensive processing so as to get away from inefficient and inherently risky batch manufacturing. Catalysis should be the norm in all chemical manufacturing and not the exception, and the catalysts themselves must have verifiably sound lifecycles and low environmental footprints including ease of recovery and reuse especially when precious metals are employed.

Bibliography

1. Clark JH (1995) Chemistry of waste minimization. Chapman and Hall, Cambridge
2. Clark JH, Rhodes CN (2000) Clean synthesis using porous inorganic solid catalysts and supported reagents. Royal Society of Chemistry, Cambridge
3. Rabo JA, Schoonover MW (2001) Early discoveries in zeolite chemistry and catalysis at Union Carbide and follow-up in industrial catalysis. Appl Catal A 222:261
4. Nagy JB, Bodart P, Hannus I, Kirics I (1998) Synthesis, characterisation and use of zeolitic materials. DecaGen Ltd., Hungary, p165
5. Sing KSW, Everett DH, Haul RAW, Moscou L, Pierotti RA, Rouquérol J, Siemieniewska T (1985) Reporting physisorption data for gas/solid systems with special reference to the determination of surface area and porosity. Pure Appl Chem 57:603
6. Kissin YV (2001) Chemical mechanisms of catalytic cracking over solid acidic catalysts: alkanes and alkenes. Catal Rev 43(1):85
7. Zhuang JQ et al (2004) Solid-state MAS NMR studies on the hydrothermal stability of the zeolite catalysts for residual oil selective catalytic cracking. J Catal 228(1):234
8. Caeiro G, Magnoux P, Lopes JM, Ribeiro FR, Menezes SMC, Costa AF, Cerqueira HS (2006) Stabilization effect of phosphorus on steamed H-MFI zeolites. Appl Catal A 314(2):160
9. Bao X et al (2005) Enhancement on the hydrothermal stability of ZSM-5 zeolites by the cooperation effect of exchanged lanthanum and phosphoric species. J Mol Struct 737(2–3):271

10. Blasco T, Corma A, Martínez-Triguero J (2006) Hydrothermal stabilization of ZSM-5 catalytic-cracking additives by phosphorus addition. J Catal 237(2):267

11. Ding W et al (2007) Understanding the enhancement of catalytic performance for olefin cracking: hydrothermally stable acids in P/HZSM-5. J Catal 248(1):20

12. Barros ZS, Zotin FMZ, Henriques CA (2007) Conversion of natural gas to higher valued products: light olefins production from methanol over ZSM-5 zeolites. Stud Surf Sci Catal 167:255

13. Lu R, Cao Z, Liu X (2008) Catalytic activity of phosphorus and steam modified HZSM-5 and the theoretical selection of phosphorus grafting model. J Nat Gas Chem 17(2):142

14. Gao X et al (2009) Modification of ZSM-5 zeolite for maximizing propylene in fluid catalytic cracking reaction. Catal Commun 10(14):1787

15. Brouwer DM, Hogeveen H (1972) Electrophilic substitutions at alkanes and in alkylcarbonium ions. Prog Phys Org Chem 9:179

16. Guisnet M, Andy P, Boucheffa Y, Gnep NS, Travers C, Benazzi E (1998) Selective isomerization of n-butenes into isobutene over aged H-ferrierite catalyst: nature of the active species. Catal Lett 50:159

17. Santacesaria E, Di Serio M, Cozzolino M, Tesser R (2004) DGMK-conference C4/C5-hydrocarbons: routes to higher value-added products, Munich

18. Seo G et al (1996) Skeletal isomerization of 1-butene over ferrierite and ZSM-5 zeolites: influence of zeolite acidity. Catal Lett 36(3–4):249

19. Mériaudeau P, Tuan VA, Le NH, Szabo G (1997) Selective isomerization of n-Butene into isobutene over deactivated H–Ferrierite catalyst: further investigations. J Catal 169(1):397

20. Asensi MA, Martínez A (1999) Selective isomerization of n-butenes to isobutene on high Si/Al ratio ferrierite in the absence of coke deposits: implications on the reaction mechanism. Appl Catal A 183:155

21. Wichterlova B et al (1999) Effect of bronsted and lewis sites in ferrierites on skeletal isomerization of n-butenes. Appl Catal A 182(2):297

22. Auerbach SM, Carrado KA, Dutta PK (2003) Handbook of zeolite science and technology. Marcel Dekker, p 481

23. Shouro D et al (2000) Mesoporous silica FSM-16 catalysts modified with various oxides for the vapor-phase Beckmann rearrangement of cyclohexanone oxime. Appl Catal A 198(1):275–282

24. Chaudhari K et al (2002) Beckmann rearrangement of cyclohexanone oxime over mesoporous Si-MCM-41 and Al-MCM-41 molecular sieves. J Mol Catal A Chem 177(2):247

25. Zhang Y et al (2005) Beckmann rearrangement of cyclohexanone oxime over Hβ zeolite and Hβ zeolite-supported boride. Catal Commun 6:53

26. Dai LX et al (1997) Development of advanced zeolite catalysts for the vapor phase Beckmann rearrangement of cyclohexanone oxime. Appl Surf Sci 121/122: 335

27. Misono M, Inui T (1999) New catalytic technologies in Japan. Catal Today 51:369

28. Izumi Y et al (2007) Development and Industrialization of the Vapor-Phase Beckmann Rearrangement Process. Bull Chem Soc Jpn 80(7):1280–1287

29. Roffia P, Leofanti G, Cesana A, Mantegazza M, Padovan M, Petrini G, Tonti S, Gervasutti P (1990) Cyclohexanone ammoximation: a break through in the 6-caprolactam production process. Stud Surf Sci Catal 55:43

30. Palkovits R, Schmidt W, Ilhan Y, Erdem-S_enatalar A, Schüth F (2009) Crosslinked TS-1 as stable catalyst for the Beckmann rearrangement of cyclohexanone oxime. Microporous Mesoporous Mater 117:228

31. Kumar R, Rao GN, Ratnasamy P (1989) Influence of the pore geometry of medium pore zeolites ZSM-5, -22, -23, -48 and -50 on shape selectivity in reactions of aromatic hydrocarbons. Stud Surf Sci Catal 49:1141

32. Sotelo JL, Uguina MA, Valverde JL, Serrano DP (1993) Kinetics of toluene alkylation with methanol over magnesium-modified ZSM-5. Ind Eng Chem Res 32:2548

33. Climent MJ, Corma A, Velty A (2004) Synthesis of hyacinth, vanilla, and blossom orange fragrances: the benefit of using zeolites and delaminated zeolites as catalysts. Appl Catal A 263(2):155

34. Climent MJ, Corma A, Garcia H, Guil-Lopez R, Iborra S, Fornés V (2001) Acid–base bifunctional catalysts for the preparation of fine chemicals: synthesis of jasminaldehyde. J Catal 197(2):385

35. Climent MJ, Corma A, Velty A, Susarte M (2000) Zeolites for the production of fine chemicals: synthesis of the fructone fragrancy. J Catal 196(2):345

36. Kantam ML et al (2005) Friedel–Crafts acylation of aromatics and heteroaromatics by beta zeolite. J Mol Catal A Chem 225(1):15

37. Andy et al (2000) Acylation of 2-methoxynaphthalene and isobutylbenzene over zeolite beta. J Catal 192(1):215

38. Heinichen HK, Holderich WF (1999) Acylation of 2-methoxynaphthalene in the presence of modified zeolite HBEA. J Catal 185(2):408

39. Casagrande M, Storaro L, Lenarda M, Ganzerla R (2000) Highly selective Friedel–Crafts acylation of 2-methoxynaphthlene catalyzed by H-BEA zeolite. Appl Catal A 201(2):263

40. Bejblova M, Zilkova N, Cejka J (2008) Transformations of aromatic hydrocarbons over zeolites. Res Chem Intermed 34(5–7):439

41. Guggenheim S, Martin RT (1995) Definition of clay and clay mineral: joint report of the aipea nomenclature and CMS nomenclature committees. Clays Clay Miner 43(2):255

42. Velde B, Meunier A (2008) The origin of clay minerals in soils and weathered rocks. Springer, Berlin Heidelberg, Chapter 1

43. Clark JH, RhodesCN (2000) Clean synthesis using porous inorganic solid catalysts and supported reagents. R Soc Chem, Chapter 3 clay materials, p 38

44. Duc M et al (2005) Sensitivity of the acid–base properties of clays to the methods of preparation and measurement: 1. Literature review. J Colloid Interface Sci 289:139

45. Brown DR, Rhodes CN (1997) Bronsted and lewis acid catalysis with ion-exchange clays. Catal Lett 45(1–2):35

46. Ravichandran J, Lakshmanan CM, Sivasankar B (1996) Acid activated montmorillonite and vermiculite clays as dehydration and cracking catalysts. React Kinet Catal Lett 59(2):301

47. Ravichandran J, Sivasankar B (1997) Properties and catalytic activity of acid-modified montmorillonite and vermiculite. Clays Clay Miner 45(6):854

48. Kloprogge JT, Duong LV, Frost RL (2005) A review of the synthesis and characterization of pillared clays and related porous materials for cracking of vegetable oil to produce biofuels. Environ Geol 47:967

49. Reddy CR et al (2007) Surface acidity study of Mn+−montmorillonite clay catalysts by FT-IR spectroscopy: correlation with esterification activity. Catal Commun 8:241

50. Reddy CR et al (2004) Synthesis of phenylacetates using aluminium-exchanged montmorillonite clay catalyst. J Mol Catal A Chem 223(1):117

51. da Silva et al (2009) Etherification of glycerol with benzyl alcohol catalyzed by solid acids. J Braz Chem Soc. 20(2):201

52. Salmon M, Zavala N, Martinez M, Miranda R, Cruz R, Cardenas J, Gavino R, Cabrera A (1994) Cyclic and linear oligomerization reaction of 3, 4, 5-trimethoxybenzyl alcohol with a bentonite-clay. Tetrahedron Lett 35(32):5797

53. Sabu KR, Sukumar R, Lalithambika M (1993) Acidic properties and catalytic activity of natural kaolinitic clays for Friedel–Crafts alkylation. Bull Chem Soc Jpn 66:3535

54. Okada S, Tanaka K, Nakadaira Y, Nakagawa N (1992) Selective Friedel–Crafts alkylation on a vermiculite, a highly active natural clay mineral with Lewis acid sites. Bull Chem Soc Jpn 65:2833

55. Laszlo P, Mathy A (1987) Catalysis of Friedel-Crafts alkylation by a montmorillonite doped with transition-metal cations. Helv Chim Acta 70(3):577

56. Clark JH, Macquarrie DJ (1997) Heterogeneous catalysis in liquid phase transformations of importance in the industrial preparation of fine chemicals. Org Process Res Dev 1:149

57. Narayanan S, Deshpande K (2000) Aniline alkylation over solid acid catalysts. Appl Catal A 199(1):1
58. Smith K, Ewart GM, El-Hiti GA, Randlesb KR (2004) Study of regioselective methanesulfonylation of simple aromatics with methanesulfonic anhydride in the presence of zeolite catalysts. Org Bimol Chem 2:3150
59. Laidlawa P, Bethell D, Brown SM, Watson G, Willock DJ, Hutchings GJ (2002) Sulfonylation of substituted benzenes using Zn-exchanged zeolites. J Mol Catal A Chem 178(1):205
60. Choudary BM, Chowdari NS, Kantam ML, Kannan R (1999) Fe(III) exchanged montmorillonite: a mild and ecofriendly catalyst for sulfonylation of aromatics. Tetrahedron Lett 40:2859
61. Choudary BM, Chowdari NS, Kantam ML (2000) Friedel–Crafts sulfonylation of aromatics catalysed by solid acids: an eco-friendly route for sulfone synthesis. J Chem Soc Perkin Trans 1:2689
62. Sharma SK, Parikh PA, Jasra RV (2007) Solvent free aldol condensation of propanal to 2-methylpentenal using solid base catalysts. J Mol Catal A Chem 278(1):135
63. Cavani F, Trifiro F, Vaccari A (1991) Hydrotalcite-type anionic clays: preparation, properties and applications. Catal Today 11:173
64. Clark JH (1994) Catalysis of organic reactions by supported inorganic reagents. VCH Publishers Inc, New York
65. Salehi P, Ali Zolfigol M, Shirini F, Baghbanzadeh M (2006) Silica sulfuric acid and silica chloride as efficient reagents for organic reactions. Curr Org Chem 10(17):2171
66. Li Z, Liu J, Gong X, Mao X, Sun X, Zhao Z (2008) Silica sulfuric acid-catalyzed expeditious environment-friendly hydrolysis of carboxylic acid esters under microwave irradiation. Chem Pap 62(6):630
67. Mobinikhaledi A, Foroughifar N, Fard MAB, Moghanian H, Ebrahimi S, Kalhor M (2009) Efficient one-pot synthesis of polyhydroquinoline derivatives using silica sulfuric acid as a heterogeneous and reusable catalyst under conventional heating and energy-saving microwave irradiation. Synth Commun 39:1166
68. Zarei A, Hajipour AR, Khazdooz L, Mirjalili BF, Chermahini AN (2009) Rapid and efficient diazotization and diazo coupling reactions on silica sulfuric acid under solvent-free conditions. Dyes Pigm 81(1):240
69. Chen X, She J, Shang Z, Wu J, Zhang P (2009) Room-temperature synthesis of pyrazoles, diazepines, β-enaminones, and β-enamino esters using silica-supported sulfuric acid as a reusable catalyst under solvent-free conditions. Synth Commun 39:947
70. Wang Y, Yuan Y, Guo S (2009) Silica sulfuric acid promotes aza-Michael addition reactions under solvent-free condition as a heterogeneous and reusable catalyst. Molecules 14:4779
71. Shobha D, Chari MA, Mukkanti K, Ahn KH (2009) Silica gel-supported sulfuric acid catalyzed synthesis of 1, 5-benzodiazepine derivatives. J Heterocycl Chem 46(5):1028
72. Li J, Meng X, Bai B, Sun M (2010) An efficient deprotection of oximes to carbonyls catalyzed by silica sulfuric acid in water under ultrasound irradiation. Ultrason Sonochem 17:14
73. Yang J, Dang N, Chang Y (2009) Silica sulfuric acid as a recyclable catalyst for a one-pot synthesis of α-aminophosphonates in solvent-free conditions. Lett Org Chem 6(6):470
74. Karam A, Gu Y, Jérôme F, Douliezb J, Barrault J (2007) Significant enhancement on selectivity in silica supported sulfonic acids catalyzed reactions. Chem Comm 22:2222
75. Dabiri M et al (2007) Silica sulfuric acid: an efficient and versatile acidic catalyst for the rapid and ecofriendly synthesis of 1,3,4-oxadiazoles at ambient temperature. Synth Commun 37:1201
76. Shaabani A, Rahmati A (2006) Silica sulfuric acid as an efficient and recoverable catalyst for the synthesis of trisubstituted imidazoles. J Mol Catal A Chem 249(1):246
77. Reddy BM, Patil MK (2009) Organic syntheses and transformations catalyzed by sulfated zirconia. Chem Rev 109(6):2185

78. Reddy BM, Sreekanth PM, Lakshmanan P (2005) Sulfated zirconia as an efficient catalyst for organic synthesis and transformation reactions. J Mol Catal A Chem 237(1):93

79. Deutsch J, Trunschke A, Müller D, Quaschning V, Kemnitz E, Lieske H (2004) Acetylation and benzoylation of various aromatics on sulfated zirconia. J Mol Catal A Chem 207(1):51

80. Deutsch J, Prescott HA, Müller D, Kemnitz E, Lieske H (2005) Acylation of naphthalenes and anthracene on sulfated zirconia. J Catal 231(2):269

81. Zane F, Melada S, Signoretto M, Pinna F (2006) Active and recyclable sulphated zirconia catalysts for the acylation of aromatic compounds. Appl Catal A 299:137

82. Hino M, Arata K (1980) Synthesis of solid superacid catalyst with acid strength of H_0?–16.04. J Chem Soc, Chem Commun (18):851

83. Paukshtis EA, Shmachkova VP, Kotsarenko NS (2000) Acidic properties of sulfated zirconia. React Kinet Catal Lett 71(2):385

84. Clark JH (2002) Solid acids for green chemistry. Acc Chem Res 35:791

85. Ratnama KJ, Reddya RS, Sekhar NS, Kantama ML, Figueras F (2007) Sulphated zirconia catalyzed acylation of phenols, alcohols and amines under solvent free conditions. J Mol Catal A Chem 276(1):230

86. Comerford JW, Clark JH, Macquarrie DJ, Breeden SW (2009) Clean, reusable and low cost heterogeneous catalyst for amide synthesis. Chem Commun 14(18):2562

87. Clark JH, Macquarrie DJ (2002) Handbook of green chemistry and technology. Blackwell Science Ltd, Chapter 13, Green Catalysts for Industry

88. Li Z et al (2007) Silica-supported aluminum chloride: a recyclable and reusable catalyst for one-pot three-component Mannich-type reactions. J Mol Catal A Chem 272(1):132

89. Choudhary VR, Jana SK, Kiran BR (2000) Highly active Si-MCM-41-supported Ga_2O_3 and In_2O_3 catalysts for friedel-crafts-type benzylation and acylation reactions in the presence or absence of moisture. J Catal 192(2):257

90. Choudhary VR, Jana SK (2002) Acylation of aromatic compounds using moisture insensitive mesoporous Si-MCM-41 supported Ga_2O_3 catalyst. Synth Commun 32(18):2843

91. Choudhary VR, Jana SK, Patil NS (2001) Acylation of benzene over clay and mesoporous Si-MCM-41 supported $InCl_3$, $GaCl_3$ and $ZnCl_2$ catalysts. Catal Lett 76(3):235

92. Choudhary VR, Patil KY, Jana SK (2004) Acylation of aromatic alcohols and phenols over $InCl_3$/montmorillonite K-10 catalysts. J Chem Sci 116(3):175

93. Derouane EG, Dillon CJ, Bethell D, Derouane-Abd Hamid SB (1999) Zeolite catalysts as solid solvents in fine chemicals synthesis: 1. catalyst deactivation in the Friedel–Crafts acetylation of anisole. J Catal 187(1):209

94. Derouane EG, Crehan G, Dillon CJ, Bethell D, He H, Derouane-Abd Hamid SB (2000) Zeolite catalysts as solid solvents in fine chemicals synthesis: 2. competitive adsorption of the reactants and products in the Friedel–Crafts acetylations of anisole and toluene. J Catal 194(2):410

95. Yadav GD, George G (2006) Friedel–Crafts acylation of anisole with propionic anhydride over mesoporous superacid catalyst UDCaT-5. Microporous Mesoporous Mater 96(1–3):36

96. Ishitani H, Naito H, Iwamoto M (2008) Fridel-Crafts acylation of anisole with carboxylic anhydrides of large molecular sizes on mesoporous silica catalyst. Catal Lett 120(1–2):14

97. Sarvari MH, Sharghi H (2005) Solvent-free catalytic Friedel-Crafts acylation of aromatic compounds with carboxylic acids by using a novel heterogeneous catalyst system: p-toluenesulfonic acid/graphite. Helv Chim Acta 88:2282

98. Wagholikar SG, Niphadkar PS, Mayadevi S, Sivasanker S (2007) Acylation of anisole with long-chain carboxylic acids over wide pore zeolites. Appl Catal A 317(2):250

99. Zarei A, Hajipour AR, Khazdooz L (2008) Friedel–Crafts acylation of aromatic compounds with carboxylic acids in the presence of P_2O_5/SiO_2 under heterogeneous conditions. Tetrahedron Lett 49:6715

100. Clark JH (1980) Fluoride ion as a base in organic synthesis. Chem Rev 80:429

101. Blass BE (2002) KF/Al_2O_3 mediated organic synthesis. Tetrahedron 58:9301

102. Handa H, Baba T, Sugisawa H, Ono Y (1998) Highly efficient self-condensation of benzaldehyde to benzyl benzoate over KF-loaded alumina. J Mol Catal A Chem 134(1–3):171

103. Kabashima H, Tsuji H, Nakatab S, Tanaka Y, Hattori H (2000) Activity for base-catalyzed reactions and characterization of alumina-supported KF catalysts. Appl Catal A 194–195:227

104. Tsuji H, Kabashima H, Kita H, Hattori H (1995) Thermal activation of KF/alumina catalyst for double bond isomerization and Michael addition. React Kinet Catal Lett 56(2):363

105. Kochkar H, Clacens JM, Figueras F (2002) Isomerization of styrene epoxide on basic solids. Catal Lett 78(1–4):91

106. Nakano Y, Niki S, Kinouchi S, Miyamae H, Igarashi M (1992) Knoevenagel reaction of malononitrile with acetone followed by double cyclization catalyzed by KF-coated alumina in aqueous solution. Bull Chem Soc Jpn 65(11):2934

107. Wang WC, Wang D, Forray C, Vaysse PJJ, Brancheko TA, Gluchowski C (1994) A convenient synthesis of 2-amino-2-oxazolines and their pharmacological evaluation at cloned human α adrenergic receptors. Bioorg Med Chem Lett 4(19):2317

108. Yamawaki J, Ando T, Hanafusa T (1981) N-Alkylation of amides and N-heterocycles with potassium fluoride on alumina. Chem Lett 1143–1146

109. Yamawaki J, Ando T (1980) Potassium fluoride on alumina as a base for crown ether synthesis. Chem Lett 9(5):533–536

110. Sawyer JS, Schmittling EA (1993) Synthesis of diaryl ethers, diaryl thioethers, and diarylamines mediated by potassium fluoride-alumina and 18-crown-6. J Org Chem 58(12):3229

111. Yadav VK, Kapoor KK (1996) KF adsorbed on alumina effectively promotes the epoxidation of electron deficient alkenes by anhydrous t-BuOOH. Tetrahedron 52:3659

112. Moghaddam FM, Bardajee GR, Veranlou ROC (2005) KF/Al$_2$O$_3$-mediated Michael addition of thiols to electron-deficient olefins. Synth Commun 35(18):2427

113. Villemin D, Alloum AB (1992) Potassium fluoride on alumina: an easy synthesis of 4-alkylidene-2-thione-1, 3-oxathiolanes from α-acetylenic alcohols. Synth Commun 22:1351

114. Villemin D, Hachemi M, Lalaoui M (1996) Potassium fluoride on alumina: synthesis of O-aryl N, N-dimethylthiocarbamates and their rearrangement into S-aryl N, N-dimethyl-thiocarbamates under microwave irradiation. Synth Commun 26(13):2461

115. Kawanami Y, Yuasa H, Toriyama F, Yoshida S, Baba T (2003) Addition of silanes to benzaldehyde catalyzed by KF loaded on alumina. Catal Commun 4:455

116. Baba T, Kato A, Yuasa H, Toriyama F, Handa H, Ono Y (1998) New Si-C bond forming reactions over solid-base catalysts. Catal Today 44:271

117. Clark JH, Cork DG, Robertson MS (1983) Fluoride ion catalysed Michael reactions. Chem Lett 12(8):1145

118. Campelo JM, Climent MS, Marinas JM (1992) Michael addition of nitromethane to 3-buten-2-one catalyzed by potassium fluoride supported on Al$_2$O$_3$, ZnO, SnO$_2$, sepiolite, AlPO$_4$, AlPO$_4$–Al$_2$O$_3$ and AlPO$_4$–ZnO. React Kinet Catal Lett 47:7

119. Kabashima H, Tsuji H, Shibuya T, Hattori H (2000) Michael addition of nitromethane to α, β-unsaturated carbonyl compounds over solid base catalysts. J Mol Catal A Chem 155(1–2):23

120. Wang SH, Wang XS, Shi DQ, Tu SJ (2003) Michael addition reaction of dimedone and chalcone catalyzed by KF/Al$_2$O$_3$. Chin J Org Chem 23(10):1146

121. Figueras F et al (2004) Effect of the support on the basic and catalytic properties of KF. J Catal 221(2):483

122. Tian DB, Zhu J, Zhu JF, Shi YX, Wang JT (2004) Michael addition of alkyl amine to α, β-unsaturated carbonyl compounds catalyzed by KF/Al2O3. Chin Chem Lett 15(8):883

123. Moghaddam FM, Bardajee GR, Taimoory SMD (2006) KF/Al$_2$O$_3$ mediated aza-Michael addition of indoles to electron-deficient olefins. Lett Org Chem 3(2):157

124. Wang X, Quan Z, Wang JK, Zhanga Z, Wang M (2006) A practical and green approach toward synthesis of N3-substituted dihydropyrimidinones: using Aza-Michael addition reaction catalyzed by KF/Al$_2$O$_3$. Bioorg Med Chem Lett 16(17):4592

125. Lenardão EJ, Ferreira PC, Jacob RG, Perin G, Leiteb FPL (2007) Solvent-free conjugated addition of thiols to citral using KF/alumina: preparation of 3-thioorganylcitronellals, potential antimicrobial agents. Tetrahedron Lett 48:6763

126. Lenardão EJ, Trecha DO, Ferreira PC, Jacob RG, Perin G (2009) Green Michael addition of thiols to electron deficient alkenes using KF/alumina and recyclable solvent or solvent-free conditions. J Braz Chem Soc 20(1):93

127. Clark JH, Farmer TJ, Macquarrie DJ (2009) The derivatization of bioplatform molecules by using KF/Alumina catalysis. ChemSusChem 2(11):1025

128. Ballinia R, Palmieri A (2006) Potassium fluoride/basic alumina as far superior heterogeneous catalyst for the chemoselective conjugate addition of nitroalkanes to electron-poor alkenes having two electron withdrawing groups in α- and β-positions. Adv Synth Catal 348(10):1154

129. Macquarrie DJ (2009) Organically modified micelle templated silicas in green chemistry. Top Catal 52(12):1640

130. Utting KA, Macquarri DJ (2000) Silica-supported imines as mild, efficient base catalysts. New J Chem 24:591

131. Motokura K, Tomita M, Tada M, Iwasawa Y (2009) Michael reactions catalyzed by basic alkylamines and dialkylaminopyridine immobilized on acidic silica-alumina surfaces. Top Catal 52:579

132. Isobe K, Hoshi T, Suzuki T, Hagiwara H (2005) Knoevenagel reaction in water catalyzed by amine supported on silica gel. Mol Divers 9(4):317

Chapter 4
Green Chemistry Metrics: Material Efficiency and Strategic Synthesis Design

John Andraos

Glossary

Atom economy	Ratio of molecular weight of target molecule to sum of molecular weights of reactants assuming a balanced chemical equation.
B/M	Ratio of sum of number of target bonds made in a synthesis plan to total number of reaction steps.
By-product of a reaction	A product formed in a reaction between reagents as a direct mechanistic consequence of producing the target product assuming a balanced chemical equation that accounts for the production of that target product.
Degree of asymmetry	A parameter determined from the shape of a synthesis tree diagram for a synthesis plan that describes the degree of skewness of a triangle whose vertices are the target product node, the origin, and the node for the last reagent along the ordinate axis.
Degree of convergence	A parameter determined from the shape of a synthesis tree diagram for a synthesis plan that describes the ratio of the angle subtended at the actual product node vertex to that at a product node vertex corresponding to the hypothetical case of all reaction substrates in a plan reacting in a single step.

This chapter was originally published as part of the Encyclopedia of Sustainability Science and Technology edited by Robert A. Meyers. DOI:10.1007/978-1-4419-0851-3

J. Andraos (✉)
Department of Chemistry, York University, 4700 Keele Street, Toronto,
ON M3J 1P3, Canada
e-mail: c1000@careerchem.com

P.T. Anastas and J.B. Zimmerman (eds.), *Innovations in Green Chemistry and Green Engineering*, DOI 10.1007/978-1-4614-5817-3_4,
© Springer Science+Business Media New York 2013

E-factor with respect to molecular weight (E_{mw})	Ratio of sum of molecular weights of by-products in a reaction to molecular weight of target product in a given reaction or synthesis plan.
E-factor with respect to mass (E_m)	The ratio of mass of total waste from all sources to mass of target product collected in a given reaction or synthesis plan.
E_{kernel}	Contribution to the total or overall E-factor with respect to mass from reaction by-products, reaction side products, and unreacted starting materials.
E_{excess}	Contribution to the total or overall E-factor with respect to mass from excess reagents.
$E_{auxiliaries}$	Contribution to the total or overall E-factor with respect to mass from auxiliary materials such as reaction solvents, work-up, and purification materials.
$E_{overall}$ (or E_m)	Ratio of mass of total waste generated from all sources to mass of target product collected for a given reaction or synthesis plan.
f^*	Fractional kernel waste contribution from target bond forming reactions.
$f(sac)$	Molecular weight fraction of reagents that absolutely do not end up in the final target molecule structure.
$f(nb)$	Fraction of number of reaction stages that do not form target bonds.
Hypsicity index	A parameter that tracks the oxidation numbers of atoms in target bond forming reactions over the course of a synthesis plan relative to their values in the target molecule.
I	Number of input materials in a synthesis plan.
Kernel mass of waste	Mass of all reagents used in a synthesis reaction or plan minus the mass of target product collected.
Kernel reaction mass efficiency	The ratio of the mass of target product collected to the sum of the stoichiometric masses of all reagents used in a chemical reaction or synthesis plan.
μ_1, First molecular weight moment	Parameter that describes the degree of building up going on from the reagent molecules toward the target molecule over the course of a synthesis plan.
Mass intensity	Ratio of mass of all materials used to mass of target product collected in a given chemical reaction or synthesis plan.
Material recovery parameter	A parameter that describes the mass consumption of all solvents and auxiliary materials used in carrying out a given chemical reaction that may be potentially recoverable by recycling.

M	Number of reaction steps in a synthesis plan.
N	Number of reaction stages in a synthesis plan.
Radial hexagon	A diagram that depicts the overall atom economy, overall reaction yield along the longest branch, overall kernel reaction mass efficiency, molecular weight fraction of reagents in whole or in part that end up in the target molecule, fraction of the kernel waste contribution from target bond forming reactions, and the degree of convergence for a given synthesis plan as a hexagon for easy visualization of the synthesis performance.
Radial pentagon	A diagram that depicts the reaction yield, atom economy, stoichiometric factor, material recovery parameter, and overall reaction mass efficiency for a given chemical reaction as a pentagon for easy visualization of the reaction performance.
Raw material cost	Sum of the costs of all reagents in a synthesis plan calculated using the basis scale in moles of the target product, assuming balanced chemical equations for all reactions in the plan and taking account of excess reagent consumption.
Reaction step	Refers to interval between a given isolated intermediate and the next consecutive isolated intermediate in a synthesis plan.
Reaction yield	Ratio of moles of target product to moles of limiting reagent for a given balanced chemical equation.
Sacrificial reactions	Nonproductive reactions in a synthesis plan that do not form target bonds appearing in the target product structure.
Side product of a reaction	A product formed in a reaction between reagents, usually undesired, that arises from a competing reaction pathway other than the one that produces the intended target product and its associated by-products.
Stoichiometric factor (SF)	A parameter that describes the total amount of excess reagents used in a given chemical reaction relative to the amounts prescribed by its stoichiometrically balanced chemical equation.
Stoichiometric coefficient	An integer appearing before the chemical formula for a reagent in a balanced chemical equation.
Synthesis tree	A diagram that describes all features of a synthesis plan, including number of steps, number of stages, number of branches, number of intermediates,

number of reagents used, and molecular weights of all chemical species.

Target bond map	A chemical structure drawing of the target product of a synthesis plan showing the target bonds made, the step numbers of each target bond, and the set of atoms that correlate directly with the corresponding reagents that ended up in the target molecule.
Target bond forming reactions	Productive reactions in a synthesis plan that result in the formation of bonds that appear in the structure of the target molecule.
Total (overall) mass of waste	The sum of all masses of materials used in synthesis plan minus the mass of the target product collected.
Total (overall) reaction mass efficiency	The ratio of the mass of target product collected to the sum of the masses of all reagents, solvents, and auxiliaries used in a given reaction or synthesis plan.
/UD/	A parameter that tracks the "oxidation length" traversed over the course of a synthesis plan equal to the sum of all the "ups" and "downs" in a hypsicity profile or bar graph beginning with the zeroth reaction stage.

Definition of the Subject and Its Importance

Over the last 2 decades, the topic of "green metrics" has grown rapidly in conjunction with the field of green chemistry. Green metrics promise to provide a rigorous, thorough, and quantitative understanding of material, energy, and cost efficiencies for individual chemical reactions and synthesis plans. Indeed, before the advent of green chemistry, good synthetic strategy and elegance were ill-defined, yet intuitive concepts couched less in quantitative terms and more by subjective ones. The quest for a reliable method of measuring material efficiency or "greenness" of a chemical reaction, synthesis, or process is of fundamental importance in the field of organic synthesis when various routes to a given target molecule are considered for selection. Such a method should be robust in its application to any kind of reaction or plan regardless of complexity. It should standardize the ranking of efficiencies of synthesis plans in an unambiguous and unbiased way. It should be used as a powerful tool in weeding out bad-performing plans quickly and identifying promising candidate plans at the drawing board stage guided by thorough literature investigations even before embarking on any experiment in the laboratory. Above all, it should facilitate the decision-making process for both chemists and non-chemists by allowing on-the-spot precise identification of potential bottlenecks in plans from easy-to-read graphics and by pointing chemists in the right direction for further optimization with confidence. In the long term, with enough examples and pattern recognition among as many plans as possible, it should be possible to understand why "good" plans are good and offer insights into what good strategy entails and how to parameterize it. *The key take home message is that optimization is an iterative exercise that compares the performances of a set of plans to a common target according to some criteria and that true optimization is achieved when the best possible values of material efficiency and synthetic elegance metrics coincide in the same plan.*

Introduction

In the period 2005–2007, algorithms were introduced to assess the material efficiencies of any reaction and any synthesis plan, and they delivered on all of the points mentioned above [1–6]. Their first and most important triumph was that they provided simple connecting relationships between key green metrics that previously were considered as independent entities. The basis of this discovery was the long forgotten law of conservation of mass pronounced by Antoine Lavoisier in 1775 that allows one to write out complete balanced chemical equations for every reaction in a synthesis plan with appropriate stoichiometric coefficients for all reactants and *all* by-products in addition to the target product of interest. The skill of balancing chemical equations has been resurrected after a long period of dormancy in the field of synthetic organic chemistry by the need to precisely identify waste composition and quantify waste production, which are essential if any real progress can be made in the newly emerging field of green chemistry whose chief aim is to minimize waste. In fact, before the advent of green metrics, waste was considered as one big lump of unwanted material, which was calculated simply as the difference between the mass of all input materials used and the mass of the desired output material. Application of green metrics forces the precise itemization of what constitutes waste so that targeted reductions of these components can be made. The mass of waste of any chemical reaction is the sum of the masses of unreacted starting materials, by-products produced as a mechanistic consequence of making the desired target product, side products produced from competing side reactions other than the intended reaction, reaction solvent, all work-up materials, and all purification materials used. Simple first-generation waste reduction strategies target the last three items in the list since they contributed the bulk of the overall mass of waste. Waste reduction strategies targeting the first three items are based on synthesis design and are necessarily more challenging to implement. The connecting green metrics are atom economy (AE) [7–9], environmental E-factor (E) [10–12], and reaction mass efficiency (RME) [13, 14]. All other metrics appearing in the literature [15–21] that pertain to material efficiency can be expressed in terms of these three and are therefore redundant. RME and E describe identically the same thing from the positive and negative points of view, respectively.

Experimental atom economy for a reaction is just another synonym for the kernel RME (reaction yield times atom economy), mass intensity (MI) is just E + 1, and mass index is the reciprocal of RME. Prior arguments that AE had a lower usefulness than RME and E were dispelled by the simple algebraic connecting relationships found; namely, AE = $1/(1 + E_{mw})$ and RME = $1/(1 + E)$, where E_{mw} and E are the E-factors based on reagent molecular weight and mass, respectively. In fact, AE is of fundamental importance in the molecular design phase of synthesis planning even before carrying out any experiments as it describes the intrinsic waste produced in a reaction or plan due to by-products arising out of the nature of the chemical reaction in question. If significant by-products are known to result from reactions in a plan at the outset, then it is inevitable that this will have an additive and cumulative effect on the production of overall waste.

This chapter summarizes the salient features of these algorithms, one for individual reactions and the other for synthesis plans. The reader is referred to prior references by the author for derivations. Numerical computations are greatly facilitated using spreadsheet programs which have been disclosed previously in a detailed investigation of the material efficiencies of the synthesis plans for oseltamivir phosphate a neuraminidase inhibitor used to treat influenza, particularly H5N1 and H1N1 [22]. Here, we illustrate how these algorithms may be implemented by showing complete worked out examples for the analysis of seven literature synthesis plans for the antidepressant pharmaceutical sertraline [23–29]. This compound is unique because its published plans show a steady progression in optimization over the period 1984–2006. This is a very rare situation in the literature where the timeline of publications shows a steady upward rise in synthesis efficiency.

Overview of Algorithms

The main theme and sequence of algorithms used in carrying out green metrics calculations on a given synthesis plan are shown in Fig. 4.1. The first thing is to access the best available literature procedures that disclose the experimental write-ups for the recipes for each reaction in the synthesis plans to the target compound of interest. Next, fully balanced chemical equations for all reactions are written down where all by-products arising from the intended mechanisms are identified and stoichiometric coefficients for all reactants and products are assigned. From these data, radial pentagons are constructed according to the method described in [5] to obtain diagrams that show the AE, reaction yield, stoichiometric factor (SF), and material recovery parameter (i.e., reaction solvent, work-up, and purification material demand) for each reaction according to the master equation given by Eq. 4.1:

$$RME = (\varepsilon)(AE)\left(\frac{1}{SF}\right)(MRP) \qquad (4.1)$$

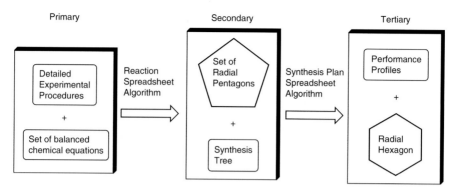

Fig. 4.1 Paradigm flowchart for determining green metrics for any synthesis plan

where

ε is the reaction yield with respect to the limiting reagent in a reaction given by the mole ratio of the target product collected and of the limiting reagent.

AE is the reaction atom economy given by the ratio of the molecular weight of the target product to the sum of molecular weights of all reagents in the balanced chemical equation.

SF is the stoichiometric factor, taking into account the use of excess reagents, and is given by

$$SF = 1 + \frac{\sum excess - masses - of - all - reagents}{\sum stoichiometric - masses - of - all - reagents} \tag{4.2}$$

MRP is the material recovery parameter, taking into account the use of auxiliary materials, and is given by

$$MRP = \frac{1}{1 + \frac{(\varepsilon)(AE)(c + s + \omega)}{m_p(SF)}} \tag{4.3}$$

where m_p is the mass of the collected target product, c is the mass of catalyst used, s is the mass of reaction solvent, and ω is the mass of all other auxiliary materials used in the work-up and purification phases of the reaction.

For ease of computation, the four factors in Eq. 4.1 are written as fractions between 0 and 1 instead of as percentages. Eq. 4.1 basically shows how RME is factored or partitioned into its four constituent contributors. The corresponding radial pentagon gives a clear visual representation of Eq. 4.1 and tells a chemist which of the parameters may be the bottlenecks in attenuating the RME value for a given reaction. The MRP factor is the strongest attenuator because the total mass of auxiliary materials far outweighs the masses of all reagents used. Hence, this was the area that was targeted in first-generation waste reduction when the 12 principles of green chemistry were announced. The radial pentagon algorithm incorporates two check calculations, one for RME using Eq. 4.1 and by simply dividing the mass of collected target product by the sum of the masses of all the input materials used in the procedure, and the other for SF using Eq. 4.2 and from SF = $(\varepsilon)(AE)/(RME)$ where RME, in this case, refers to the reaction mass efficiency calculated when all auxiliary materials are assumed to be recovered, that is, not committed to waste so that MRP is set to 1.

A huge bonus from the radial pentagon analysis is that it picks up all errors that appear in literature procedures. The most common are incorrect reporting of reaction yields, masses of ingredients, and moles of ingredients. Unfortunately, there is less care in the reporting of these key parameters by authors in experimental sections of research papers than expected. Moreover, reviewers and editors of journals do not routinely check for the accuracy of these details, particularly if they are buried in supplementary material, as they naively assume that authors have assumed that responsibility. Patent documents are notorious for such errors as they are often made with intent. Practicing chemists know all too well that such mistakes come to light when someone else embarks on duplicating the reported experiment. The biggest drawback in determining the true RME for a reaction is the lack of full disclosure of all masses of auxiliary materials used in a given procedure. Of particular note are the amounts of solutions used in work-up washes and extractions, mass of drying agent used, and masses of chromatographic materials (solid supports and eluents). If these problems remain unchecked, the successful wide implementation of green chemistry practices will be severely hindered as only cursory guesses and assumptions can be made in such analyses. With missing data, the best that can be accomplished to is to put upper bounds on RME values and lower bounds on E values.

The key set of parameters from the radial pentagon analysis for each reaction that are used as direct inputs in the second spreadsheet algorithm to assess the material efficiency of the entire synthesis plan are as follows: molecular weights of reagents and by-products, molecular weight of product, reaction yield, stoichiometric factor, total mass of excess reagents, mole scale of limiting reagent, mass of reaction solvent, and masses of all other auxiliaries. The use of the second spreadsheet algorithm is greatly facilitated by the construction of a synthesis tree for the plan according to the treatments described in [3, 5] that gives an exact count of the number of reaction stages, reaction steps, input materials, and branches in the plan. It is an excellent bookkeeping and proofreading tool when reading research papers describing synthesis plans. A great advantage of using synthesis trees is that they give a visual representation of an entire plan in one simple diagram. Of particular note is the ease of analyzing convergent plans with multiple branches. Green metrics are determined from a synthesis tree diagram by reading it in the reverse sense from the target product node toward the input reagent nodes. In this way, the mole scales of all reagents are normalized relative to the target product, which is the basis scale for the entire plan. The method uses a simple connect-the-dots approach to determine the appropriate chain of reaction yields required to determine the mole scale of each reagent in a plan originating from any branch. It makes sense to follow the reverse approach since the target product node is the only one that is common to all branches in the tree diagram.

The algorithm for the computation of overall AE, RME, and E for any synthesis plan of any degree of complexity is given by a sequence of computations beginning with the last step ($j = n$) and working toward the first step ($j = 1$) and is described elsewhere [22]. The overall E-factor may be partitioned into its three components as shown explicitly below:

$$E_{total} = E_{kernel} + E_{excess} + E_{auxiliaries} \qquad (4.4)$$

where,

$$E_{kernel} = \frac{1}{p_n} \sum_j \left(\frac{1}{\prod\limits_k^{n \to j} \varepsilon_k} \right) \left(\frac{p_j}{(AE)_j} \right) \left[1 - \varepsilon_j (AE)_j \right] \qquad (4.5)$$

$$E_{excess} = \frac{1}{p_n} \sum_j \left(\frac{1}{\prod\limits_k^{n \to j} \varepsilon_k} \right) \left(\frac{p_j}{(AE)_j} \right) \left[(SF)_j - 1 \right] \qquad (4.6)$$

$$E_{auxiliaries} = \frac{1}{p_n} \sum_j \left(\frac{1}{\prod\limits_k^{n \to j} \varepsilon_k} \right) \left(\frac{c_j + s_j + \omega_j}{x_j^*} \right) \qquad (4.7)$$

Symbol definitions:

x is number of moles of target product in synthesis plan, where x is set to 1 mole.

$\prod\limits_k^{n \to j} \varepsilon_k$ is the multiplicative chain of reaction yields connecting the target product node to the reactant nodes for step j as per synthesis tree diagram read from right to left (i.e., in the direction $n \to j$).

p_j is the molecular weight of product of step j.

ε_j is the reaction yield with respect to limiting reagent for step j.

$(AE)_j$ is the atom economy for step j.

$(SF)_j$ is the stoichiometric factor for step j that accounts for excess reagents used in that step.

$(c_j + s_j + \omega_j)$ is the sum of masses of auxiliary materials used in step j, namely the mass of catalyst, mass of reaction solvent, and mass of all other post-reaction materials used in the work-up and purification phases.

x_j^* is the experimental mole scale of limiting reagent in step j as reported in an experimental procedure.

The summation runs over the n reaction steps. A step begins with an isolated intermediate and ends with the following next isolated intermediate.

The relative magnitudes of these three components break down into the following order: $E_{auxiliary} >>> E_{excess} > E_{kernel}$. For synthesis plans, the hierarchy of contributor metrics that control overall material efficiency according to RME and E at the kernel molecular design stage is in descending order of the number of steps (n), reaction yields (ε), and atom economies (AE). When taking into account all materials used, the recovery and/or minimization of auxiliary materials in the work-up and purification phases plays the largest role in minimizing waste and, hence, the true overall magnitude of RME and E for a synthesis plan. Key assumptions implicit in the use of the algorithm described above are that an intermediate product collected in a given step is entirely committed as a reagent in the next step so that a true mass throughput can be assessed, and that reaction yield performances are invariant with mole scale, something that can only be checked by experiment but is a necessary assumption in implementing such an analysis. The results of synthesis plan algorithm are depicted graphically as a radial hexagon by direct analogy to the radial pentagon analysis. The key six parameters chosen are each fractions ranging from 0 to 1. For the material efficiency performance, there is overall AE, overall yield along the longest branch, and overall reaction mass efficiency. For the strategy efficiency performance, there is molecular weight fraction of reagents ending up as part of the structure of the final target product, the fraction of kernel waste arising from target bond reactions, and degree of convergence. The outer perimeter represents a perfectly green plan. The more the resultant actual hexagon is distorted toward the center, the worse is the plan. Again, strengths and weaknesses in a plan may be seen at once. In addition, the algorithm produces bar graphs that depict profiles of percentage of kernel and percentage of true waste distributions, atom economy, reaction yield, kernel reaction mass efficiency, cumulative mass of waste produced, hypsicity (oxidation level tracking), and target bond reactions per reaction stage. Bottlenecks in the plan may be spotted at once and then appropriate action taken to force optimization in the desired direction.

Strategy Metrics

Parameters that describe and track synthesis strategy for a given plan include: (1) target bond structure maps, (2) target bond forming reaction profiles from which the number of target bonds per reaction step is determined, (3) molecular weight fraction of sacrificial reagents used, and (4) hypsicity (oxidation level) index. A target bond structure map is simply a drawing of the target product structure showing which target bonds are made and at what reaction step.

A target bond is denoted with a heavy connecting line, and its associated circled reaction step number is given alongside. From such a diagram, it is possible to trace the origin of each atom in the target structure back to the associated atoms in the reagents used. In effect, the set of reagent atoms is mapped onto the set of atoms comprising the target structure. From such a map, it is possible to determine the molecular weight fraction of sacrificial reagents whose atoms never get incorporated in the final target structure according to

$$f(\text{sac}) = 1 - \frac{\sum MW_{\text{reagents in whole or in part ending up in target product}}}{\sum MW_{\text{all reagents}}}$$

$$= 1 - \frac{(AE)_{\text{overall}} \sum MW_{\text{reagents in whole or in part ending up in target product}}}{MW_{\text{target product}}}$$

Sacrificial reagents include those that serve as protecting groups, those that change the electronic states of key atoms so that skeletal building bond forming reactions are possible, those that are used to control stereochemistry, those that are used in substitution reactions to switch poor leaving groups into better ones, and those that are reducing or oxidizing agents that are used in *subtractive* redox reactions, that is, where oxygen atoms or hydrogen atoms are *removed* from a structure. The condition concerning redox reactions is important as it is *additive* redox reactions that are desirable to reduce f(sac) since they contribute oxygen or hydrogen atoms to the target structure. It is obvious that the ultimate goal is to minimize the magnitude of f(sac). One can deduce readily from the above equation that when the overall atom economy is unity, that is, when all atoms in all of the reagents used end up in the target structure, the two sums in the first part of the above equation become equal to each other, and hence, f(sac) is equal to zero. Also, from the target structure map, it is possible to construct a bar graph of the number of target bonds made versus the reaction step count. Gaps in such bar graphs coincide with the use of sacrificial reagents in those steps and bars indicate productive steps. The number of target bonds made per step may be used as an indicator of synthetic efficiency and elegance since good synthesis strategies are characterized by fewer steps and the accomplishment of more target bonds made per step.

The target structure map also provides the set of atoms in the target structure that are involved in bond-making and bond-breaking processes throughout the synthesis. These atoms are precisely the ones connected by the heavy lines in the target structure map. It is possible to trace the oxidation numbers of these atoms from the target structure back to the progenitor reagent used via the intervening intermediate products. In a manner similar to the determination of the molecular weight first moment building up parameter, the difference between the oxidation number of an atom in an intermediate structure and that same atom in the final target product is determined for each atom involved in this special set of atoms as a function of reaction stage. This idea of tracking the changes in oxidation state, or hypsicity, (Gk: *hypsos*, meaning level or height) of key atoms involved in bond-making and bond-breaking steps was introduced by Hendrickson [30]. He proposed that good synthesis plans aim for the *isohypsic* condition, which is characterized by a zero net change in oxidation state of all atoms of starting materials and intermediates involved until the target product is reached. This can be achieved by designing synthesis plans that eliminate redox reactions entirely as it is consistent with the conclusion that such reactions are to be minimized in a plan because they are the most material inefficient class of chemical reaction and therefore contribute to significant attenuations in kernel and global RMEs. If these cannot be avoided due to practical considerations, then the next best thing to achieve the isohypsic condition is to strategically sequence redox reactions in such a way that for every increase in oxidation level of an atom occurring in a step, it is matched by a concomitant decrease in oxidation level of equal magnitude in the next step, or vice versa. This cuts down on the accumulation of excess gains or losses in oxidation level of atoms, as the case may be, in starting materials and intermediates with respect to the oxidation levels of those atoms in the final target molecule over the course of the synthesis. This second option of achieving the isohypsic condition is purely due to an algebraic cancellation of all the "ups" and "downs" in oxidation state changes and, hence, will coincide with appreciably higher kernel E-factor values and lower kernel RME values for such plans since generally redox reactions are least material efficient. The first option, however, will coincide with lower kernel E-factor values and higher kernel RME values since redox reactions would be completely eliminated.

Formally, we may define a hypsicity index, HI, as

$$
HI = \frac{\sum\limits_{stages,j}\left[\sum\limits_{atoms,i}\left[(Ox)_{stage,j}^{atom,i}-(Ox)_{stage,N}^{atom,i}\right]\right]}{N+1} = \frac{\sum\limits_{stages,j}\Delta_j}{N+1}
$$

where (Ox) represents the relevant oxidation number of an atom. If HI is zero, then the synthesis is *isohypsic*. If HI is positive valued, then to get to the target molecule a net reduction is required over the course of the synthesis since an accumulated gain in oxidation level has resulted. Such a condition is termed *hyperhypsic*, by analogy with the term *hyperchromic*, which describes increases in intensities of absorption bands in spectroscopy. Conversely, if HI is negative valued, then to get to the target molecule a net oxidation is required over the course of the synthesis since an accumulated loss in oxidation level has resulted. Such a condition is termed *hypohypsic*, again by analogy with the term *hypochromic*. It is important to note that changes in oxidation number can occur for atoms in reactions that are not formally classified as reductions or oxidations with respect to the substrate of interest. A good example of this is the Grignard reaction which is classified as a carbon–carbon bond forming reaction and yet involves a formal oxidation with respect to magnesium in the preparation of the Grignard reagent. Another is electrophilic aromatic substitution, which begins with an oxidation state of -1 for the ArC-H carbon atom, which then increases to $+1$ when hydrogen is substituted for chlorine, for example. The hypsicity index therefore accounts for all such changes in oxidation numbers of atoms regardless of the reaction type.

The following sequence of steps may be followed to determine HI for a synthesis plan:

1. Enumerate atoms in the target structure that are only involved in the building up process from corresponding starting materials according to the structure map. This set of atoms defines those that are involved in bonding changes occurring in the relevant reaction steps.
2. Work backwards intermediate by intermediate to trace the oxidation numbers of the above set of atoms back to original starting materials as appropriate following the reaction stages back to the zeroth stage.
3. For each key atom, i, in each reaction stage, j, determine the difference in oxidation number of that atom with respect to what it is in the final target structure. Hence,

$$(Ox)_{stage,j}^{atom,i} - (Ox)_{stage,N}^{atom,i}$$

4. Sum the differences determined in step (3) over all key atoms in stage j. This yields the term $\sum\limits_{atoms,i} \left[(Ox)_{stage,j}^{atom,i} - (Ox)_{stage,N}^{atom,i} \right] = \Delta_j$.
5. Finally, take the sum $\sum\limits_{stages,j} \Delta_j$ over the number of stages and divide by N + 1

accounting for the extra zeroth reaction stage.

From the above discussion, for a material efficient synthesis plan, it is not a sufficient condition to just aim for an HI value of zero. What matters is that as many of the atoms as possible in oxidizing and reducing agents end up in the target structure. These can only arise from oxidation and reduction reactions that are of the *additive* type where oxygen and hydrogen atoms are added to a structure in a reaction and remain there until the final target structure is reached, and not of the *subtractive* types as described earlier in the discussion of sacrificial reagents. So, it is possible to have HI values that are strongly positive or negative and still be material efficient so long as those key atoms are incorporated as part of the final target structure. It is preferable to have a hypsicity profile that exhibits either a steadily increasing or a steadily decreasing oxidation level rather than an undulating one. Increases and decreases in oxidation levels parallel the types of redox reactions employed in a plan. This will of course depend on the oxidation levels of atoms in the selected reagents used throughout the synthesis. Hence, careful correlation of HI values and overall AE, kernel RME, and f(sac) values need to be made to understand the synergy between material efficiency and oxidation level changes. It is impossible to make inferences solely on the basis of the magnitude of the HI. One needs to examine the shape or distribution of the bar graph that tracks the changes in oxidation level as function of reaction stage for a given synthesis plan.

An Example: Sertaline Plans

Seven synthesis plans for sertraline were examined – four industrial (Pfizer G1, Pfizer G2, Pfizer G3, and Gedeon Richter) and three academic (Buchwald, Lautens, and Zhao). Table 4.1 summarizes the material efficiency metric performances, Table 4.2 summarizes the E-factor breakdown, and Table 4.3 summarizes the strategy efficiency metric performances. From these data, it appears that the overall best material efficient synthesis plan is that of Zhao since it has the highest overall yield, atom economy, and overall kernel RME. Consequently, it produces the least kernel and overall masses of waste. It also uses the least number of input materials. The Pfizer plans show a nice progression from G1 to G3 with a sevenfold decrease in overall E-factor from G1 to G2 and a further 2.5-fold decrease from G2 to G3. The Zhao, Pfizer G3, Gedeon Richter, and Buchwald have the least number of steps at five. The Buchwald and Pfizer G3 plans have all of their kernel waste coming from target bond

Table 4.1 Summary of material efficiency metrics for the seven plans to produce sertraline ranked in descending order according to percentage of kernel reaction mass efficiency

Plan	Type	N^a	n^b	I^c	$f*^d$	Overall yield (%)	AE (%)	Kernel RME (%)
Zhao[e]	linear	5	5	9	0.541	35.0	43.0	18.1
Pfizer G3[f]	linear	5	5	14	1	25.0	39.0	11.2
Gedeon Richter[f]	linear	5	5	11	0.639	20.5	36.6	9.4
Buchwald[e]	linear	5	5	12	1	19.2	38.9	8.6
Pfizer G2[f]	linear	8	8	17	0.894	8.7	34.9	4.3
Lautens[e]	convergent	11	13	28	0.389	14.1	7.1	1.9
Pfizer G1[f]	linear	10	10	18	0.731	1.8	26.7	1

[a]Number of reaction stages
[b]Number of reaction steps
[c]Number of input reagents
[d]Fractional kernel waste contribution from target bond forming reactions
[e]Target is (+)-sertraline
[f]Target is (+)-sertraline hydrochloride

Table 4.2 Summary of mass of waste and E-factor breakdown for seven plans to produce sertraline ranked in ascending order[a]

Plan	E_{kernel}	E_{excess}	$E_{auxiliary}$	E_{total}	Kernel mass of waste (kg)	Total mass of waste (kg)
Zhao[b]	4.54	89.21	64.64	158.38	1.4	48.4
Pfizer G3[c]	7.92	55.21	226.23	289.76	2.7	99.2
Gedeon Richter[c]	9.59	42.07	339.53	391.19	3.3	133.9
Buchwald[b]	10.56	90.25	352.24	453.05	3.2	138.6
Pfizer G2[c]	22.34	163.56	539.67	725.56	7.6	248.4
Lautens[b]	51.99	114.58	2424.98	2591.55	15.9	792.8
Pfizer G1[c]	103.95	350.30	4536.58	4990.82	35.6	1708.6

[a]Basis is 1 mole target product
[b]Target is (+)-sertraline
[c]Target is (+)-sertraline hydrochloride

Table 4.3 Summary of strategy efficiency metrics for the seven plans to produce sertraline

Plan	$\mu 1$ (g/mol)	β	δ	f(sac)	HI	B/M	f(nb)	|UD|
Zhao[a]	−60.98	0.738	0.435	0.495	+1	1.20	0.33	8
Pfizer G3[b]	−96.77	0.781	0.508	0.540	+2	1.60	0	12
Gedeon Richter[b]	−58.69	0.734	0.476	0.521	+1.33	1.40	0.20	8
Buchwald[a]	−85.98	0.760	0.486	0.358	+2.33	1.40	0	12
Pfizer G2[b]	−72.83	0.839	0.426	0.562	+1.56	1.00	0.25	12
Lautens[a]	−15.81	0.776	0.471	0.764	+1	0.92	0.45	24
Pfizer G1[b]	−50.04	0.848	0.389	0.579	+1.64	0.90	0.30	10

[a]Target is (+)-sertraline
[b]Target is (+)-sertraline hydrochloride

forming reactions; whereas, the Lautens plan has most of its kernel waste coming from sacrificial reactions. The Lautens plan also has the worst overall atom economy. In terms of E-factor contributors, the Zhao plan has the least E-auxiliary whereas the Pfizer G3 plan has the least E_{kernel} and the Gedeon Richter plan has the least E_{excess} contributions. The overall most wasteful Pfizer G1 plan has the highest contributors in all three categories. From a strategy perspective, it is less clear-cut to decide the overall best plan. The Pfizer G3, Buchwald, and Zhao plans show the greatest degree of building up from starting materials toward product throughout the course of their syntheses. The Gedeon Richter plan has the least degree of asymmetry. The Pfizer G3 plan has the highest degree of convergence. The Pfizer G3 plan has highest number of target bonds made per reaction stage. The Pfizer G3 and Buchwald plans have all reaction stages producing at least one target bond; whereas, almost half the reaction stages in the Lautens plan do not produce target bonds. The Zhao and Gedeon Richter plans have the least number of oxidation changes throughout the course of their synthesis plans, whereas, the Lautens plan has the most.

Figures 4.2 and 4.3 show the synthesis schemes for the best overall performing Zhao and Pfizer G3 plans. The Zhao plan involves a Friedel–Crafts alkylation in step 1 to give a racemic product, an asymmetric ketone reduction in step 2 with a chiral proline organocatalyst to give an optically active trans-tetralol intermediate, alcohol oxidation in step 3, imination in step 4, and catalytic hydrogenation in step 5. The Pfizer G3 plan involves a Friedel–Crafts acylation in step 1, a tandem reduction-lactonization sequence in step 2, a tandem lactone ring opening Friedel–Crafts acylation in step 3 to create the 1-tetralone ring system, a telescoped imination-hydrogenation-diastereomeric salt resolution sequence in step 4, and, finally, isolation of the hydrochloride salt of the correct cis-amine product in step 5. The corresponding synthesis tree diagrams are shown in Fig. 4.4. Both plans are linear.

Figures 4.5 and 4.6 show the radial pentagons for each reaction in the Zhao and Pfizer G3 plans. For the Zhao plan, the bottlenecks are the low atom economy in step 1, the low yield in step 2, the excess reagent usage in steps 1, 2, and 4, and the high auxiliary material usage in steps 1, 3, and 4. For the Pfizer G3 plan, the bottlenecks are the low atom economy in step 1, the low yield in step 4, the high excess reagent usage in steps 1 to 3, and the high auxiliary material consumption in steps 3 to 5.

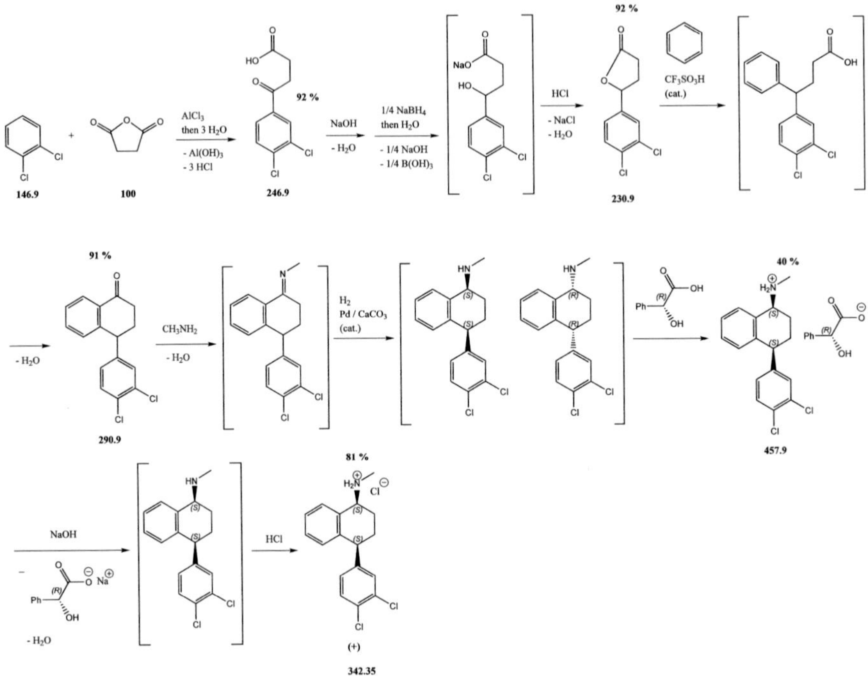

Fig. 4.2 Pfizer G3 synthesis plan

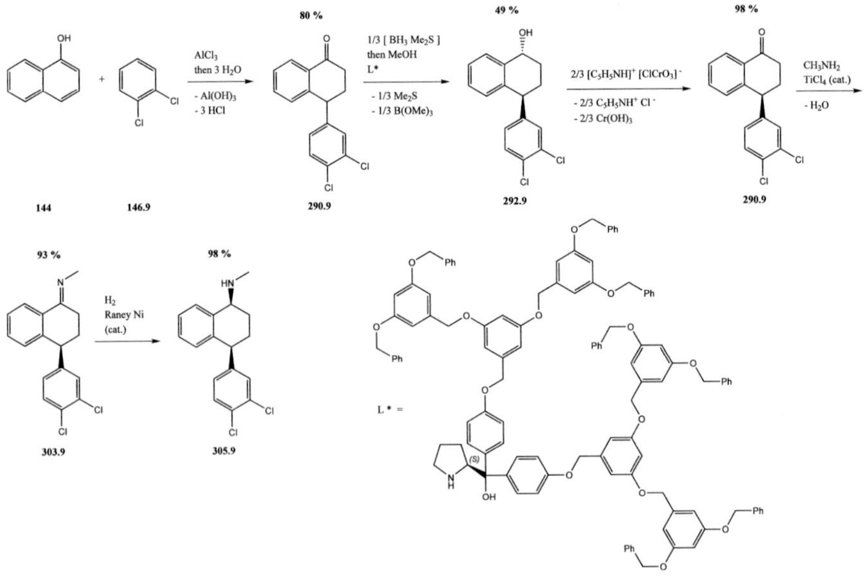

Fig. 4.3 Zhao synthesis plan

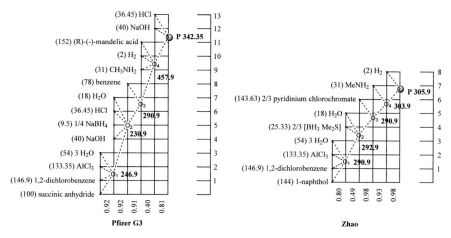

Fig. 4.4 Synthesis trees for Pfizer G3 and Zhao plans

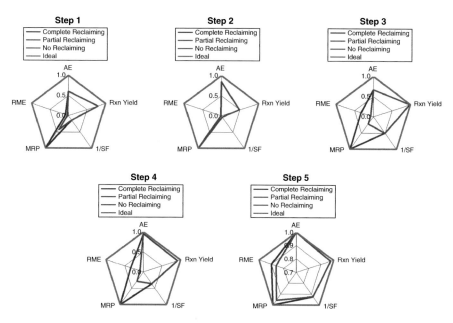

Fig. 4.5 Radial pentagons for Zhao plan

The numerical data presented in Tables 4.1–4.3 are conveniently displayed graphically in Fig. 4.7, which shows the radial hexagons for all seven plans. It is easily observed that the overall best performing Zhao plan has the least distorted hexagon with the largest area compared to the other plans. The Lautens plan has the least area. Figures 4.8 and 4.9 show the kernel and total waste distribution profiles for each plan. The kernel waste distributions account for wastes coming from

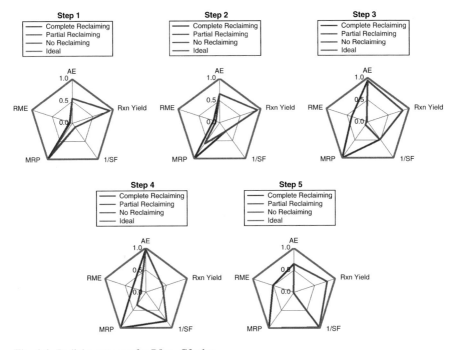

Fig. 4.6 Radial pentagons for Pfizer G3 plan

reaction by-products, side products, unreacted starting materials, and excess reagents. The total waste distributions add on the contributions from auxiliary materials. Reaction steps producing significant amounts of waste from auxiliary materials may be directly correlated with their corresponding MRP performances from their radial pentagons. On comparing both kinds of distributions for each plan, it is observed that the Zhao plan is the only one that has consistently the same shape. Every synthesis plan is made up of reactions that either make target bonds found in the final product or not. In terms of optimization, the goal is to minimize both the global waste and the proportion arising from sacrificial non-productive reactions at the same time. A disastrous situation to avoid is coming up with a synthesis plan that produces a lot of waste, most of which originating from sacrificial reactions such as redox adjustments and excessive use of protecting groups. Figure 4.10 shows the kernel mass of waste profile for all sertaline plans as a function of the global waste produced and the waste contributions from target bond forming and sacrificial reactions. The lengths of the bars correlate directly with the overall kernel E-factors and each bar, in turn, is subdivided into these two broad groups of reactions. The molecular weight building up profiles are given in Fig. 4.11. It is easily observed that the Buchwald and Zhao plans exhibit a high degree of building up since no intermediate along their reaction sequences has a molecular weight exceeding that of the target product sertraline. From the hypsicity profiles shown in Fig. 4.12, it is observed that the Buchwald, Pfizer

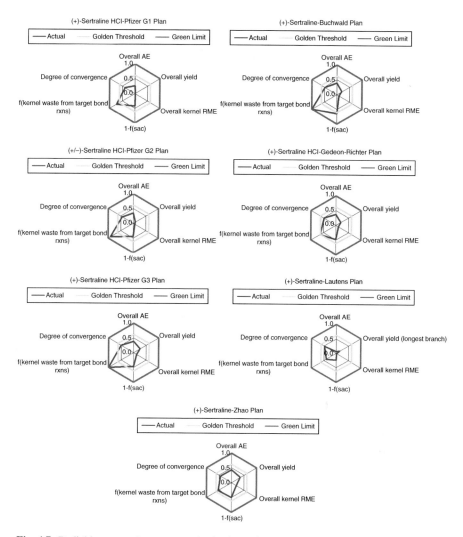

Fig. 4.7 Radial hexagons for seven synthesis plans of sertraline

G2, and Pfizer G3 are hyperhypsic with steadily decreasing oxidation steps which is consistent with the additive reduction reactions employed in their plans. It is obvious that the Lautens plan has the most oxidation number changes compared to the other plans. The target bond profiles and maps for all plans are shown in Figs. 4.13 and 4.14. These profiles confirm the conciseness and, therefore, the strategic efficiencies of the Buchwald and Pfizer G3 plans since there are no gaps in their respective bar graphs. From Fig. 4.14, it is observed that the plans break down into four distinct strategies in constructing the 1,2,3,4-tetrahydro-naphthalene ring system: (1) the Pfizer G2, Pfizer G3, and Buchwald plans

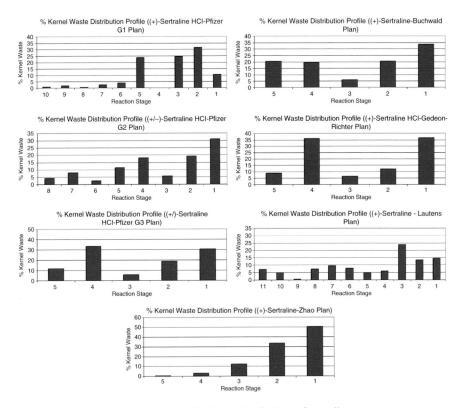

Fig. 4.8 Kernel waste distribution profiles for synthesis plans of sertraline

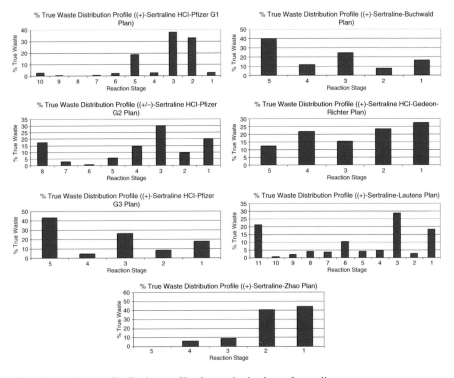

Fig. 4.9 Total waste distribution profiles for synthesis plans of sertraline

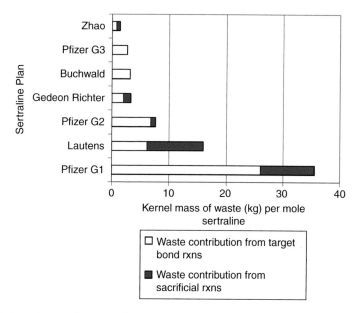

Fig. 4.10 Kernel mass of waste profile for synthesis plans showing contributions from target bond reactions and sacrificial reactions

involve [4 + 2] cycloadditions via Friedel–Crafts acylations; (2) the Gedeon Richter and Zhao plans begin with 1-naphthol as starting material so the ring frame is already set; (3) the Pfizer G1 plan involves building up the saturated ring by chain extension in step 2 via an aldol condensation and then a [6 + 0] ring closure in step 5 via an intramolecular Friedel–Crafts acylation; and (4) the Lautens plan involves a unique ring-constructing sequence via a Diels–Alder [2.2.1] adduct, which then undergoes ring opening via an asymmetric reduction as shown in Fig. 4.15. Though the Lautens strategy clearly demonstrates ingenuity and novelty, it unfortunately failed to translate this attribute into an overall material efficient plan that could compete with the other industrial plans beginning from the third-generation feedstock 1-naphthol available in two steps from the coal tar product naphthalene. The main reasons for this are the high number of reaction steps (11 versus 5) and the high proportion of sacrificial nontarget bond forming reactions involved (see Table 4.1). This ugly mismatch between demonstration of novelty in a few steps and overall low-material-efficiency performance in a plan occurs unfortunately all too often. Hence, the achievement of an optimum synergy between the two is indeed hard to achieve within the context of realizing a truly global "green" and hence cost-effective synthesis plan. However, fruitful opportunities for making new reaction discoveries and linking existing reactions in new combinations always exist and will help to make this goal a reality as synthetic organic chemistry continues to expand (Fig. 4.16).

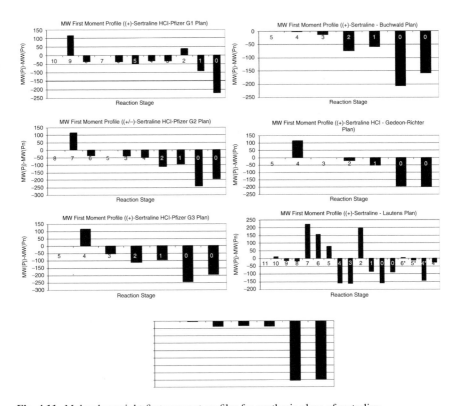

Fig. 4.11 Molecular weight first moment profiles for synthesis plans of sertraline

Summary of Optimization

Characteristics of a "good" synthesis plan that merge both optimum material and strategy efficiencies are given below.

Material Efficiency

1. Minimize overall waste by first identifying major waste contributors such as solvent, work-up, and purification materials used and then minimizing them where possible.
2. Waste contribution from target bond forming reactions should exceed that from sacrificial reactions.
3. Maximize overall reaction yield along longest branch.

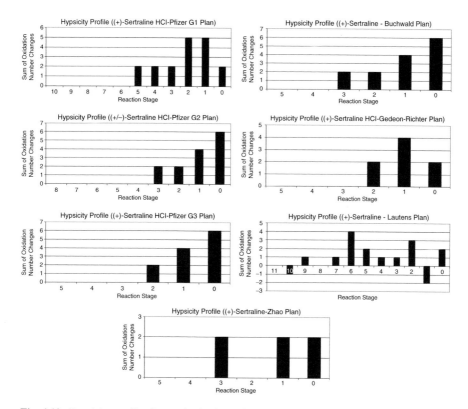

Fig. 4.12 Hypsicity profiles for synthesis plans of sertraline

4. Maximize overall reaction mass efficiency so that kernel RME $\geq 60\%$ for each reaction in a synthesis plan.
5. Maintain SF as close to one as far as possible to reduce impact of excess reagent consumption.

Strategy Efficiency

1. Number of reaction steps that form target bonds should exceed number of reaction steps that are sacrificial reactions.
2. Minimize overall number of reaction steps.
3. Maximize overall atom economy so that AE $\geq 60\%$ for each reaction in a synthesis plan.
4. Maximize degree of convergence.

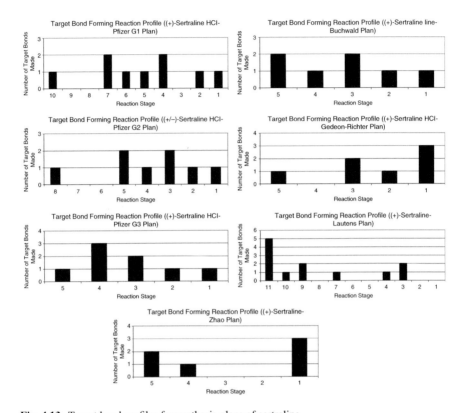

Fig. 4.13 Target bond profiles for synthesis plans of sertraline

5. Maximize the number of direct synthesis–type reactions such as carbon–carbon and non-carbon–carbon bond forming reactions, additive redox reactions, multicomponent, tandem, domino, or cascade reactions, strategic rearrangements particularly with respect to ring constructions, and non-sacrificial substitution reactions.

6. Minimize the number of indirect synthesis–type reactions such as eliminations or fragmentations, subtractive redox reactions, and nonstrategic rearrangements.

7. Minimize fractional contribution of sacrificial reagents.

8. Minimize degree of asymmetry.

9. Design plans with large negative molecular weight first moments so that plans begin from low molecular weight starting materials and progressively build up toward the molecular weight of the target molecule with minimal overshoots above the molecular weight of the target product.

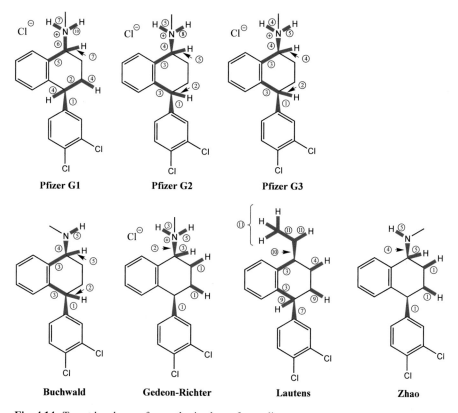

Fig. 4.14 Target bond maps for synthesis plans of sertraline

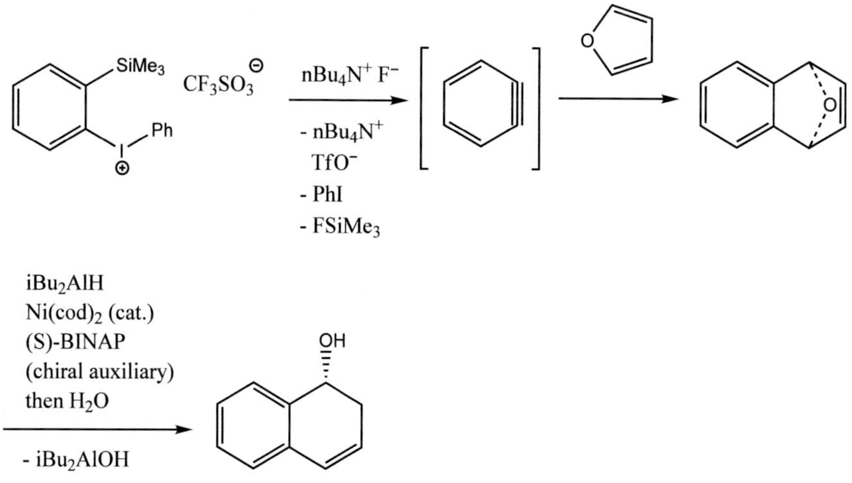

Fig. 4.15 Lautens ring construction strategy

Buchwald, Pfizer G2, Pfizer G3

step 3

Pfizer G1

step 2   step 5

Lautens

step 3

Gedeon Richter, Zhao

Fig. 4.16 Graphs showing ring construction strategies

10. With respect to redox economy: aim for HI = 0 by eliminating all redox reactions, or aim for HI > 0 with a steadily decreasing hypsicity profile using additive-type reduction reactions, or aim for HI < 0 with a steadily increasing hypsicity profile using additive-type oxidation reactions.

Thresholds and Probabilities

From the connecting relationships $RME = \frac{1}{1+E}$ and $AE = \frac{1}{1+E_{mw}}$, it is possible to set a threshold for "greenness" for individual reactions at the kernel level (excluding all auxiliary materials) as $AE \geq 0.618$ so that $AE \geq E_{mw}$, and $RME \geq 0.618$ so that $RME \geq E$, respectively. This threshold which exactly corresponds to the golden ratio equal to $\phi = \frac{\sqrt{5}-1}{2}$ is found by equating RME and E in the former expression and AE and E_{mw} in the latter. This is the justification for putting a 60% minimum cutoff for both AE and RME in the optimization recommendations given in section 6. The implication of this from a marketing point of view is that aiming for "green" chemistry is really aiming for "golden" chemistry. If individual reactions fail to meet this criterion at the kernel level, then there is no hope that they will meet it when all auxiliary materials are taken into account.

For an individual reaction, the kernel RME is given by $RME = (\varepsilon)(AE)$, which is a reduced form of Eq. 4.1 when SF = 1 (no excess reagents) and MRP = 1 (all auxiliaries recovered or eliminated). If a threshold value of ϕ is set for the kernel RME, then it is possible to define the probability of achieving this target RME given a reaction's AE and range of possible reaction yield performances. For any given chemical reaction, the magnitude of AE is fixed while the reaction yield is in principle variable over the full range 0–100%. Applying the inequality $RME = (\varepsilon)(AE) \geq 0.618$ implies that $(AE) \geq \frac{0.618}{\varepsilon}$. Figure 4.17 shows plots of AE versus reaction yield under two scenarios, depending on whether a reaction's atom economy, $(AE)^*$, is below or above the threshold value of 0.618. The curved line represents the equation $AE = \frac{0.618}{\varepsilon}$. The region above this line satisfies the condition $RME > 0.618$, and the region below the line satisfies the condition $RME < 0.618$. If a reaction's AE is below 0.618, then the probability of it achieving an RME above 0.618 is zero no matter how high the reaction yield is. This is obvious because the horizontal line does not intersect the curve $AE = \frac{0.618}{\varepsilon}$. However, if a reaction's AE is above 0.618, then the probability of it achieving this goal is given by the length of the bolded line segment, or $p = 1 - \frac{0.618}{(AE)^*}$. So, for example, if a reaction has $(AE)^* = 0.8$, then $p = 0.23$ or 23%. The best possible scenario would yield a probability of only $p = 38\%$ for a reaction with 100% AE. This low probability may be amplified considerably if we consider a minimum cutoff for the reaction yield as well that is better than 0%. This means case II shown in Fig. 4.17 may be further subdivided into two scenarios given by the conditions $0 < \varepsilon_{min} < \frac{0.618}{(AE)^*}$ and $\frac{0.618}{(AE)^*} < \varepsilon_{min} \leq 1$

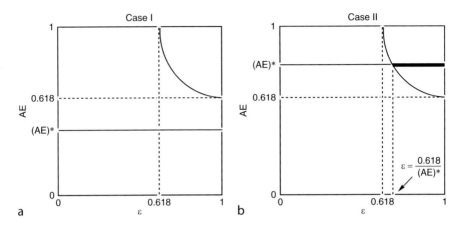

Fig. 4.17 Plots of AE versus reaction yield according to AE $= \phi/\varepsilon = 0.618/\varepsilon$ showing the two possible cases: (**a**) when (AE)$^* \leq 0.618$ and $0 < \varepsilon_{min} \leq 1$; and (**b**) when $0.618 < $ (AE)$^* \leq 1$ and $0 < \varepsilon_{min} \leq 1$

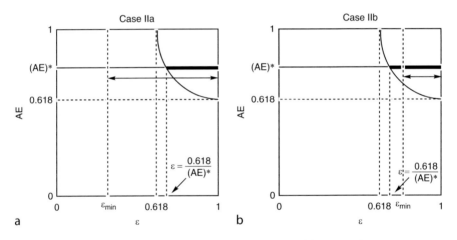

Fig. 4.18 Plots of AE versus reaction yield according to AE $= \phi/\varepsilon = 0.618/\varepsilon$ showing the two possible cases: (**a**) $0.618 < $ (AE)$^* \leq 1$ and $0 < \varepsilon_{min} < \frac{0.618}{(AE)*}$; and (**b**) when $0.618 < $ (AE)$^* \leq 1$ and $\frac{0.618}{(AE)*} < \varepsilon_{min} \leq 1$

as shown in Fig. 4.18. In the former case, $p = \left(1 - \frac{0.618}{(AE)*}\right)/(1 - \varepsilon_{min})$ and in the latter, $p = 1$. So, if a reaction proceeds with a minimum yield of 70% and it has an AE of 80%, then the probability of achieving an RME of 0.618 improves over threefold from $p = 0.23$ to $p = 0.76$. If the minimum yield is no worse than $0.618/0.8 = 0.77$ (or 77%), then it is certain that the goal condition is attainable. These simple calculations clearly illustrate the importance of achieving both atom economies and reaction yields as high as possible in order to increase the chances of achieving even a modest target threshold RME of 61.8%. In a nutshell, synthesis plans composed of reactions with high atom economies and high yields working together have the best chance of achieving high kernel reaction mass efficiencies.

Future Directions

The future success of applying green chemistry thinking and metrics to synthesis design and optimization will depend on:

1. Making use of multicomponent and tandem one-pot reactions as central theme reactions in synthesis planning
2. Disclosing all material and energy consumption parameters for each reaction in a given plan (especially masses of chromatographic solvents, drying agents, work-up wash solutions, and reactant gases) and clamping down on careless and intentional errors in publications, particularly experimental sections and supplementary materials of journal articles, with respect to mass and mole amounts of materials used and reaction yields so that true assessments of RME and E-factors can be made
3. Implementing the radial pentagon spreadsheet algorithm as a standard proofreading tool for reviewers of synthesis papers to trap errors described in (2) and help authors in improving the standard and reliability of their reported procedures
4. Mandating the reporting of green metrics as proof of greenness and part of the standard protocol in reports of synthesis plans in the literature, especially if plans are advertised as "green-er" than prior published plans and hence reduce false claims of "greenness"
5. Accepting that optimization is a continuous ongoing iterative exercise and that ranking of plans is the inevitable consequence of metrics analysis
6. Accepting that optimization is a multivariable problem and that claims of achievements of optimization to a given target molecule are legitimate when the magnitudes of all variables in the set of plans considered are appropriately maximized and minimized and appear in the same plan
7. Realizing that when a new target molecule is desired to be made with no prior published guidance available, it is necessary to go through poor performing plans before hitting on the "right" one
8. Realizing that quantitative assessment of synthesis plans has an important role in deciding which may be good candidate plans to pursue
9. Changing the well-worn paradigm of designing synthesis plans around a fixed type of reaction just because of its novelty or because it honors the discoverer of that reaction to one that uses material and synthetic efficiency as the uppermost constraints in the choice of the set of reactions ultimately selected for a plan

10. Reaching a consensus on defining the set of first- and second-generation feedstock compounds that form a common basis of starting materials for all synthetic target molecules so that synthesis plans to any given target may be traced back to these and then ranked in an unbiased way
11. Ceasing to divorce the goals of green chemistry from those of elegant synthesis design and cost-efficient process chemistry
12. Creating a searchable structure database based on target bond mapping so that ring construction strategies may be encoded and as an aid to ensure novelty in planning a synthesis to a given target using retrosynthetic analysis

Some caveats to bear in mind when carrying out green metrics analyses include:

1. Generalizations about relative efficiencies of linear versus convergent routes, short plans versus longer ones, or stereoselective versus racemic with resolution must be taken with caution as there exist plenty of examples in the literature when counterintuitive results occur because gains made in one set of parameters are lost in others.
2. The synthesis plans of specialized catalysts or solvents used, such as chiral catalysts, ligands, and ionic liquids, must also be worked out using separate synthesis tree diagrams with appropriate mole-scaling factors as part of the overall assessment of material efficiency to a given target molecule.
3. When using the radial pentagon analysis for single reactions, one must be aware of determining correct mole amounts for reagents that are reported in procedures as solutions given in terms of weight percent and properly assigning the role of each material used in the right place in the template spreadsheet.
4. No claims of greenness can be made if only one plan exists for a given target molecule.
5. Claims of greenness cannot be made based on one criterion, such as the use of a "green" solvent for example.
6. Once green metrics analyses are done on a set of literature procedures to a given target molecule and it is found in the ranking process that optimized parameters are scattered over a number of plans, it is imperative that the next disclosed plan should demonstrate improvements over the best reported prior plan.

Bibliography

Primary Literature

1. Andraos J (2005) Unification of reaction metrics for green chemistry: applications to reaction analysis. Org Process Res Dev 9:149–163
2. Andraos J (2005) Unification of reaction metrics for green chemistry II: evaluation of named organic reactions and application to reaction discovery. Org Process Res Dev 9:404–431
3. Andraos J (2005) On using tree analysis to quantify the material, input energy, and cost throughput efficiencies of simple and complex synthesis plans and networks: towards

a blueprint for quantitative total synthesis and green chemistry. Org Process Res Dev 10:212–240

4. Andraos J, Izhakova J (2006) Perspectives on the application of green chemistry principles to total synthesis design. Chimica Oggi/The Int J Ind Chem 24(6, Suppl.):31–36

5. Andraos J, Sayed M (2007) On the use of "green" metrics in the undergraduate organic chemistry lecture and laboratory to assess the mass efficiency of organic reactions. J Chem Educ 84:1004–1010

6. Andraos J (2007) Gauging material efficiency. Can Chem News 59(4):14–17

7. Trost BM (1991) The atom economy – a search for synthetic efficiency. Science 254:1471–1477

8. Trost BM (2002) On inventing reactions for atom economy. Acc Chem Res 35:695–705

9. Trost BM (1995) Atom economy. A challenge for organic synthesis - homogeneous catalysis leads the way. Angew Chem Int Ed 34:259–281

10. Sheldon RA (1994) Consider the environmental quotient. Chem Tech 24(3):38–47

11. Sheldon RA (2000) Atom utilisation, E factors and the catalytic solution. CR Acad Sci Paris Sér IIc Chim 3:541–551

12. Sheldon RA (2001) Atom efficiency and catalysis in organic synthesis. Pure Appl Chem 72:1233–1246

13. Curzons AD, Constable DJC, Mortimer DN, Cunningham VL (2001) So you think your process is green, how do you know? – using principles of sustainability to determine what is green - a corporate perspective. Green Chem 3:1–6

14. Constable DJC, Curzons AD, Freitas dos Santos LM, Geen GR, Hannah RE, Hayler JD, Kitteringham J, McGuire MA, Richardson JE, Smith P, Webb RL, Yu M (2001) Green chemistry measures for process research and development. Green Chem 3:7–9

15. Steinbach A, Winkenbach R (2000) Choose processes for their productivity. Chem Eng April: 94–104

16. Constable DJC, Curzons AD, Cunningham VL (2002) Metrics to "green" chemistry – which are the best? Green Chem 4:521–527

17. Eissen M, Metzger JO (2002) Environmental performance metrics for daily use in synthetic chemistry. Chem Eur J 8:3580–3585

18. Eissen M, Hungerbühler K, Dirks S, Metzger J (2003) Mass efficiency as metric for the effectiveness of catalysts. Green Chem 5:G25–G27

19. Metzger JO, Eissen M (2004) Concepts on the contribution of chemistry to a sustainable development - renewable raw materials. CR Acad Sci Paris Sér IIc Chim 7:569–581

20. Eissen M, Mazur R, Quebbemann HG, Pennemann KH (2004) Atom economy and yield of synthesis sequences. Helv Chim Acta 87:524–535

21. van Aken K, Strekowski L, Patiny L (2006) EcoScale, a semi-quantitative tool to select an organic preparation based on economical and ecological parameters. Beilstein J Org Chem 2. doi:10.1186/1860-5397-2-3

22. Andraos J (2009) Global green chemistry metrics analysis algorithm and spreadsheets: evaluation of the material efficiency performances of synthesis plans for oseltamivir phosphate (Tamiflu) as a test case. Org Process Res Dev 13:161–185

23. Welch WM, Kraska AR, Sarges R, Koe BK (1984) Nontricyclic antidepressant agents derived from cis- and trans-1-amino-4-aryltetraline. J Med Chem 27:1508–1515

24. Quallich GJ, Williams MT, Friedmann RC (1999) Friedel-Crafts synthesis of 4-(3, 4-dichlorophenyl)-3, 4-dihydro-1(2 H)-naphthalenone, a key intermediate in the preparation of the antidepressant sertraline. J Org Chem 55:4971–4973

25. Lautens M, Rovis T (1999) Selective functionalization of 1, 2-dihydronaphthalenols leads to a concise, stereoselective synthesis of sertraline. Tetrahedron 55:8967–8976

26. Yun J, Buchwald SL (2000) Efficient kinetic resolution in the asymmetric hydrosilylation of imines of 3-substituted indanones and 4-substituted tetralones. J Org Chem 65:767–774

27. Vukics K, Fodor T, Fischer J, Fellegvári I, Lévai S (2002) Improved industrial synthesis of antidepressant sertraline. Org Process Res Dev 6:82–85

28. Taber GP, Pfisterer DM, Colberg JC (2004) A new and simplified process for preparing N-[4-(3, 4-dichlorophenyl)-3, 4-dihydro-1(2 H)-naphthalenylidene]methanamine and a telescoped process for the synthesis of (1 S-cis)-4-(3, 4-dichlorophenol)-1, 2, 3, 4-tetrahydro-N-methyl-1-naphthalenamine mandelate: key intermediates in the synthesis of sertaline hydrochloride. Org Process Res Dev 8:385–388
29. Wang G, Zheng C, Zhao G (2006) Asymmetric reduction of substituted indanones and tetralones catalyzed by chiral dendrimer and its application to the synthesis of (+)-sertraline. Tetrahedron Asymm 17:2074–2081
30. Hendrickson JB (1971) A systematic characterization of structures and reactions for use in organic synthesis. J Am Chem Soc 93:6847–6854

Books and Reviews

Abdel-Magid FA (2004) Chemical process research: the art of practical organic synthesis. American Chemical Society, Washington

Anastas PT, Warner JC (1998) Green chemistry: theory and practice. Oxford University Press, Oxford

Anderson NG (2000) Practical process research and development. Academic, San Diego

Baran PS, Maimone TJ, Richter JM (2007) Total synthesis of marine natural products without using protecting groups. Nature 446:404–408

Bertz SH, Sommer TJ (1993) Applications of graph theory to synthesis planning: complexity, reflexivity, and vulnerability. In: Hudlicky T (ed) Organic synthesis: theory and applications, vol 2, JAI Press. Greenwich, CT, pp 67–92

Burns NZ, Baran PS, Hoffmann RW (2009) Redox economy in organic synthesis. Angew Chem Int Ed 48:2854–2867

Calvo-Flores FG (2009) Sustainable chemistry metrics. ChemSusChem 2:905–919

Carey JS, Laffan D, Thomson C, Williams MT (2006) Analysis of the reactions used for the preparation of drug candidate molecules. Org Biomol Chem 4:2337–2347

Carlson R (1992) Design and optimization in organic synthesis. Elsevier, Amsterdam

Cornforth JW (1993) The trouble with synthesis. Aust J Chem 46:157–170

Eissen M (2001) Bewertung der Umweltverträglichkeit organisch-chemischer Synthesen. PhD thesis, Universität Oldenburg

Fuchs PL (2001) Increase in intricacy – a tool for evaluating organic synthesis. Tetrahedron 57:6855–6875

Hendrickson JB (1977) Systematic synthesis design. 6. Yield analysis and convergency. J Am Chem Soc 99:5439–5450

Hoffmann RW (2006) Protecting-group-free synthesis. Synlett 3531–3541

Lapkin A, Constable DJC (2008) Green chemistry metrics: measuring and monitoring sustainable processes. Wiley, Chichester

Lee S, Robinson G (1995) Process development: fine chemicals from grams to kilograms. Oxford University Press, Oxford

Newhouse T, Baran PS, Hoffmann RW (2009) The economies of synthesis. Chem Soc Rev 38:3010–3021

Nicolaou KC, Vourloumis D, Winssinger N, Baran PS (2000) The art and science of total synthesis at the dawn of the twenty-first century. Angew Chem Int Ed 39:44–122

Nicolaou KC (2003) Perspectives in total synthesis: a personal account. Tetrahedron 59:6683–6738

Nicolaou KC, Snyder SA (2004) The essence of total synthesis. Proc Nat Acad Sci USA 101:11929–11936

Nicolaou KC, Edmonds DJ, Bulger PG (2006) Cascade reactions in total synthesis. Angew Chem Int Ed 45:7134–7186

Orru RVA, de Greef M (2003) Recent advances in solution-phase multicomponent methodology for the synthesis of heterocyclic compounds. Synthesis 1471–1499

Posner GH (1986) Multicomponent one-pot annulations forming three to six bonds. Chem Rev 86:831–844

Qiu F (2008) Strategic efficiency – the new thrust for synthetic organic chemists. Can J Chem 86:903–906

Seebach D (1990) Organic synthesis –where now? Angew Chem Int Ed 29:1320–1367

Serratosa F (1990) Organic chemistry in action: the design of organic synthesis. Elsevier, Amsterdam

Sheldon RA (1997) The E factor: fifteen years on. Green Chem 9:1273–1283

Sheldon RA (2008) E factors, green chemistry and catalysis: an odyssey. Chem Commun 3352–3365

Smit WA, Bochkov AF, Caple R (1998) Organic synthesis: the science behind the art. Royal Society of Chemistry, Cambridge

Snieckus V (1999) Optimization in organic synthesis. Med Res Rev 19:342–347

Tietze LF (1996) Domino reactions in organic synthesis. Chem Rev 96:115–136

Tietze LF, Modi A (2000) Multicomponent domino reactions for the synthesis of biologically active natural products and drugs. Med Chem Rev 20:304–322

Ugi I, Dömling A, Hörl W (1994) Multicomponent reactions in organic chemistry. Endeavour New Ser 18(3):115–122

Ugi I, Dömling A, Werner B (2000) Since 1995 the new chemistry of multicomponent reactions and their libraries, including their heterocyclic chemistry. J Heterocycl Chem 37:647–658

Ugi I (2001) Recent progress in the chemistry of multicomponent reactions. Pure Appl Chem 73:187–191

Weber L, Illgen K, Almstetter M (1999) Discovery of new multi-component reactions with combinatorial methods. Synlett 366–374

Weber L (2002) Multi-component reactions and evolutionary chemistry. Drug Discov Today 7:143–147

Weber L (2002) The application of multi-component reactions in drug discovery. Curr Med Chem 9:1241–1253

Wender P, Miller BL (1993) Toward the ideal synthesis: connectivity analysis and multibond-forming processes. In: Hudlicky T (ed) Organic synthesis: theory and applications, vol 2, JAI Press. Greenwich, Connecticut, pp 27–65

Zhang TY (2006) Process chemistry: the science, business, logic, and logistics. Chem Rev 106:2583–2595

Zhu J, Bienyamé H (2005) Multicomponent reactions. Wiley, Weinheim

Chapter 5
Green Chemistry with Microwave Energy

Rajender S. Varma

Glossary

Green chemistry	Green chemistry is the broad discipline that encompasses the design of chemical processes and products that eliminate or reduce the generation and use of hazardous substances. It applies across the life cycle, including the design, manufacture, and use of a chemical product.
Microwaves	Microwaves (0.3–300 GHz) lie in the electromagnetic radiation spectrum between radiowave (Rf) and infrared (IR) frequencies with relatively large wavelengths and are a form of energy and not heat. This nonionizing radiation, incapable of breaking chemical bonds, is a form of energy that manifests itself as heat through interaction with the polar medium.
Sustainability	Literally meaning to "maintain," "support," or "endure" the concept of sustainability calls for policies and strategies that meet society's present needs without compromising the ability of future generations to meet their own needs.

This chapter was originally published as part of the Encyclopedia of Sustainability Science and Technology edited by Robert A. Meyers. DOI:10.1007/978-1-4419-0851-3

R.S. Varma (✉)
Sustainable Technology Division, National Risk Management Research Laboratory,
U.S. Environmental Protection Agency, 26 West Martin Luther King Drive,
Cincinnati, OH 45268, USA
e-mail: Varma.Rajender@epa.gov

P.T. Anastas and J.B. Zimmerman (eds.), *Innovations in Green Chemistry and Green Engineering*, DOI 10.1007/978-1-4614-5817-3_5,
© Springer Science+Business Media New York 2013

Definition of the Subject

Green chemistry utilizes a set of 12 principles that reduces or eliminates the use or generation of hazardous substances in the design, manufacture, and applications of chemical products [1]. This newer chemical approach protects the environment by inventing safer and eco-friendly chemical processes that prevent pollution "at source" rather than cleaning up "end-of-the-pipe" by-products and pollutants generated by traditional synthesis. The diverse nature of our chemical universe promotes a need for various greener strategic pathways in our quest to attain sustainability. The synthetic chemical community has been under increased pressure to produce, in an environmentally benign fashion, the myriad of chemical entities required by society in relatively short spans of time. This is especially true for the pharmaceutical and fine chemical industries. Among others, one of the best options is to accelerate these synthetic processes by using microwave (MW)-assisted chemistry techniques in conjunction with safer reaction media. The efficient use of the MW heating approach for the synthesis of a wide variety of organics and nanomaterials in aqueous and solvent-free media is discussed in this entry, including the sustainable application of recyclable and reusable nano-catalysts.

Introduction

The emerging area of green chemistry emphasizes minimum hazard as the performance criteria while designing new chemical processes. Rather than remediation, which involves cleaning up of waste after it has been produced, the main objective is to avoid waste generation in the first place. The desired approach requires new environmentally benign syntheses, catalytic methods, and chemical products that are "benign by design" [1]. One of the thrust areas for achieving this target is to explore alternative expeditious reaction conditions and eco-friendly reaction media to accomplish the desired chemical transformations with minimized by-products or waste as well as eliminating the use of conventional organic solvents. Consequently, several newer strategies have appeared, such as solvent-free (dry media) reactions with [2, 3] and without microwave irradiation [4].

The nonclassical heating technique using microwaves, termed as the "Bunsen burner of the 21st century," is rapidly becoming popular for expeditious chemical syntheses and transformations [5]. This entry summarizes noteworthy greener methods that use MWs that have resulted in the development of sustainable synthetic protocols for drugs and fine chemicals. A brief account of the author's own experiences in MW-assisted organic transformations involving benign alternatives, such as solid-supported reagents [2, 4], and greener reaction media, such as aqueous, ionic liquid, and solvent-free, for the synthesis of various heterocycles [3], oxidation-reduction reactions, coupling reactions, and some name reactions is included [5–9].

Microwaves are a nonionizing form of radiation energy that are not strong enough to break chemical bonds but transfer energy selectively to various substances. Some materials (such as hydrocarbons, glass, and ceramics) are nearly transparent to microwaves and therefore behave as good insulators in a MW oven since they are heated only to a very limited extent. Metals reflect MW; molecules with dipole moment (many types of organic compounds) and salts absorb MW energy directly. Microwaves couple directly with molecules in the reaction mixture with rapid rise in temperature; dipole rotation and ionic conduction are the two most important fundamental mechanisms for transfer of energy from microwaves to the molecules being heated. Polar molecules try to align themselves with the rapidly changing electric field of the MW and the coupling ability is determined by the polarity of the molecules. Therefore, there are some significant differences in terms of thermal gradient between the conventional chemical reactions in the liquid phase and the same reactions conducted under MW irradiation.

When a liquid reaction mixture is subjected to conventional heating, the walls of the vessel are directly heated but the reaction mixture receives thermal energy by conduction/convection. The temperature of the reaction mixture cannot be higher than the temperature of the vessel walls. In contrast, MW energy permeates through glass vessels directly and is available for absorption by molecules in the reaction mixture. Consequently, it is possible for these molecules to be at higher energy levels and the contents of the flask to be at a higher bulk temperature in a few minutes. Another important feature of microwaves is that they penetrate several centimeters into a liquid; in contrast, radiant heat (e.g., from infrared rays) raises the temperature of only the surface layer. A reaction mixture under MW irradiation is therefore at a higher temperature in the middle than at the surface. Accordingly, the temperature profile of the reaction mixture can be quite different depending on whether there is conventional heating or MW-induced energy transfer.

Reactions Under Solvent-Free Conditions (With or Without Any Support)

Although the initial surface-mediated chemical transformation dates back to 1924 [10], it was not until almost half a century later that the technique received well-deserved attention as attested by several books [11, 12, 13], reviews, and account articles [14, 15]. Heterogeneous reactions facilitated by supported reagents on inorganic oxide surfaces have received special attention in recent years. The use of MW irradiation techniques for the acceleration of organic reactions had a profound impact on these heterogeneous reactions since the appearance of initial reports on the application of microwaves for chemical synthesis in polar solvents [16, 17]. The approach has now matured into a useful technique for a variety of applications in organic synthesis and functional group transformations, as is testified by a large number of books [18, 19] and review articles on this theme [2–5, 20–23].

The reactions appear to occur at relatively low bulk temperature although higher localized temperatures may be reached during MW irradiation. Unfortunately, accurate recording of temperature has not been made in the majority of such studies. The situation has now improved as a result of the availability of commercial MW systems. The recyclability of some of these solid mineral supports renders them into environmentally friendlier "green" protocols.

Since the initial report [24], a large number of MW-promoted solvent-free protocols have been illustrated for a wide variety of useful chemical transformations, such as cleavage (protection/deprotection), condensation, rearrangement reactions, oxidation, reduction, and the synthesis of several heterocyclic compounds on mineral supports [2–5, 20–23]. A range of industrially significant chemical precursors, such as enones, imines, enamines, nitroalkenes, and heterocyclic compounds, has been obtained in a relatively environmentally friendlier manner [2–5, 20–23]. A vast majority of these solvent-free reactions has been performed using an unmodified household MW oven or commercial MW equipment usually operating at 2,450 MHz in open glass containers with neat reactants. The general procedure involves simple mixing of neat reactants with the catalyst, their adsorption on mineral or "doped" supports, and subjecting the reaction mixture to MW irradiation.

It is clear that this solventless approach addresses the problems associated with waste disposal of solvents that are used severalfold in chemical reactions, thus minimizing or avoiding the excess use of chemicals and solvents.

Protection–Deprotection Reactions

Although inherently wasteful, protection–deprotection reaction sequences constitute an integral part of organic syntheses, such as the preparation of monomers, fine chemicals, and precursors for pharmaceuticals. These reactions often involve the use of acidic, basic, or hazardous reagents, and toxic metal salts [25]. The solventless MW-accelerated cleavage of functional groups provides an attractive alternative to conventional deprotection reactions.

Deacylation Reactions

The utility of recyclable alumina as a viable support surface for deacylation reaction was reported by Varma and his colleagues wherein the orthogonal deprotection of alcohols [26] and regeneration of aryladehydes from the corresponding diacetates [27] is possible under solvent-free conditions on neutral alumina surface using MW irradiation (Scheme 5.1).

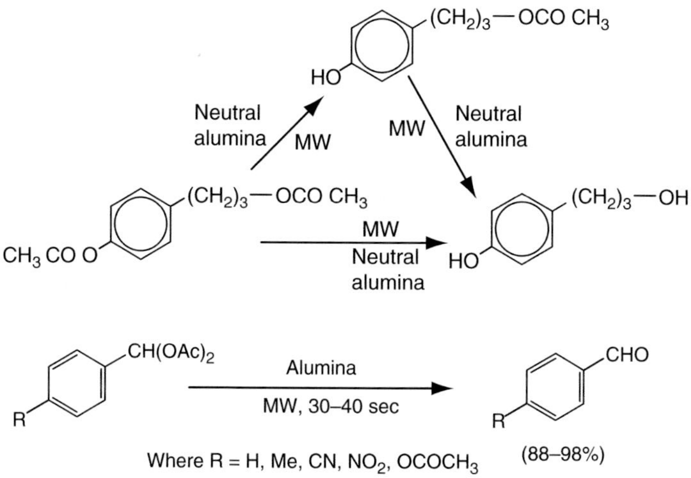

Scheme 5.1 Deacylation of protected alcohols and phenols and regeneration of aldehydes from aldehyde diacetates on neutral alumina

Debenzylation of Carboxylic Esters

An efficient solvent-free debenzylation process for the cleavage of carboxylic esters on alumina surface was developed by Varma and coworkers [28]. By changing the surface characteristics of the solid support from neutral to acidic, the cleavage of 9-fluorenylmethoxycarbonyl (Fmoc) group and related protected amines was achieved. The cleavage of *N*-protected moieties, however, required the use of basic alumina and a MW irradiation time of 12–13 min at ~130–140°C.

Desilylation Reactions

Tertiary-butyldimethylsilyl (TBDMS) ether derivatives of alcohols can be rapidly cleaved to regenerate the corresponding hydroxy compounds on alumina using MW irradiation [29]. This approach circumvents the use of corrosive fluoride ions that is normally employed for cleaving such silyl protecting groups.

Thionation Reactions: Thioketones, Thiolactones, Thioamides, Thionoesters, and Thioflavonoids

The conversion of carbonyl compounds to the corresponding thio analogues is especially useful under solventless conditions and circumvents the use of conventional phosphorous pentasulfide under basic conditions, hydrogen sulfide in the presence of acid, or Lawesson's reagent. Using the MW approach, no acidic or basic medium is used and the carbonyl compounds are simply admixed with neat

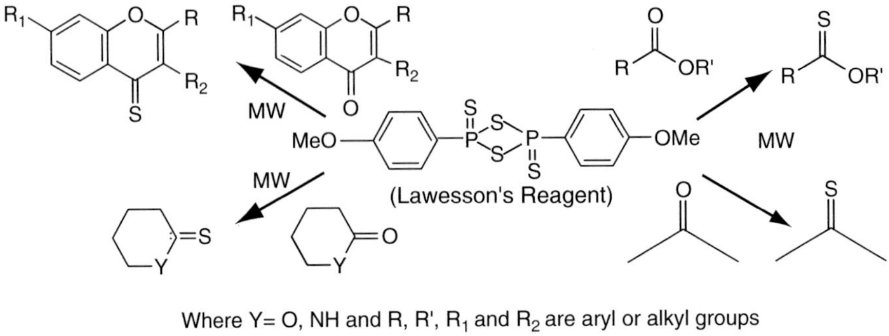
MW
MeO
OMe
(Lawesson's Reagent)
MW
MW
Where Y= O, NH and R, R', R1 and R2 are aryl or alkyl groups

Scheme 5.2 Solvent-free synthesis of thioketones, thiolactones, thioamides, and thionoesters by Lawesson's reagent. From Varma and Kumar (Ref. [30])

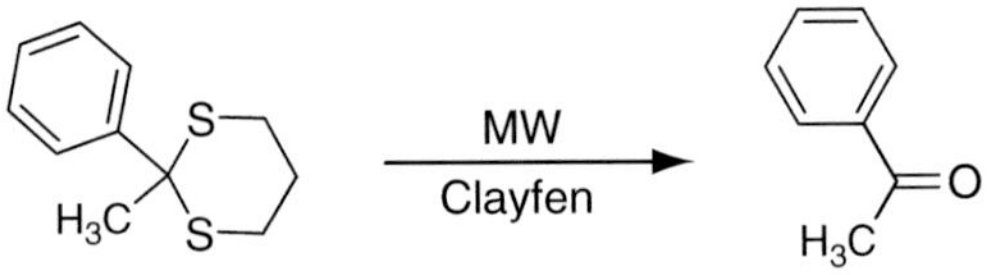
MW
Clayfen

Scheme 5.3 Clayfen-catalyzed dethioacetalization

Lawesson's reagent (0.5 equiv.). This benign approach is general and is applicable to the high yield conversion of ketones, flavones, isoflavones, lactones, amides, and esters to the corresponding thio analogs (Scheme 5.2). The protocol uses comparatively a much smaller amount of Lawesson's reagent and avoids the use of large excesses of dry hydrocarbon solvents, such as benzene, xylene, triethylamine, or pyridine, that are conventionally used [30].

Dethioacetalization

Thio acetals and ketals derived from aldehydes and ketones were rapidly deprotected by clayfen within seconds under solvent-free conditions (Scheme 5.3), thus avoiding the use of excess solvents and toxic oxidants commonly employed in the dethioacetalization process [31].

Oxidation Reactions

Oxidation of Alcohols

The oxidation of alcohols to carbonyl compounds is an important transformation in organic synthesis. The use of supported reagents has gained popularity because of

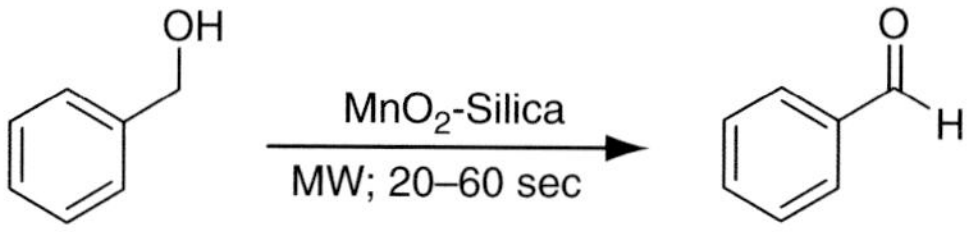

Scheme 5.4 MnO$_2$-silica-catalyzed oxidation of alcohols

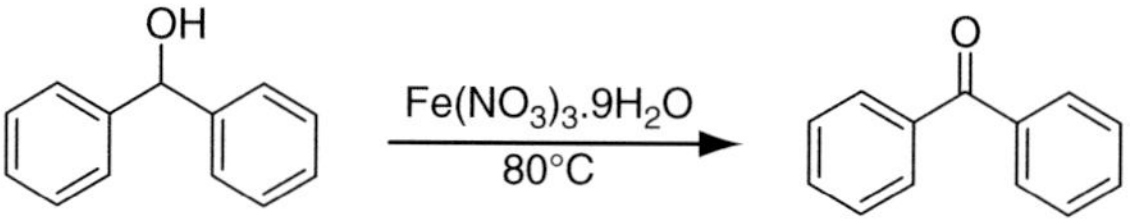

Scheme 5.5 Solventless oxidation of alcohols to carbonyl compounds

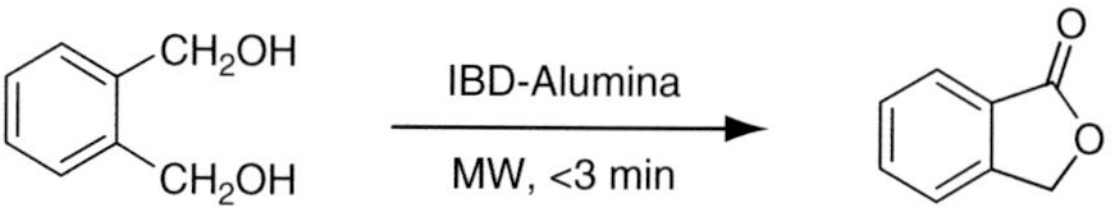

Scheme 5.6 IBD-alumina catalyzed oxyhyperiodination of alcohols

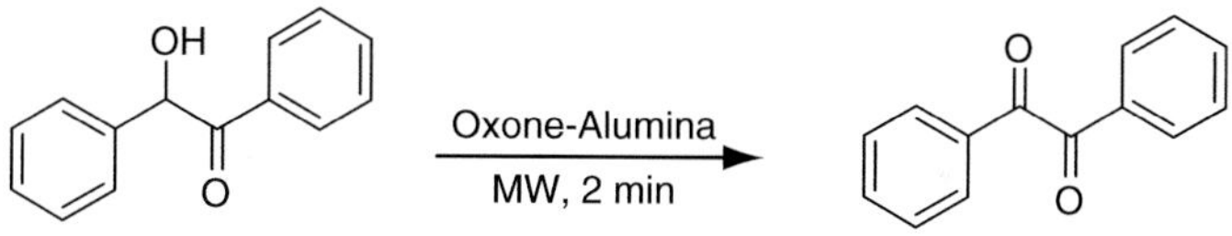

Scheme 5.7 Oxone-alumina catalyzed oxidation of benzoin

the improved selectivity, reactivity, and associated ease of manipulation. Alcohols can be rapidly and selectively oxidized to the corresponding carbonyl compounds by silica-supported active manganese dioxide (Scheme 5.4) under solvent-free conditions using microwaves [32].

This oxidation protocol can also be catalyzed by ferric nitrate under solvent-free conditions [33], without any solid support (Scheme 5.5).

The nonmetallic oxidant, iodobenzene diacetate (IBD) on alumina, accomplished the oxidation of alcohols under solvent-free MW irradiation conditions and entails only the mixing of the neat alcohols with 1.1 equivalents of IBD doped on neutral alumina (Scheme 5.6); this rapid procedure avoids the over-oxidation of alcohols to carboxylic acids [34].

The solid-state oxidation of α-hydroxyketones to generate vicinal ketones (benzils) is possible using Oxone® supported on wet alumina (Scheme 5.7). The protocol circumvents the major limitations of Oxone®, namely, the low solubility in organic solvents and the inherent danger of combustibility upon heating with solvents [35].

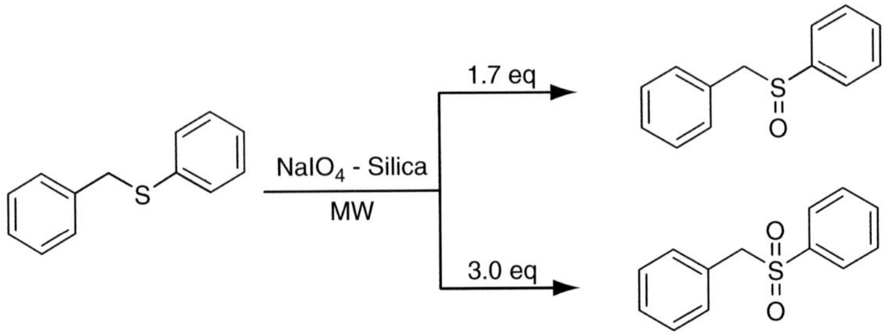

Scheme 5.8 NaIO$_4$-silica catalyzed selective oxidation of sulfides

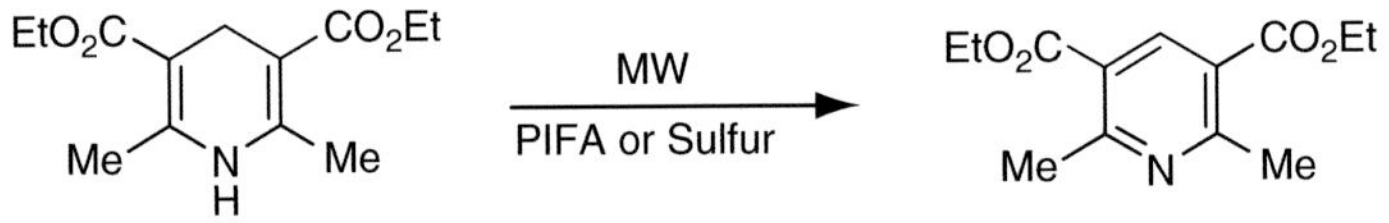

Scheme 5.9 PIFA or sulfur-catalyzed oxidation of 1,4-dihydropyridines

Oxidation of Sulfides

Selective oxidation of sulfides to sulfoxides and sulfones is possible [36] using wet silica-supported sodium periodate under MW irradiation (Scheme 5.8). A unique feature of this protocol is its applicability to long-chain aliphatic sulfides which are usually insoluble in polar solvents and are difficult to oxidize.

Oxidation of Dihydropyridines

Solventless oxidation of 1,4-dihydropyridines to pyridines was effected by phenyliodine(III) bis(trifluoroacetate) (PIFA) at room temperature or elemental sulfur under MW irradiation conditions (Scheme 5.9). Dealkylation at the 4-position in the cases of ethyl, isopropyl, and benzyl-substituted dihydropyridine derivatives with PIFA was circumvented by an alternative general procedure using elemental sulfur which provides pyridines in good yields [37].

Reduction Reactions

Reduction of Carbonyl Compounds

A manipulatively simple and rapid method for the reduction of carbonyl compounds was developed [38] by Varma et al. under solvent-free conditions using NaBH$_4$-Alumina and MW irradiation (Scheme 5.10a). When sodium

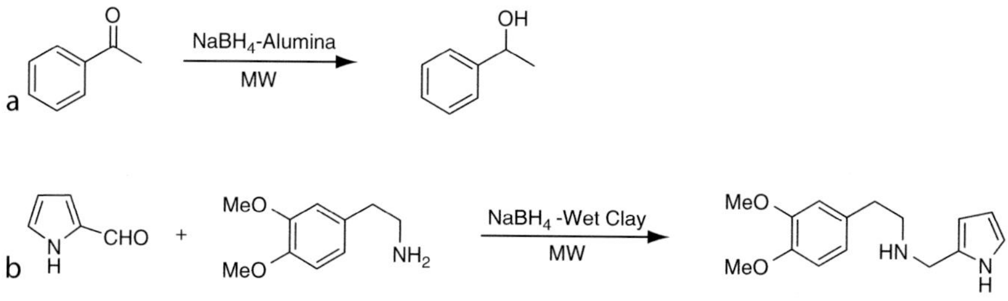

Scheme 5.10 (a) NaBH$_4$-catalyzed reduction of carbonyl compounds and (b) Reductive-amination of carbonyl compounds

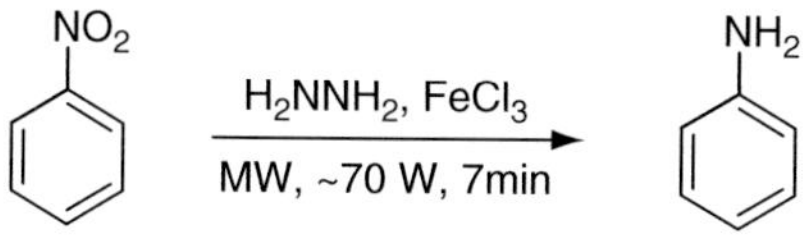

Scheme 5.11 Reduction of nitro compounds to amines

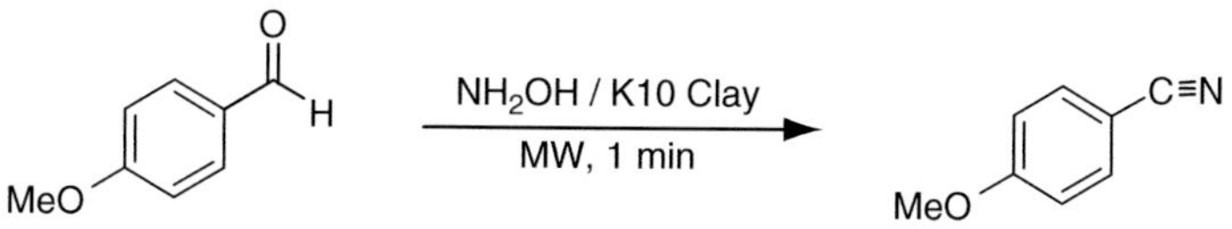

Scheme 5.12 Hydroxylamine-clay-catalyzed synthesis of aromatic nitriles

borohydride on wet clay was used as support, the reductive-amination of carbonyl compounds was efficiently accomplished under MW irradiation (Scheme 5.10b) [39]. The reactions involving Schiff's bases generated from cyclohexanone and aniline and aliphatic aldehydes and amines, however, require a relatively longer time for completion.

Reduction of Nitro Compounds

Varma and coworkers developed a solvent-free MW reduction protocol that leads to a facile preparation of aromatic amines from the corresponding nitro compounds [40] with hydrazine hydrate supported on solid materials such as alumina, silica gel, and clays (Scheme 5.11).

Similarly, hydroxylamine supported on clay can convert arylaldehydes into nitriles (Scheme 5.12) [41].

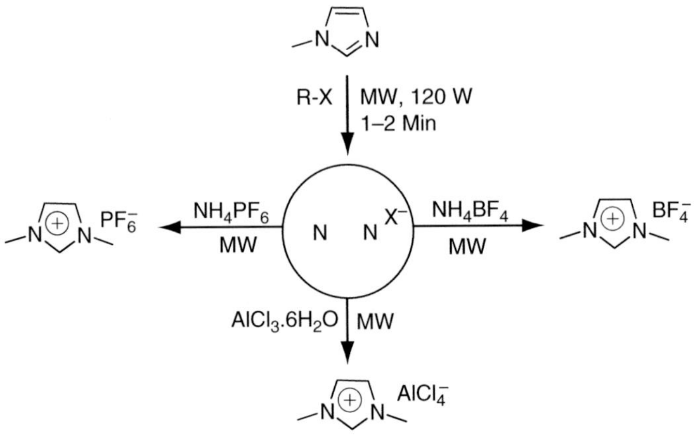

Scheme 5.13 MW-assisted synthesis of ionic liquids

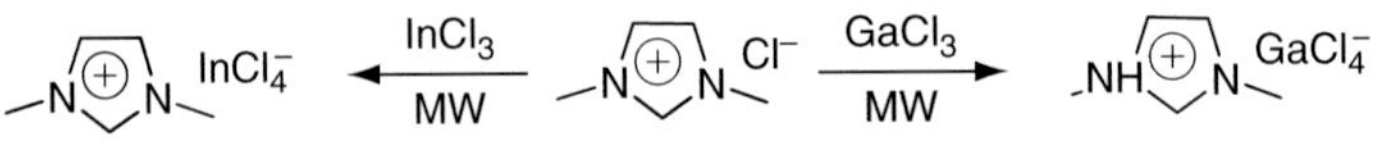

Scheme 5.14 MW-assisted synthesis of In and Ga containing ionic liquids

MW-Assisted Synthesis of Ionic Liquids and Their Application as Catalysts

Ionic liquids (ILs) have been the focus of attention due to their potential in a variety of commercial applications, such as electrochemistry, heavy metal ion extraction, phase transfer catalysis, and polymerization, and as substitutes for conventional volatile organic solvents (Rogers and Seddon 2002) [42, 43]. Additional environmentally friendly attributes of these ILs include negligible vapor pressure, potential for recycling, compatibility with various organic compounds and organometallic catalysts, and ease of separation of products from reactions.

Solvent-Free Synthesis of Ionic Liquids

Ionic liquids, being polar and ionic in character, couple with MW irradiation very efficiently and are, therefore, ideal MW-absorbing candidates for enhancing chemical reactions. The first efficient preparation of 1,3-dialkylimidazolium halides via MW irradiation was developed by Varma et al. (Scheme 5.13) [44, 45]. The reaction time was reduced to minutes and the method avoids the excessive use of organic solvents as the reaction medium. These syntheses can also be ideally carried out using ultrasound under solvent-free conditions [46].

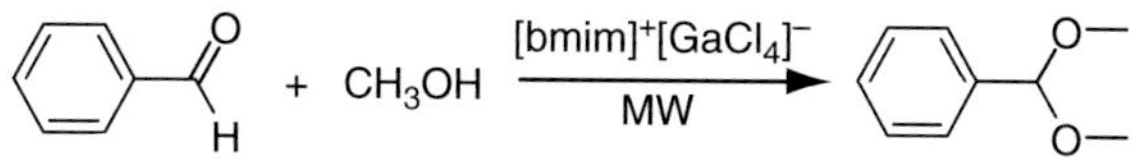

Scheme 5.15 Ionic liquid-catalyzed protection of carbonyl compounds

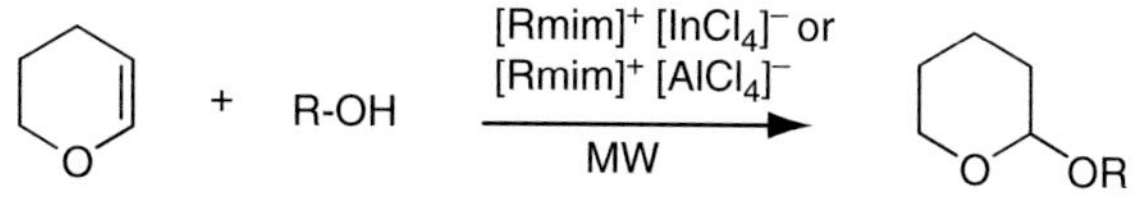

Scheme 5.16 Ionic liquid-catalyzed THP-protection of alcohols

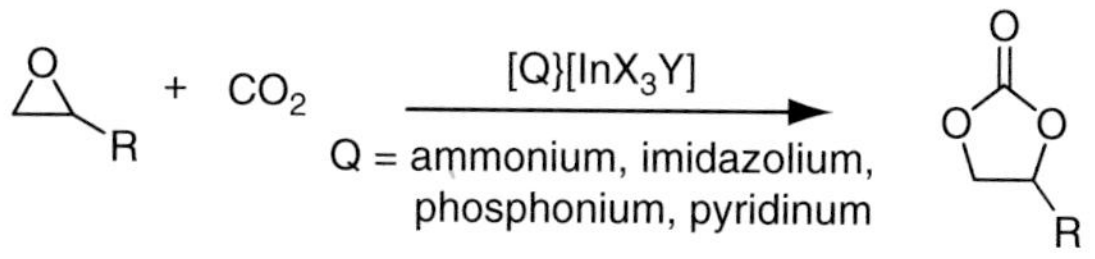

Scheme 5.17 Synthesis of cyclic carbonates catalyzed by tetrahaloindate(III)-based ionic liquids

Metal-bearing classes of ILs, [Rmim][InCl$_4$] and [bmim][GaCl$_4$], were prepared using a solvent-free MW procedure (Scheme 5.14) and utilized as catalysts [47, 48]. This approach is much faster, efficient, and eco-friendly as it does not use any organic solvent.

Synthetic Application of ILs as Catalysts

Ionic liquids have emerged as a new class of green solvents for chemical processes and transformations (Rogers and Seddon 2002). Their polarity renders them good solvents for various organic reactions and catalyzes, including the dissolutions of renewable materials such as cellulose [49].

1-Butyl-3-methylimidazolium tetrachlorogallate, [bmim][GaCl$_4$], has been used as an active catalyst for the efficient acetalization of aldehydes (Scheme 5.15) [48].

Imidazolium-based tetrachloroaluminates ([Rmim][AlCl$_4$]) [50] and tetrachloroindate ([Rmim][InlCl$_4$]) [47] have been used as recyclable catalysts for the efficient protection of alcohols to form tetrahydropyranyl (THP) derivatives (Scheme 5.16).

The coupling reaction of carbon dioxide (CO$_2$) with epoxides generates the five-membered cyclic carbonates which are precursors for polymeric materials such as polyurethanes and polycarbonates, serve as aprotic polar solvents, and are utilized as intermediates in the production of pharmaceutical and fine chemicals. The catalytic amounts of tetrahaloindate(III)-based ILs are found to exhibit the highest catalytic activities for the synthesis of cyclic carbonates (Scheme 5.17) [51].

Reactions in Aqueous Medium

Solvent-free reactions undoubtedly minimize the environmental impacts resulting from the use of solvents in chemical production but there are limitations in view of

Fig. 5.1 Effect of microwaves on the reaction mixture in polar medium

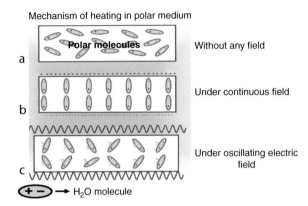

the poorly understood heat- and mass-transfer issues under these conditions [2]. Fluorous solvents, ionic liquids [52], aqueous systems [7], and supercritical carbon dioxide have emerged as alternatives in this movement. Water, which is naturally abundant and can be contained because of its relatively higher vapor pressure, appears to be a sustainable alternative because of its nontoxic, noncorrosive, and nonflammable nature [6–8].

In addition to the isolation of products from the aqueous medium, the insolubility of most of the organic substrates in water is a challenge. However, some of these issues can be overcome by using the MW heating technique wherein water is rapidly heated to high temperatures enabling it to act like a pseudo-organic solvent. Two key mechanisms, namely, dipolar polarization and ionic conduction of water molecules (Fig. 5.1), come into play upon irradiation of a reaction mixture in an aqueous medium by microwaves, invoking two distinguishing effects:

(a) *Specific microwave effect*: the dipole–dipole-type interaction of the dipolar water molecules and reactants with the electric field component of MW (Fig. 5.1b), resulting in what is described by some as a specific MW (nonthermal) effect which is a controversial and debatable issue among chemists. Kappe et al. have shown that this effect is mainly due to thermal phenomenon and is thus not nonthermal [53].

(b) *Thermal effect*: the dielectric heating that ensues from the tendency of dipoles of water molecules and reactants to follow the rapidly alternating electric fields and induce energy dissipation in the form of heat through molecular friction and dielectric loss (Fig. 5.1c).

It has been observed that various organic reactions can be conducted in an aqueous medium using MW irradiation without using any phase-transfer catalyst (PTC). This is because water at higher temperature behaves as a pseudo-organic solvent, as the dielectric constant decreases substantially and an ionic product increases the solvating power toward organic molecules to be similar to that of acetone or ethanol [9].

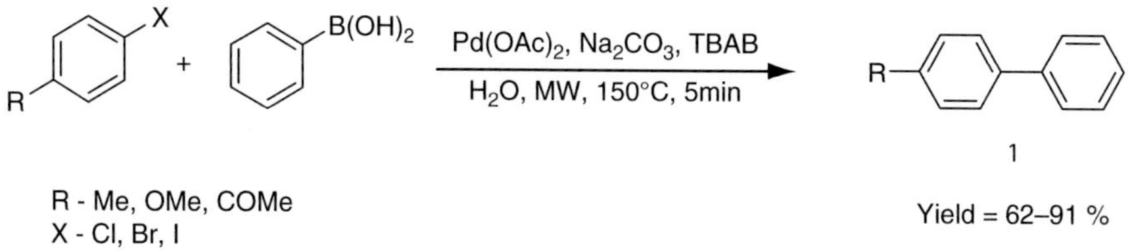

Scheme 5.18 Aqueous MW-assisted Suzuki reaction

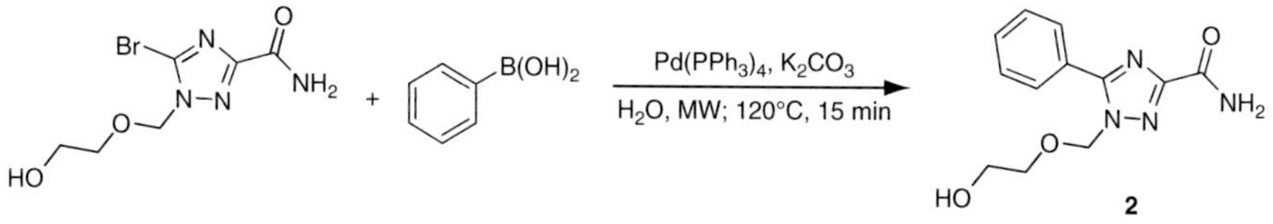

Scheme 5.19 Aqueous MW-assisted Suzuki reaction of triazole

Although MW-assisted reactions in organic solvents have developed rapidly for a variety of applications in drug discovery [54] and organic synthesis [7], the focus has now shifted to the more environmentally benign methods, which use greener solvents such as polyethylene glycol (PEG) [55] and water [6–8]. Some illustrative recent synthetic reactions in aqueous reaction medium using MW irradiation are summarized below.

Coupling Reactions (Suzuki, Heck, and Sonogashira)

Among the heterogeneous palladium (Pd)-catalyzed reaction systems, MW-assisted coupling reaction in aqueous medium is a "greener" choice by chemists. Suzuki reaction has been studied in aqueous medium using MW irradiation (Scheme 5.18) for the preparation of various biaryl derivatives [1] from aryl halide and phenylboronic acid [56] in presence of tetra butyl ammonium bromide (TBAB); the poor yield for aryl chlorides as compared to other halides could be improved by conducting the reaction using simultaneous cooling technique in conjugation with MW heating.

The versatility of aqueous MW-assisted Suzuki reaction has also been demonstrated in the synthesis of natural products and heterocyclic compounds [57].

5-Aryltriazole acyclonucleosides [2] bearing aromatic groups on the triazole ring have been synthesized via the Suzuki coupling reaction in aqueous solution (Scheme 5.19) [58]. This high-yield coupling method directly afforded the product and involved no protection and deprotection steps, thus providing a convenient one-step procedure for 5-aryltriazole acyclonucleosides.

Benzothiazole-based Pd(II) complexes 3 and 4 have been found to be highly active catalysts in Suzuki and Heck cross-coupling reactions of aryl chlorides and

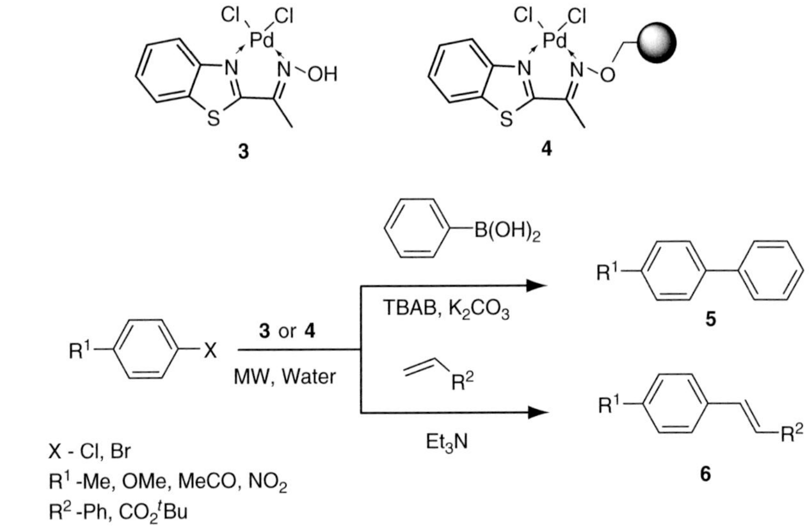

Scheme 5.20 Aqueous MW Suzuki and Heck reactions using benzothiazole-based Pd(II) complexes

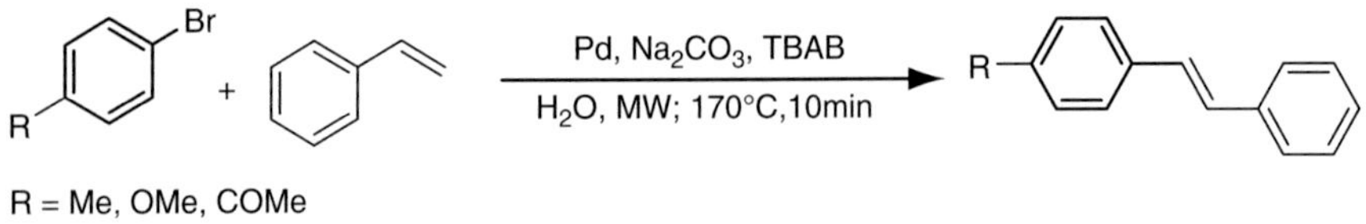

R = Me, OMe, COMe

Scheme 5.21 Aqueous MW-assisted Heck reaction

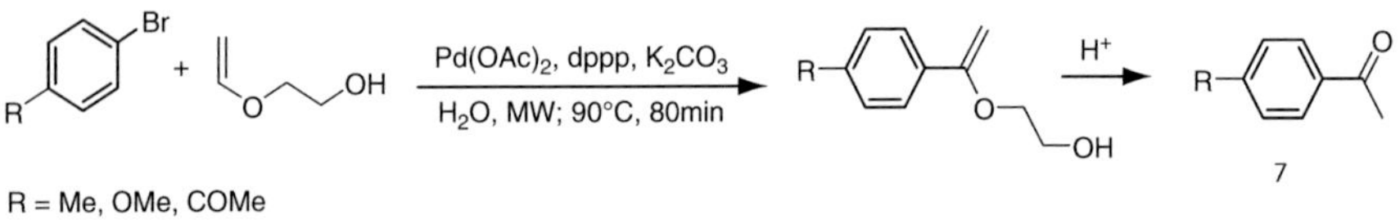

R = Me, OMe, COMe

Scheme 5.22 Aqueous MW-expedited internal Heck reaction

bromides with olefins and arylboronic acids under aqueous MW irradiation conditions. The immobilized catalyst 4 was found to have higher durability compared with its mobilized counterpart 3 (Scheme 5.20) [59].

Arvela and Leadbeater performed the Heck coupling reaction in water using MW heating (Scheme 5.21), with Pd catalyst concentrations as low as 500 ppb [60].

Larhed et al. have recently reported highly regio-selective and fast Pd(0)-catalyzed internal R-arylation of ethylene glycol vinyl ether with aryl halides in aqueous medium (Scheme 5.22) in presence of 1,3-bis(diphenylphosphino)propane (dppp) [61].

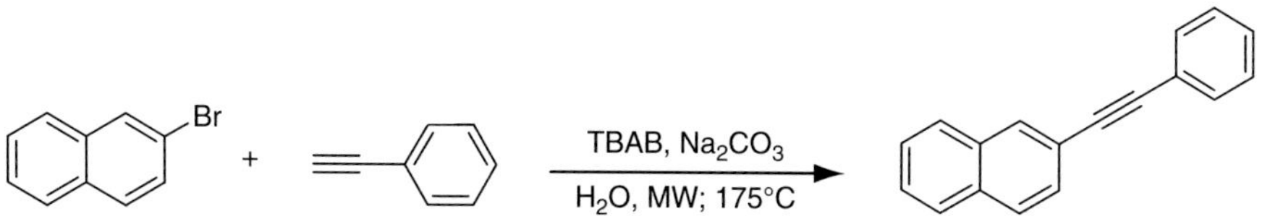

Scheme 5.23   Aqueous MW-assisted Sonogashira reaction

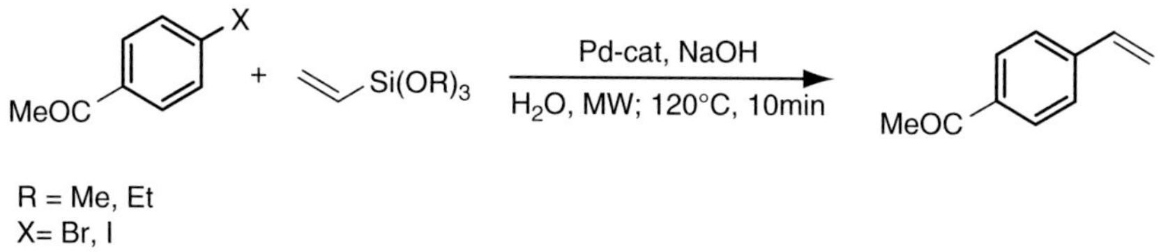

R = Me, Et
X= Br, I

Scheme 5.24   Aqueous MW-assisted Hiyama reaction

Aryl bromides and iodides were efficiently converted to corresponding acetophenones [7] in high yields in water using ethylene glycol vinyl ether as the olefin and potassium carbonate as the base. This Pd(0)-catalyzed method is advantageous as no heavy metal additives or ionic liquids are necessary. The reaction proceeded cleanly without any noticeable byproduct formation and avoided the need for inert atmosphere. MW irradiation has proven to be beneficial in activation of aryl chlorides toward the internal Heck arylation.

An aqueous Sonogashira coupling reaction was reported by Eycken et al. under MW irradiation (Scheme 5.23). This reaction precludes the need for copper(I) or any transition-metal phosphane complex, thus overcoming the problem of toxicity and air sensitivity of transition-metal complexes, as well as the use of expensive phosphane ligands [62].

Hiyama Cross-Coupling Reaction

The relatively benign and stable organosilanes have emerged as useful entities for cross-coupling reactions. This is exemplified by the Hiyama cross-coupling reactions between vinylalkoxysilanes and aryl halides which were promoted by aqueous sodium hydroxide under fluoride-free conditions and carried out using MW irradiation in aqueous medium (Scheme 5.24) [63].

Stille Reaction

Stille reaction between organo-tin and aryl halide was achieved using aqueous MW chemistry by van der Eycken et al. (Scheme 5.25) [64].

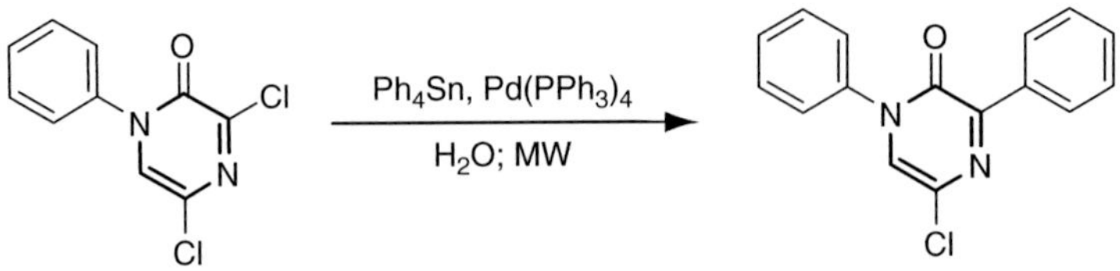

Scheme 5.25 Aqueous MW-assisted Stille reaction

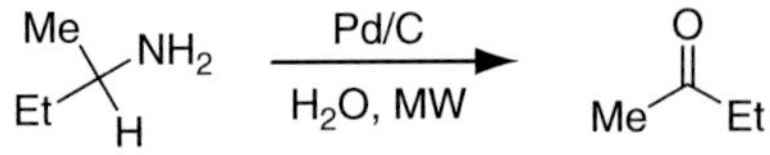

Scheme 5.26 MW-assisted aqueous retro-reductive amination reaction

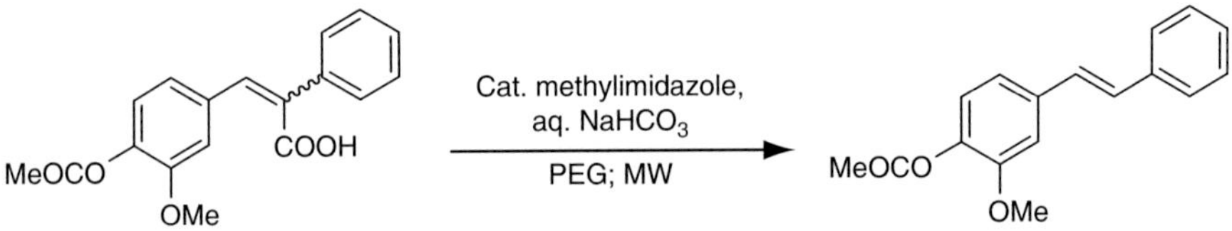

Scheme 5.27 Decarboxylation of cinnamic acid in aqueous medium using microwaves

Transformation of Amines to Ketones

Although the selective transformation of amines to ketones is a common biological process, conversion of amines to ketones by chemical means are rather limited. A retro-reductive, MW-assisted, Pd-catalyzed amination reaction for direct conversion of amines to ketones in water was accomplished by Olah et al. (Scheme 5.26). This expeditious reaction proceeds smoothly without any heavy metal–based oxidant or volatile organic solvents [65].

Decarboxylation of Cinnamic Acids

A metal-free protocol for decarboxylation of substituted α-phenylcinnamic acid derivatives in aqueous media was developed by Sinha et al. [66], wherein a catalytic amount of methylimidazole in aqueous $NaHCO_3$ in polyethylene glycol (PEG) under MW irradiation conditions furnished the corresponding para/ortho hydroxylated (E)-stilbenes (Scheme 5.27). This is clearly a clean alternative to the conventional multistep methods that often involve the use of toxic quinoline and a copper salt combination. The critical role of water in facilitating the decarboxylation highlights the synthetic utility of water-mediated organic transformations.

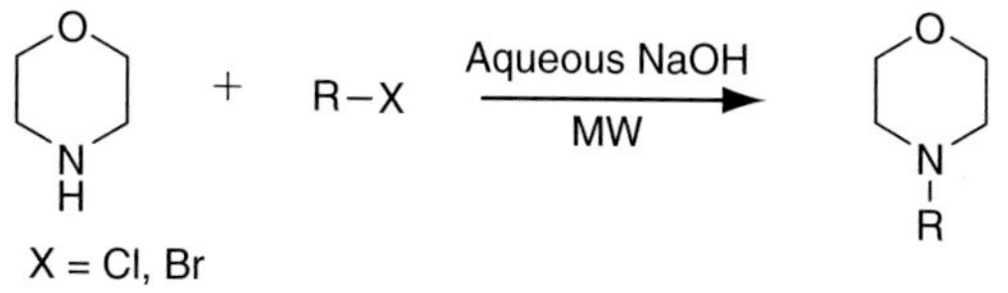

Scheme 5.28 *N*-Alkylation in basic water using MW irradiation

Reactions in Near-Critical Water

Dallinger and Kappe examined several MW-assisted organic reactions in near-critical water (NCW) in the temperature range of 270–300°C. The hydrolysis of esters or amides, the hydration of alkynes, Diels–Alder cycloadditions, pinacol rearrangements, and the Fischer indole synthesis were successfully performed in MW-generated NCW without the addition of an acid or base catalyst [67]. This study demonstrated that it is technically feasible to perform MW synthesis in water on scales from 15 to 400 ml at temperatures of up to 300°C and 80 bars of pressure in a multimode MW reactor.

Synthesis of Heterocycles

Heterocyclic compounds are a special class among pharmaceutically significant natural products and synthetic compounds [68, 69]. The remarkable ability of heterocyclic nuclei to serve both as biomimetics and reactive pharmacophores has provided a unique value as traditional key elements for numerous drugs. Conventional organic synthesis is too slow to satisfy the demand for generation of small molecules and this void has been filled by combinatorial and automated chemistry to meet the increasing requirement of new compounds for drug discovery, expediently [70]. The efficiency of MW flash-heating has resulted in dramatic reductions in reaction times from days and hours to minutes and seconds [5, 6, 9, 71–73].

Nitrogen Heterocycles

Nitrogen heterocycles are abundant in nature and are of great significance to life because their structural subunits exist in many natural products such as vitamins, hormones, antibiotics, and alkaloids, as well as herbicides, dyes, and pharmaceuticals [68].

An expedient *N*-alkylation of nitrogen heterocycles was reported in aqueous media under MW irradiation conditions (Scheme 5.28) [74].

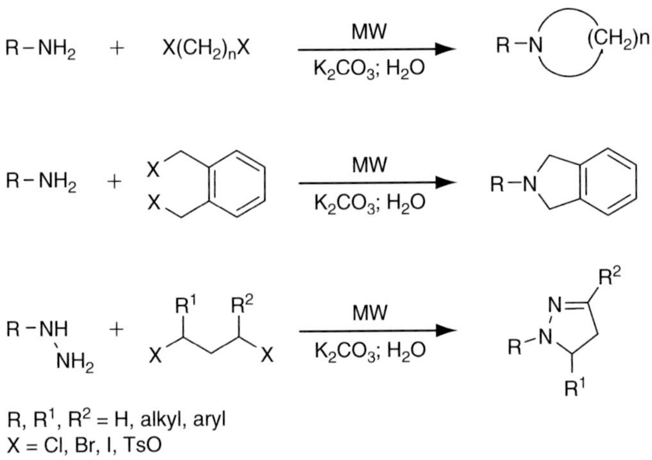

Scheme 5.29 Synthesis of *N*-heterocycles in aqueous media using MW irradiation

The double *N*-alkylation of primary amines and hydrazine derivatives (Scheme 5.29) in carbonated water using microwaves provided access to the synthesis of nitrogen-containing heterocycles, namely, substituted azetidines, pyrrolidines, piperidines, azepanes, *N*-substituted 2,3-dihydro-1H-isoindoles, 4,5-dihydropyrazoles, pyrazolidines, and 1,2-dihydrophthalazines. The readily available alkyl dihalides (or ditosylates) thus provide a facile entry to important classes of building blocks in natural products and pharmaceuticals [75–77].

This MW-accelerated general approach shortened the reaction time significantly and utilized readily available amines and hydrazines with alkyl dihalides or ditosylates to assemble two C–N bonds in a simple S_N2-like sequential heterocyclization experimental protocol which has never been fully realized under conventional reaction conditions. The strategy avoids multistep reactions, functional group protection/deprotection sequences, and eliminates the use of expensive phase transfer and transition metal catalysts. The experimental observations are consistent with the mechanistic postulation wherein the polar transition state of the reaction is favored by MW irradiation with respect to the dielectric polarization nature of MW energy transfer. In large-scale experiments, the phase separation of the desired product in either solid or liquid form from the aqueous media can facilitate product purification by simple filtration or decantation instead of tedious column chromatography, distillation, or extraction processes, which reduces the usage of volatile organic solvents [77].

A direct Grignard type of addition of alkynes to in situ generated imines, from aldehyde and amines, was catalyzed by CuBr and provides a rapid and solvent-free approach access to propargylamines in excellent yields (Scheme 5.30) [78].

Direct synthesis of these cyclic ureas using a MW-assisted protocol that proceeds rapidly in the presence of ZnO were reported (Scheme 5.31) [79].

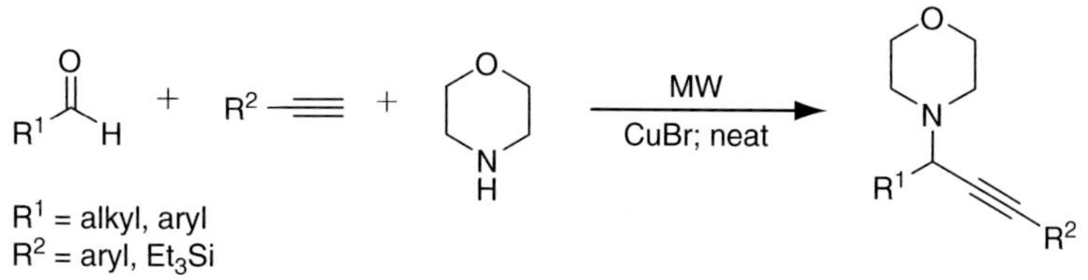

Scheme 5.30 CuBr-catalyzed solvent-free synthesis of propargylamines

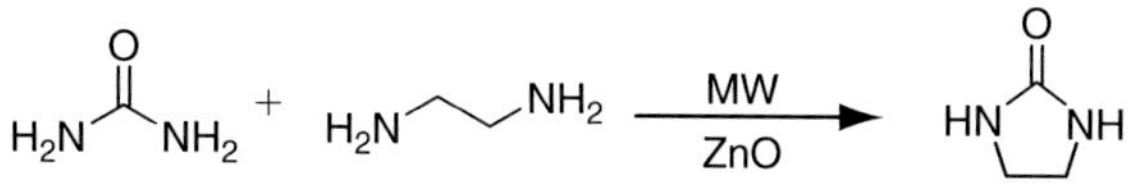

Scheme 5.31 ZnO-catalyzed MW synthesis of imidazolidine-2-one

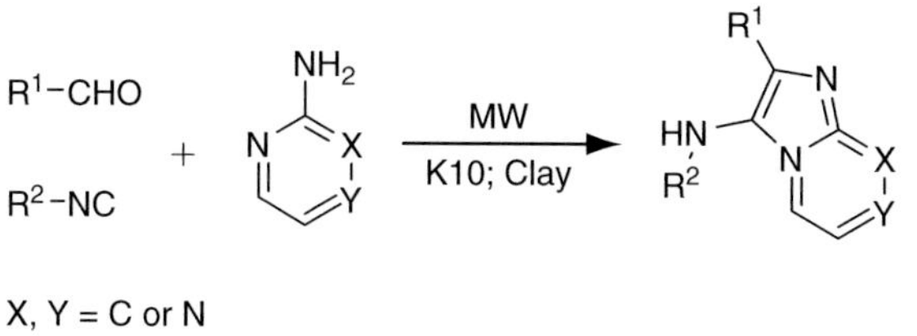

X, Y = C or N

Scheme 5.32 Clay-catalyzed solvent-free synthesis of annulated *N*-heterocycles

The imidazo[1,2-a] annulated nitrogen heterocycles bearing pyridine, pyrazine, and pyrimidine moities constitute a class of biologically active compounds that can be assembled via atom-economic, one-pot condensation of aldehydes, amines, and isocyanides (three-component Ugi reaction) using a MW approach (Scheme 5.32) in the presence of recyclable montmorillonite K-10 clay under solvent-free conditions [80].

Dihydropyrimidinones are an important class of biologically active organic compounds that were synthesized under solvent-free conditions [81] or by an environmentally benign Biginelli protocol using polystyrene sulfonic acid (PSSA) as a catalyst (Scheme 5.33). A very simple eco-friendly isolation procedure entails the filtration of the precipitated products [82].

Oxygen-Heterocycles

Heterocycles bearing oxygen atom are important classes of building blocks in organic synthesis and have attracted the attention of medicinal chemists over the years [69].

The solvent-free synthesis of 2-aminosubstituted isoflav-3-enes was developed, which can be carried out in one pot using microwaves via the in situ generation of enamines and their subsequent reactions with salicylaldehydes (Scheme 5.34) [83].

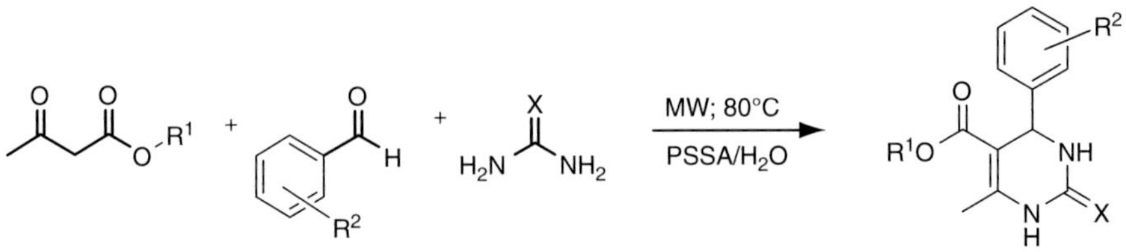

Scheme 5.33 Biginelli reaction using microwaves in aqueous medium

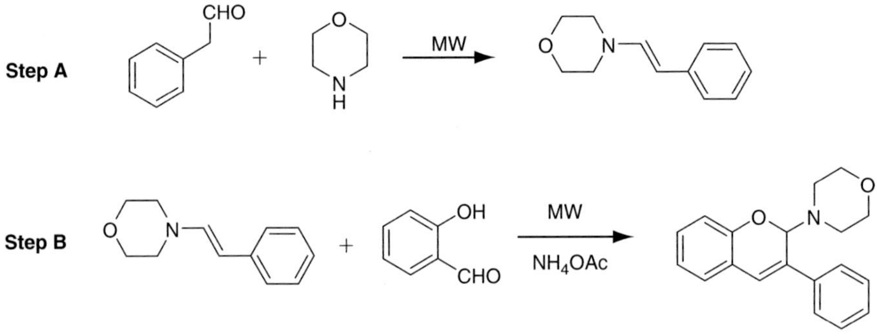

Scheme 5.34 One-pot solventless synthesis of isoflav-3-enes

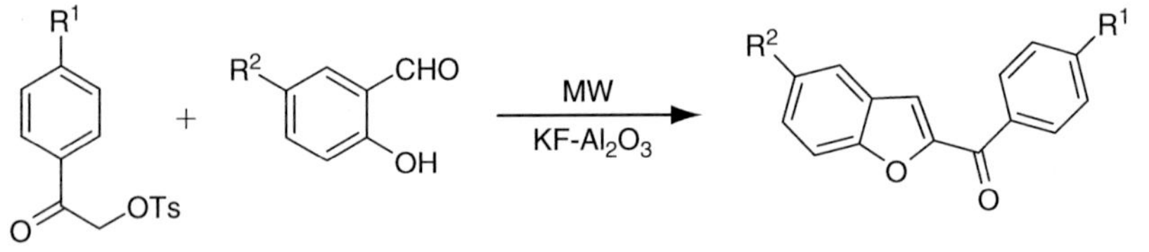

Scheme 5.35 MW assisted synthesis of 2-aroylbenzo[*b*]furans

This environmentally friendly procedure does not require azeotropic removal of water using large excesses of aromatic hydrocarbon solvents for the generation of enamines or the activation of the catalyst.

The expeditious solvent-free syntheses of 2-aroylbenzo[*b*]furans were developed from readily accessible α-tosyloxyketones and mineral oxides in processes that are accelerated by exposure to microwaves (Scheme 5.35) [84].

Dioxane rings, common structural motifs in numerous bioactive molecules, can be assembled via tandem bis-aldol reaction of ketones with paraformaldehyde in aqueous media catalyzed by PSSA under MW irradiation conditions to produce 1,3-dioxanes (Scheme 5.36) [85].

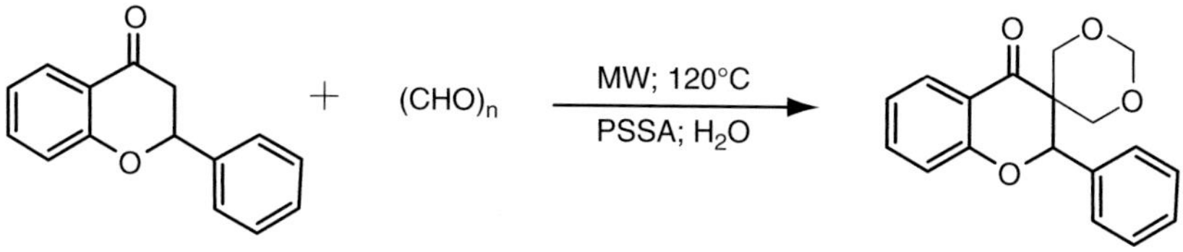

Scheme 5.36 PSSA-catalyzed synthesis of 1,3-dioxanes in aqueous media

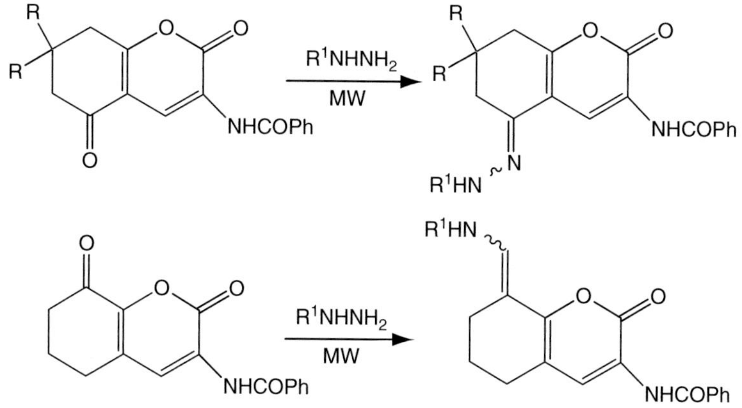

Scheme 5.37 Synthesis of heterocyclic hydrazones under solvent-free conditions (Ref. [85])

Heterocyclic Hydrazones

The first example of a reaction between two solids, under solvent-free and catalyst-free environment, was accomplished by Varma et al. The reaction of neat 5- or 8-oxobenzopyran-2(1H)-ones with a variety of aromatic and heteroaromatic hydrazines provides rapid access to several synthetically useful heterocyclic hydrazones (Scheme 5.37) [86].

An aqueous protocol for the synthesis of heterocyclic hydrazones using PSSA as a catalyst was also developed by this group recently (Scheme 5.38). The protocol entails the simple filtration as the product isolation step in a reaction that proceeds in the absence of any organic solvent under MW irradiation [87].

Sulfur Heterocycles

Synthesis of thiazoles often involves utilization of lachrymatory starting materials, phenacyl halides, and hazardous reagents under drastic conditions. Varma and coworkers accomplished the synthesis of 1,3-thiazoles (which are not easily accessible under classical heating conditions) in excellent yields [83] from thioamides and α-tosyloxyketones catalyzed by montmorillonite K-10 clay, (Scheme 5.39).

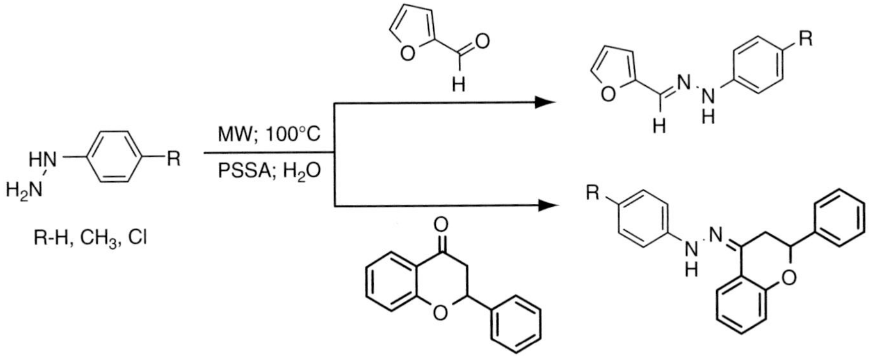

Scheme 5.38 Synthesis of hydrazone derivatives of furaldehyde and flavanone in water

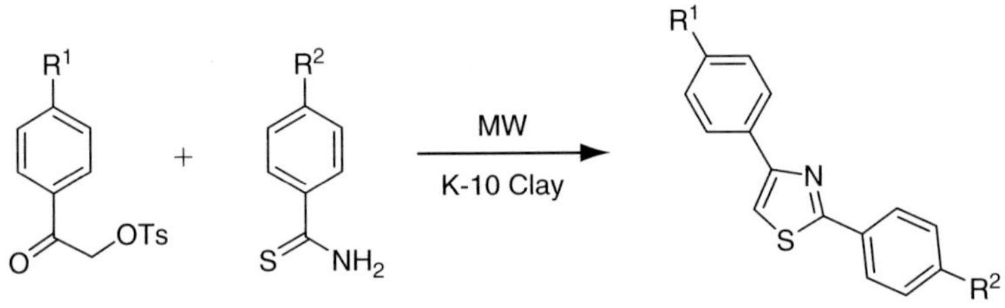

Scheme 5.39 MW-assisted solventless synthesis of 1,3-thiazoles

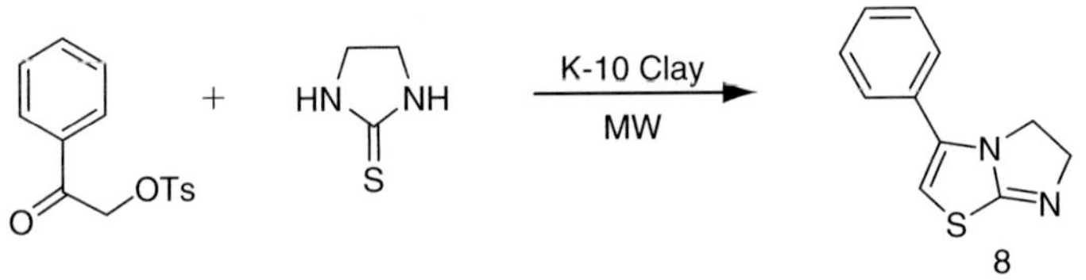

Scheme 5.40 MW-assisted solventless synthesis of bridgehead-thiazoles

The strategy was extended to a concise synthesis of bridgehead 3-aryl-5,6-dihydroimidazo[2,1-*b*][1,3]thiazoles [8], which are difficult to obtain, requiring heating over an extended period of time, and uses α-haloketones or α-tosyloxyketones under strongly acidic conditions. The solventless mixing of α-tosyloxyketones with thioamides in the presence of montmorillonite K-10 clay and exposing the reactants to MW irradiation for 3 min affords the substituted bridge-head thiazoles (Scheme 5.40) [83].

A novel one-pot solventless synthesis of 1,3,4-oxadiazoles and 1,3,4-thiadiazoles via condensation of acid hydrazide and triethyl orthoalkanates under MW irradiation was recently developed (Scheme 5.41) [88].

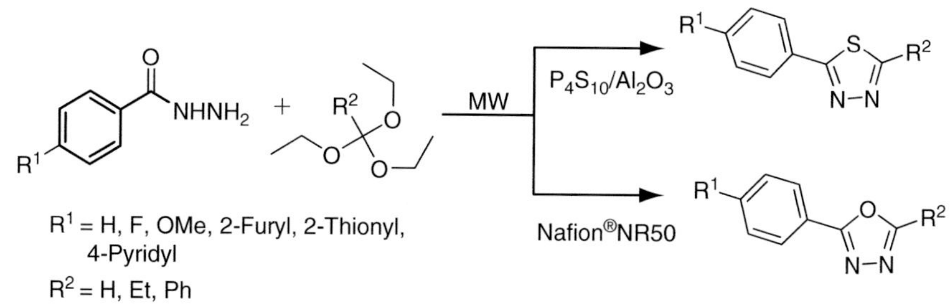

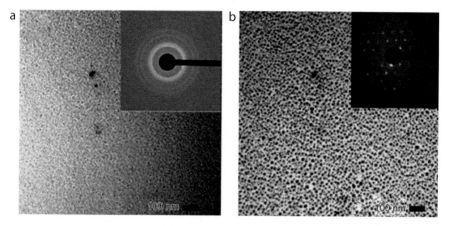

Scheme 5.41 One-pot solventless synthesis of 1,3,4-oxadiazoles and 1,3,4-thiadiazoles

Fig. 5.2 Au nanoparticles synthesized with (**a**) glucose and (**b**) sucrose under MW irradiation conditions (Ref. [88])

Synthesis of Nanomaterials

The use of MW irradiation as an efficient, environmentally friendly, and economically viable heating method for the production of nanomaterials has increased in recent years. Synthesis of gold (Au) nanoparticles with various morphologies using sugar as reducing agent under MW heating was reported by Nadagouda and Varma [89]. The authors reported the synthesis of nanoparticles (Fig. 5.2), prism, hexagons, and rods using a wide variety of sugars (D-glucose, sucrose, mannose, etc.) as reducing agents. The highly dispersed nanoparticles are produced in the size range of 10–20 nm in less than a minute. When low concentration of sugar is used, hexagonal, triangular, and rod-shaped particles with submicron sizes are obtained.

Baruwati and Varma [90] reported the MW-assisted synthesis of Au nanoparticles along with other noble metal nanoparticles using red wine and red grape pomace extract as a single source of reducing and capping agent; highly dispersed nanoparticles are produced in the size range 5–20 nm (Fig. 5.3) and only

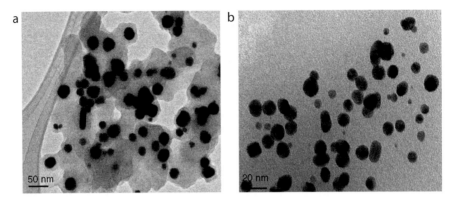

Fig. 5.3 Au nanoparticles synthesized using (**a**) red wine and (**b**) red grape pomace at MW power level 50 W, time 1 min (Ref. [89])

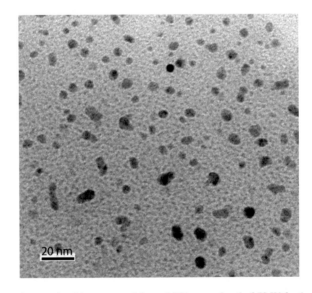

Fig. 5.4 TEM micrograph of Ag nanoparticles at MW power level of 50 W for 1 min (Ref. [90])

45–60 s reaction time is required. This general method produces silver (Ag), Pd, platinum (Pt), and iron (Fe) in the particle range 5–20 nm. In contrast, white wine produces highly agglomerated nanoparticles because of lack of polyphenolics which are present in red grape pomace as well as in red wine.

Highly dispersed Ag nanoparticles in the size range 5–10 nm were prepared by Varma et al. [91] using aqueous MW approach within a minute utilizing a completely benign and ubiquitous tripeptide, glutathione (Structure 1). Normally, crystalline Ag nanoparticles (Fig. 5.4) are obtained and the effect of MW power on the morphology of ensuing Ag nanoparticles has also been investigated for this green and sustainable procedure which is adaptable for the synthesis of Pd, Pt, and Au nanoparticles

Chemical structure of tripeptide, Glutathione

Greener Synthesis of Nanomaterials

The development of solution-based controlled synthesis of nanomaterials via a bottom-up approach often uses toxic reducing and capping agents and some dispersants. Greener alternatives, especially using a biomimetic approach, are now possible wherein benign entities such as vitamin B_1 [92], vitamin B_2 [93], vitamin C [94], tea polyphenols [95], simple sugars [89], and PEG [96] can generate nanoparticles. These nanoparticles can be cross-linked under the influence of microwaves to form nanocomposites with cellulose [97] or polyvinyl alcohol (PVA) [98].

Vitamin B_1 was used for one-step synthesize of Pd nanobelts, nanoplates, and nanotrees without using any special capping agents (Fig. 5.5) [92]. Depending upon the Pd concentration, Pd nanoparticles crystallized in various shapes and sizes. At lower Pd concentration, plate-like shape was obtained. The Pd plates were grown on a single Pd nanorod backbone mimicking the leaf-like structures. However with increase in Pd concentration, formation of tree-like structures was observed. Upon further increase in concentration, Pd nanoplates become thicker by vertically aligning themselves to form ball-like shape and this general protocol can be extended to prepare other noble nanomaterials such as Au and Pt. The Pd nanoparticles showed excellent catalytic activity for several C–C bond-forming reactions such as Suzuki, Heck, and Sonogashira reactions under MW irradiation conditions [92].

The control of the size and morphology of nanostructures to tailor the physical and chemical properties is an important aspect in nanoscience. Recently, Varma et al. designed a convenient method for the synthesis of metal oxides with 3D nanostructures [99, 100] which are obtainable from hydrolysis of inexpensive starting materials in water without using any reducing or capping reagent. This economical and environmentally sustainable synthetic concept could ultimately enable the fine-tuning of material responses to magnetic, electrical, optical, and mechanical stimuli. Well-defined morphologies, including octahedron, sphere, triangular rod, pine, and hexagonal snowflake with particles in the size range of 100–500 nm were obtained (Fig. 5.6). Nano-ferrites were then functionalized and

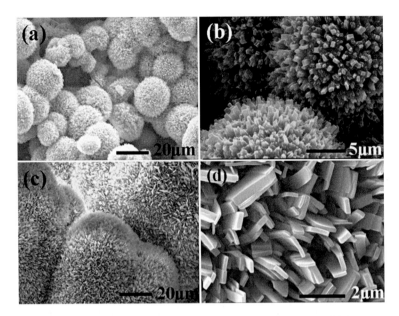

Fig. 5.5 SEM images of Pd nanoparticles generated using vitamin B$_1$ (Ref. [91])

Fig. 5.6 MW-assisted synthesis of metal oxides with well-defined morphologies (Ref. [99])

coated with Pd metal, which catalyzed various C–C coupling and hydrogenation reactions with high yields. In addition, the effortless recovery and increased efficiency, combined with the inherent stability of this catalyst, rendered the method sustainable [99, 100]. In view of these unique morphologies, synthesized nanomaterials will have significant applications in biomedical science and catalysis.

MW-Assisted Synthesis of Quantum Dots (QD) in Aqueous Medium

During the last decade, synthesis of high-quality semiconductor nanocrystals popularly referred to as quantum dots, QD, has been a subject of intense research because of their size-dependent properties. These colloidal semiconductor nanoparticles are much superior compared to the conventional dye molecules in terms of flexible photoexcitation, sharp photoemission, and superb resistance to photobleaching [101]. Their optical properties could be manipulated by changing the size and composition to meet specific wavelength requirements [102] and they have found application in quantum-dot lasers, optoelectronics, nonlinear optical devices, solar cells, and bio-tagging [103–108]. Synthesis of water-soluble ZnSe nanocrystals in aqueous medium under MW conditions has been reported by Qian et al. [109]. These nanocrystals are water soluble, have high crystallinity, and their photoluminescence (PL) quantum yield ranges up to 17%. These properties are marked improvements when compared to ZnSe QDs prepared by conventional aqueous synthesis method. The method has eliminated the use of expensive, environmentally unfriendly reagents such as trioctylphosphine (TOP), tributylphosphine (TBP), and trioctylphosphine oxide (TOPO).

ZnS nanoballs were synthesized in saturated water solutions under MW irradiation conditions by Zhao et al. [110] under ambient air; the ensuing products were highly crystalline and about 300 nm in diameter. CdSe, PbSe, and $Cu_{2-x}Se$ nanoparticles were prepared under MW irradiation conditions by Zhu et al. [111] wherein $CdSO_4$, $Pb(Ac)_2$, and $CuSO_4$ reacted with Na_2SeSO_3 in water in the presence of complexing agents: potassium nitrilotriacetate ($N(CH_2\text{-}COOK)_3$-NTA) for CdSe and PbSe or triethanol amine for $Cu_{2-x}Se$ in a MW refluxing system. Although the method is simple and adequate for producing CdSe nanoparticles in 4–5 nm range, PbSe, and $Cu_{2-x}Se$ nanoparticles were found to be bigger in size in the range 30–80 nm and often agglomerated. The method also describes how different phases of CdSe could be obtained by varying the MW heating times. A 10 min irradiation led to the formation of CdSe in the cubic (sphalerite) phase, while after 30 min, CdSe obtained is in the hexagonal cadmoselite phase.

A one-pot synthetic method for the synthesis of CdSe/ZnS core/shell QDs using MW radiation was reported by Schumacher et al. [112] and is based on the addition of a water-soluble Zn^{2+} complex, $Zn(NH_3)_4^{2+}$ to a solution containing CdSe initial nanocrystals and 3-mercaptopropionic acid (MPA). Subsequent MW heating for less than 2 h generated high-quality CdSe/ZnS-based QDs possessing good photoluminescent quantum yield (13%) and biocompatibility.

A seed-mediated approach for rapid synthesis of high-quality alloyed quantum dots (CdSe–CdS) in aqueous phase by MW irradiation was reported by Qian et al. [113]. Initially, CdSe seeds were first formed by the reaction of NaHSe and Cd^{2+}, and then alloyed quantum dots (CdSe–CdS) were rapidly generated by the release of sulfide ions from 3-mercaptopropionic acid as sulfide source with MW irradiation.

Magnetic Nanoparticles

Recently, magnetic nanoparticles have emerged as viable alternatives to conventional materials, as a readily available, robust, high-surface-area heterogeneous catalyst support [114]. Post-synthetic surface modification of magnetic nanoparticles controls desirable chemical functionality and enables the generation of catalytic sites on the surfaces of resulting nano-catalyst. Their insoluble

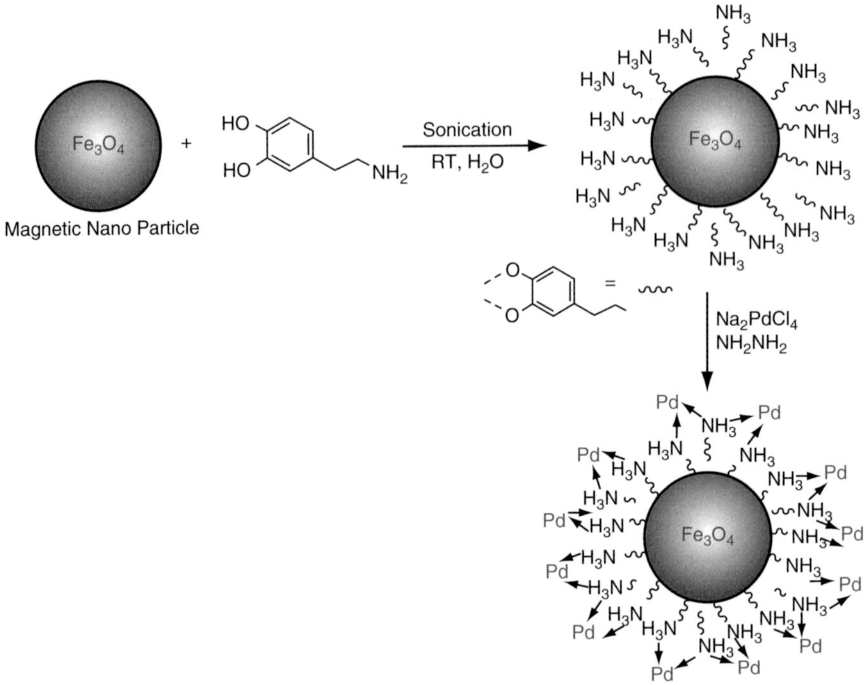

Scheme 5.42 Dopamine functionalized nanoferrite-Pd-catalyst. From Polshettiwar and Varma (Ref. [119])

character together with paramagnetic nature enables effortless separation of these nano-catalysts from the reaction mixture using an external magnet, which eliminates the necessity of catalyst filtration. These novel nano-catalysts bridge the gap between homogeneous and heterogeneous catalysis, thus preserving the desirable attributes of both the systems.

This concept was recently explored for the development of other metal catalysts [99, 100, 115–119]. Varma and coworkers developed a convenient synthesis of nano-ferrite-supported Pd catalyst from inexpensive starting materials in water (Scheme 5.42) [115, 116]. This catalyst catalyzes the oxidation of alcohols and olefins with high turnover numbers and excellent selectivity. Also, being magnetically separable, this approach eliminates the requirement of catalyst filtration after completion of the reaction, which is an additional sustainable attribute of this oxidation protocol.

MW-Assisted Nano-Catalysis in Water

Chemists have been under intense pressure to develop newer methods, which are expeditious and environmentally benign. One of the better alternatives is the use of nano-catalysis in conjunction with MW heating technology. The efficiency of MW heating has resulted in dramatic reductions in reaction times, reduced from days to minutes, which is very significant in process chemistry for the expedient generation of organics and nanomaterials [5, 6, 121–125].

Naturally abundant water is a good alternative because of its nonflammable, nontoxic, and noncorrosive nature, which may help reduce the dependence of chemists on hazardous solvents [6–8, 67, 74–77, 85, 126, 127]. Additionally, water can be contained because of its relatively lower vapor pressure when compared to organic solvents, thus rendering it a sustainable alternative. Interestingly, the combination of MW and aqueous medium has shown excellent benefits such as shorter reaction times, homogeneous in-core heating, and enhanced yields and selectivity [6–8, 67, 74–77]. In addition to these microwave "thermal effects" and "nonthermal effects" [128], there are additional benefits of using microwaves for nano-catalyzed aqueous protocols as described below:

1. Selectivity toward water
   Microwave heating depends on the composition and structure of molecules (i.e., their dielectric properties) and this property can facilitate selective heating. Microwaves initiate the rapid and intense heating of polar molecules such as water, while nonpolar molecules do not absorb the radiation and are not heated. Loupy and Varma [121], Strauss [122], and Larhed [129] demonstrated that this selective heating can be exploited to develop a high yield rapid MW protocol using a two phase (polar–nonpolar) solvent system. The advantageous use of water in the MW-assisted processes, especially without the use of phase-transfer catalysts [121], has been well demonstrated [6–8, 67, 74–77].

2. Selectivity toward catalyst
   Selective heating can be exploited in heterogeneous catalysis methods as demonstrated in MW-assisted rapid molybdenum-catalyzed allylic reactions by Larhead and his coworkers [130] and in the case of oxidation of alcohol using Magtrieve™ by Bogdal et al. [131]. These authors established that the polar catalyst absorbed extra energy and heated at a higher temperature than the overall reaction temperature, thus making the process more energy efficient.

3. Nano-catalysts serve as susceptors
   Susceptors are materials that efficiently absorb MW irradiation and transfer the generated thermal energy to molecules in the vicinity that are weak MW absorbers. Although transmission of the energy occurs through conventional mechanisms, MW heating is more rapid than conventional heating. Kappe [132] and Leadbeater [133] used silicon carbide and ionic liquid, respectively, as susceptors and established that addition of these materials in the reaction mixture enhanced its overall capacity to absorb microwaves and significantly reduced the required MW energy. The use of these materials as susceptors in the reaction mixture, however, adds to the overall cost of the protocol. Ideally, if suitably designed nanomaterials can play a dual role of catalyst and susceptor, then the advantageous attributes can be enjoyed without the need of any additional material as a susceptor.

4. Nano-catalyst stability

MW-assisted reactions are often fast and consequently the residence time of nano-catalysts at this elevated temperature is kept to a minimum. Catalytic processes with short reaction times thus safeguard the catalyst from deactivation and decomposition, thereby increasing the overall efficiency of the catalyst and the entire method.

It appears that this approach of unifying MW technique with nano-catalysis and benign water (as a reaction medium) can offer an extraordinary synergistic effect with greater potential than these three individual components in isolation. To illustrate the concept of this green and sustainable approach, some representative protocols are presented below.

Ruthenium Hydroxide Nano-Catalyst in MW-Assisted Hydration of Nitriles in Water Amides have been generally prepared by the hydration of nitriles, with catalysis by strong acids and bases which produces several by-products including carboxylic acids. Under the influence of strong reagents and harsh conditions, however, sensitive functional groups on molecules could not be kept intact, thus decreasing the selectivity of the reaction protocol. Heterogeneous catalyst systems have been developed to overcome these drawbacks associated with homogeneous processes. The limited turnover numbers of these protocols and reusability of the catalyst are continuing challenges. A hydration method in pure water improved the reaction conditions and product yield [134], but it used expensive ruthenium (Ru) complexes as catalysts and required traditional workup using volatile organic solvents for isolation purposes. A green and sustainable pathway was developed using ruthenium hydroxide nano-catalyst under aqueous MW conditions [135] wherein nano-$Ru(OH)_x$ was prepared in two steps. Magnetic nanoparticles were functionalized post-synthetically [100, 117] via sonication of nano-ferrites with dopamine in aqueous medium, followed by the addition of ruthenium (Ru) chloride and subsequent hydrolysis using sodium hydroxide solution (Scheme 5.43).

The nano-$Ru(OH)_x$ catalyst exhibited high activity for hydration of a range of activated, inactivated, and heterocyclic nitriles in water medium and the reactions rates were not influenced by the nature of the substituents on the benzonitrile molecules. This protocol with high catalytic activity displayed excellent chemoselectively and neither an electronic effect nor the position of the substituents influenced the reaction rate. In the hydration of the benzonitrile-containing dioxole ring, the reaction proceeded only at the cyano group to afford the corresponding amide, while keeping the ring intact (Scheme 5.44). Therefore, this protocol could be very useful in the total synthesis of drug molecules, where the selective hydration of a nitrile group to an amide is a requirement, without influencing other sensitive functional groups.

After completion of the reaction and when the stirring is stopped, the reaction mixture turns clear and catalyst is deposited on the magnetic bar because of the paramagnetic nature of the nano-$Ru(OH)_x$. The catalyst is conveniently removed using an external magnet, thus avoiding even a filtration step. After separation of

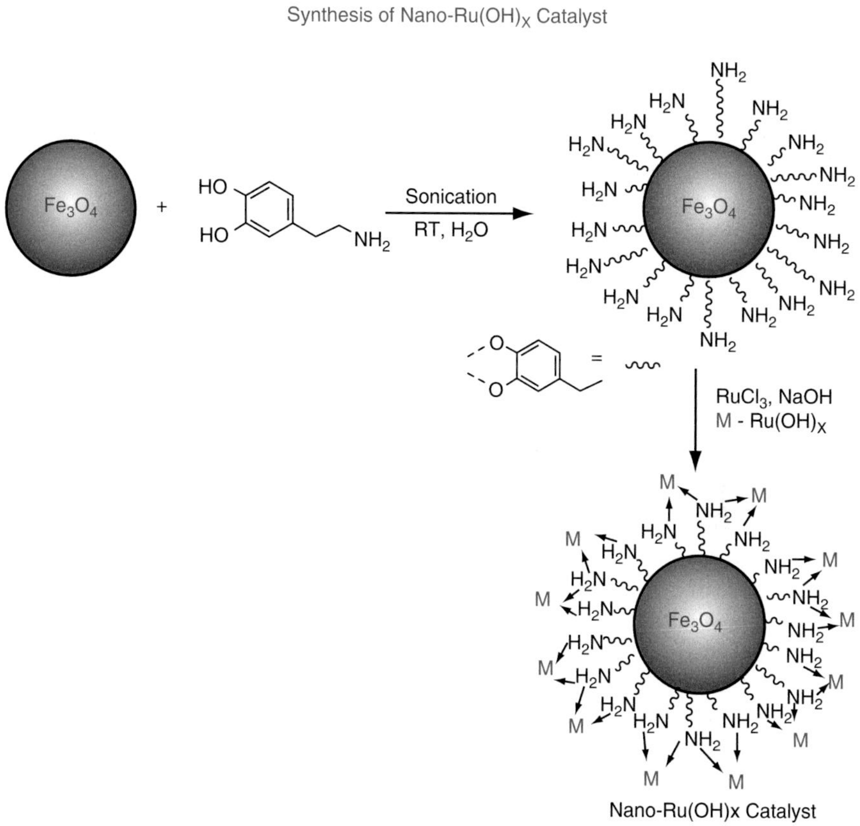

Synthesis of Nano-Ru(OH)$_X$ Catalyst

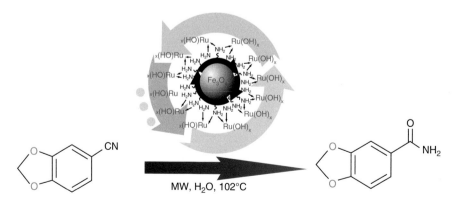

Scheme 5.43 Magnetically separable nano-Ru(OH)$_x$ catalyst. From Polshettiwar and Varma (Ref. [119])

Scheme 5.44 Hydration of nitrile using nano-Ru(OH)$_x$ catalyst. From Polshettiwar and Varma (Ref. [119])

catalyst, the clear reaction mixture is cooled slowly and crystals of benzamides with acceptable purity are precipitated. The complete operation is conducted in pure aqueous medium and no organic solvents are used during the reaction or in the workup steps.

Glutathione-Based Nano-organocatalyst for Aqueous MW-Assisted Synthesis of Heterocycles Organocatalysis has been a very active area of research during the past decade and this metal-free approach has attracted universal interest. Although a wide range of reactions has been successfully introduced using this strategy, most of these transformations are generally conducted in organic solvents. In aqueous protocols, it was observed that the addition of water often accelerated the organocatalyst-mediated reaction, making the overall protocol efficient and eco-friendly [136–138]. However, most of these methods use small amounts of water as reaction medium and excessive amounts of volatile organic solvents are used during the workup, which unfortunately defeats the central idea of reducing the environmental burden of organic contaminants [139].

These drawbacks were successfully circumvented in a green and sustainable manner using glutathione-based nano-organocatalyst under aqueous MW conditions [140–142]. Glutathione, a tripeptide consisting of glutamic acid, cysteine, and glycine units, is a ubiquitous antioxidant present in human and plant cells. The use of glutathione as an active catalytic moiety is preferred due to its benign nature as well as the presence of the highly active thiol group, which can be used for attachment to solid support (ferrites). The catalyst is conveniently prepared by sono-chemical covalent anchoring of glutathione molecules via coupling of its thiol group with the free hydroxyl groups of ferrite surfaces (Scheme 5.45).

The successful use of this glutathione-based nano-organocatalyst approach was demonstrated by Varma et al. for the synthesis of a series of pyrrole heterocycles by Paal–Knorr reaction under aqueous MW conditions. It showed excellent catalytic activity and several amines reacted with tetrahydro-2,5-dimethoxyfuran to produce the respective pyrrole derivatives in good yields (Scheme 5.46) [140–142].

Using this strategy, various hydrazines and hydrazides were found to react efficiently with 1,3-diketones thus affording the desired pyrazoles in good yields (Scheme 5.47) [140–141]. All these reactions proceed efficiently in aqueous medium and get completed in less than 20 min under MW irradiation conditions. This general approach has recently been extended to the homocoupling of boronic acids in aqueous medium under the influence of microwaves [142].

Separation of the catalyst and final product from the reaction mixture is one of the most vital aspects of synthetic protocols. Catalyst recovery occurs generally via filtration or extractive isolation of products, both of which require excessive amounts of organic solvents. However, in the aforementioned protocols, within a few seconds after stirring is stopped, catalyst gets deposited on the magnetic bar and can be easily removed using an external magnet.

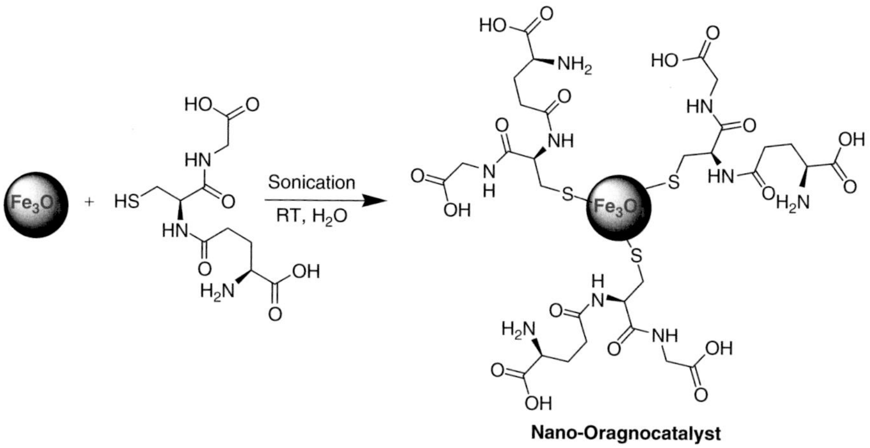

Scheme 5.45 Nano-ferrite functionalization using glutathione. From Polshettiwar and Varma (Ref. [119])

R - alkyl, aryl, heterocyclic

Yield = 72 – 92 %

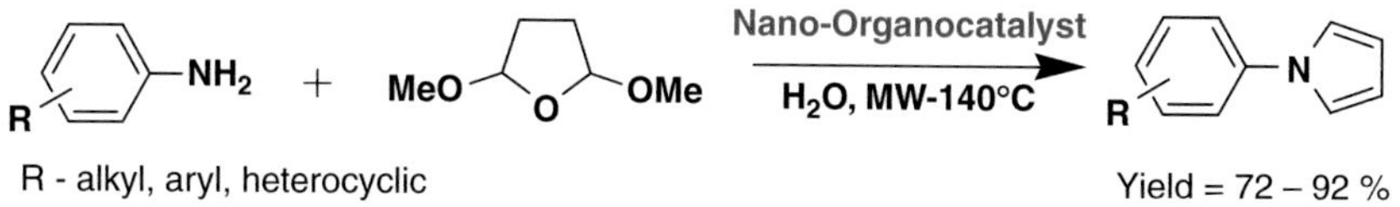

Scheme 5.46 Paal–Knorr reactions using nano-organocatalyst. From Polshettiwar and Varma (Ref. [119])

R^1 = Me, OEt
X = H, Et, Cl

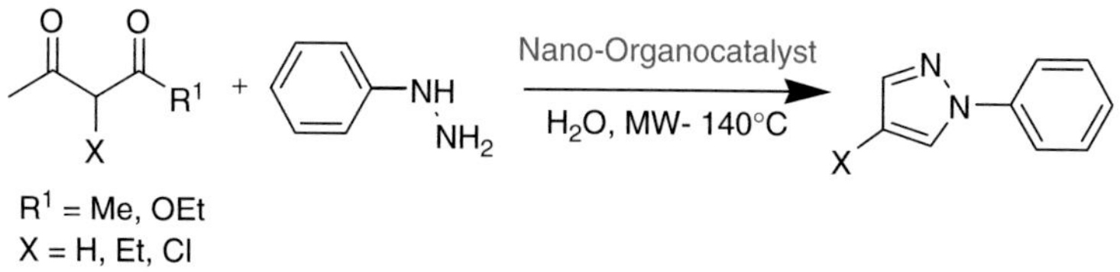

Scheme 5.47 Synthesis of pyrazoles using nano-organocatalyst. From Polshettiwar and Varma (Ref. [119])

In most of the experiments, after completion of the reactions, the phase separation of the desired product from the aqueous medium occurs; this facilitates the isolation of synthesized heterocycles by simple decantation, without using any volatile organic solvents during the reaction or during product workup (Schemes 5.46, 5.47).

Future Directions

The use of various alternative pathways in green chemistry domain such as solvent-free synthesis and transformations [143–145], mechanochemical reactions by grinding [146–149], reactions in benign solvents like water [150–153], polyethylene glycol (PEG) [154, 155], and eco-friendly generated ionic liquids [44–48] and their utility in catalysis [156–161] can be augmented nicely by the use of microwave irradiation heating technique.

The "greener" production of nanoparticles [162] with relatively attractive toxicological profile [163, 164] and their enhanced utility in catalysis [115, 116, 135, 140–142] and environmental remediation [165–167] is garnering attention.

Nano-catalysts mimic homogeneous (easily accessible, high surface area) as well as heterogeneous (stable, easy to handle and isolate) catalyst systems. Nano-catalysts possessing a paramagnetic core thus allow rapid and selective chemical transformations with excellent product yield coupled with the ease of catalyst separation and recovery. Among these options, nanocatalyst-catalyzed transformations in aqueous reaction medium are one of the ideal solutions for the development of green and sustainable protocols. However, execution of many organic reactions in water is not simple due to the inherent limitation of solubility of nonpolar reactants in polar aqueous medium, which can be assisted by using MW irradiation conditions. Thus, a combination of benign water medium, nonconventional MW heating, and nano-catalyst seems to be the optimum pathway to develop the next generation of highly efficient processes [120].

Disclaimer The views expressed in this article are those of the author and do not necessarily reflect the views and policies of the US Environmental Protection Agency. The use of trade names does not imply endorsement by the US Government.

Bibliography

Primary Literature

1. Anastas PT, Warner JC (2000) Green chemistry: theory and practice. Oxford University Press, Oxford
2. Varma RS (1999) Solvent-free organic syntheses using supported reagents and microwave irradiation. Green Chem 1:43–55
3. Varma RS (1999) Solvent-free syntheses of heterocycles using microwave irradiation. J Heterocyclic Chem 36:1565–1571
4. Varma RS (2000) Clay and clay-supported reagents in organic synthesis. Tetrahedron 58:1235–1255
5. Polshettiwar V, Varma RS (2008) Microwave-assisted organic synthesis and transformations using benign reaction media. Acc Chem Res 41:629–639

6. Polshettiwar V, Varma RS (2008) Aqueous microwave chemistry: a clean and green synthetic tool for rapid drug discovery. Chem Soc Rev 37:1546–1557
7. Li C-J, Chen L (2006) Organic chemistry in water. Chem Soc Rev 35:68–82
8. Varma RS (2007) Clean chemical synthesis in water. Org Chem Highlight. http://www.organic-chemistry.org/Highlights/2007/ 01February.shtm
9. Strauss CR, Trainor RW (1995) Developments in microwave-assisted organic chemistry. Aust J Chem 48:1665–1692
10. Anonymous (1924) Using chemical reagents on porous carriers. Akt –Ges Fur Chemiewerte Brit Pat 231: 901 [Chem Abst (1925) 19: 3571]
11. Laszlo P (1987) Preparative chemistry using supported reagents. Academic, San Diego
12. Smith K (1992) Solid supports and catalyst in organic synthesis. Ellis Hardwood, Chichester
13. Clark JH (1994) Catalysis of organic reactions by supported inorganic reagents. VCH, New York
14. McKillop A, Young KW (1979) Organic synthesis using supported reagents – Part I & Part II. Synthesis 401–422 and 481–500
15. Cornelis A, Laszlo P (1985) Clay-supported copper(II) and iron(III) nitrates: novel multi-purpose reagents for organic synthesis. Synthesis 100:909–918
16. Gedye R, Smith F, Westaway K, Humera A, Baldisera L, Laberge L, Rousell J (1986) The use of microwave ovens for rapid organic synthesis. Tetrahedron Lett 27:279–282
17. Giguere RJ, Bray TL, Duncan SM, Majetich G (1986) Application of commercial microwave ovens to organic synthesis. Tetrahedron Lett 27:4945–4948
18. Varma RS (2002) Advances in green chemistry: chemical syntheses using microwave irradiation. AstraZeneca Research Foundation India, Bangalore [85 Reaction schemes, ~300 references]
19. Varma RS (2002) Organic synthesis using microwaves and supported reagents. In: Loupy A (ed) Microwaves in organic synthesis, Chapter 6. Wiley-VCH, New York, pp 181–218
20. Pillai UR, Sahle-Demessie E, Varma RS (2002) Environmentally friendlier organic transformations on mineral supports under non-traditional conditions. J Mater Chem 12:3199–3207
21. Varma RS (2001) Solvent-free accelerated organic syntheses using microwaves. Pure Appl Chem 73:193–198
22. Perreux L, Loupy A (2001) A tentative rationalization of microwave effects in organic synthesis according to the reaction medium, and mechanistic considerations. Tetrahedron 57:9199–9223
23. Loupy A, Petit A, Hamelin J, Texier-Boullet F, Jacquault P, Mathe D (1998) New solvent-free organic synthesis using focused microwaves. Synthesis 1998:1213–1234
24. Gutierrez E, Loupy A, Bram G, Ruiz-Hitzky E (1989) Inorganic solids in "dry media" an efficient way for developing microwave irradiation activated organic reactions. Tetrahedron Lett 30:945–948
25. Greene TW, Wuts PGM (1991) Protective groups in organic synthesis, 2nd edn. Wiley, New York
26. Varma RS, Varma M, Chatterjee AK (1993) Microwave-assisted deacetylation on alumina: a simple deprotection method. J Chem Soc Perkin Trans –1 999–1000
27. Varma RS, Chatterjee AK, Varma M (1993) Alumina-mediated deacetylation of benzaldehyde diacetates. A simple deprotection method. Tetrahedron Lett 34:3207–3210
28. Varma RS, Chatterjee AK, Varma M (1993) Alumina-mediated microwave thermolysis: a new approach to deprotection of benzyl esters. Tetrahedron Lett 34:4603–4606
29. Varma RS, Lamture JB, Varma M (1993) Alumina-mediated cleavage of t-butyldimethylsilyl ethers. Tetrahedron Lett 34:3029–3032
30. Varma RS, Kumar D (1999) Microwave-accelerated solvent-free synthesis of thioketones, thiolactones, thioamides thionoesters and thioflavonoids. Org Lett 1:697–700
31. Varma RS, Saini RK (1997) Solid state dethioacetalization using clayfen. Tetrahedron Lett 38:2623–2624

32. Varma RS, Saini RK, Dahiya R (1997) Active manganese dioxide on silica: oxidation of alcohols under solvent-free conditions using microwaves. Tetrahedron Lett 38:7823–7824
33. Namboodiri VV, Polshettiwar V, Varma RS (2007) Expeditious oxidation of alcohols to carbonyl compounds using iron (III) nitrate. Tetrahedron Lett 48:8839–8842
34. Varma RS, Dahiya R, Saini RK (1997) Iodobenzene diacetate on alumina: rapid oxidation of alcohols to carbonyl compounds in solventless system using microwaves. Tetrahedron Lett 38:7029–7032
35. Varma RS, Dahiya R, Kumar D (1998) Solvent-free oxidation of benzoins using oxone® on wet alumina under microwave irradiation. Molecules Online 2:82–85
36. Varma RS, Saini RK, Meshram HM (1997) Selective oxidation of sulfides to sulfoxides and sulfones by microwave thermolysis on wet silica-supported sodium periodate. Tetrahedron Lett 38:6525–6528
37. Varma RS, Kumar D (1999) Solid state oxidation of 1,4-dihydropyridines to pyridines using phenyliodine(III) bis(trifluoroacetate) or elemental sulfur. J Chem Soc Perkin Trans −1 1755–1757
38. Varma RS, Saini RK (1997) Microwave-assisted reduction of carbonyl compounds in solid state using sodium borohydride supported on alumina. Tetrahedron Lett 38:4337–4338
39. Varma RS, Dahiya R (1998) Sodium borohydride on wet clay: solvent-free reductive amination of carbonyl compounds using microwaves. Tetrahedron 54:6293–6298
40. Vass A, Dudas J, Toth J, Varma RS (2001) Solvent-free reduction of aromatic nitro compounds with alumina-supported hydrazine under microwave irradiation. Tetrahedron Lett 42:5347–5349
41. Varma RS, Naicker KP (1998) Hydroxylamine on clay: a direct synthesis of nitriles from aromatic aldehydes using microwaves under solvent-free conditions. Molecules Online 2: 94–96
42. Welton T (2004) Ionic liquids in catalysis. Coord Chem Rev 248:2459–2477
43. Rogers RD, Seddon KN, Volkove S (2002) Green Industrial applications of ionic liquids. NOTO science, series
44. Varma RS, Namboodiri, VV (2001) An expeditious solvent-free route to ionic liquids using microwaves. Chem. Commun 643–644
45. Namboodiri VV, Varma RS (2002) An improved preparation of 1, 3-dialkylimidazolium tetrafluoroborate ionic liquids using microwaves. Tetrahedron Lett 43:5381–5383
46. Namboodiri VV, Varma RS (2002) Solvent-free sonochemical preparation of ionic liquids. Org Lett 4:3161–3163
47. Kim YJ, Varma RS (2005) Microwave-assisted preparation of imidazolium-based tetrachloroindate (III) and their application in the tetrahydropyranylation of alcohols. Tetrahedron Lett 46:1467–1469
48. Kim YJ, Varma RS (2005) Microwave-assisted preparation of 1-butyl-3-methylimidazolium tetrachlorogallate and its catalytic use in acetal formation under mild conditions. Tetrahedron Lett 46:7447–7449
49. Swatloski RP, Spear SK, Holbrey JD, Rogers RD (2002) Dissolution of cellulose with ionic liquids. J Am Chem Soc 124:4974–4975
50. Varma RS, Namboodiri VV (2002) Microwave-assisted preparation of dialkylimidazolium tetrachloroaluminates and their use as catalysts in the solvent-free tetrahydropyranylation of alcohols and phenols. Chem Commun 342–343
51. Kim YJ, Varma RS (2005) Tetrahaloindate(III)-based ionic liquids in the coupling reaction of carbon dioxide and epoxides to generate cyclic carbonates: H-bonding and mechanistic studies. J Org Chem 70:7882–7891
52. Plechkova NV, Seddon KR (2008) Applications of ionic liquids in the chemical industry. Chem Soc Rev 37:123–150
53. Herrero MA, Kremsner JM, Kappe CO (2008) Nonthermal microwave effects revisited – on the importance of internal temperature monitoring and agitation in microwave chemistry. J Org Chem 73:36–47

54. Polshettiwar V, Varma RS (2007) Greener and sustainable approaches to the synthesis of pharmaceutically active heterocycles. Curr Opin Drug Discov Devel 10:723–737

55. Chen J, Spear SK, Huddleston JG, Rogers RD (2005) Polyethylene glycol and solutions of polyethylene glycol as green reaction media. Green Chem 7:64–82

56. Leadbeater NE, Marco M (2003) Rapid and amenable Suzuki coupling reaction in water using microwave and conventional heating. J Org Chem 68:888–892

57. Crozet MD, Castera-Ducros C, Vanelle P (2006) An efficient microwave-assisted Suzuki cross-coupling reaction of imidazo [1, 2-a] pyridines in aqueous medium. Tetrahedron Lett 47:7061–7065

58. Zhu R, Qu F, Queleverb G, Peng L (2007) Direct synthesis of 5-aryltriazole acyclonu-cleosides via Suzuki coupling in aqueous solution. Tetrahedron Lett 48:2389–2393

59. Dawood KM (2007) Microwave-assisted Suzuki–Miyaura and Heck–Mizoroki cross-coupling reactions of aryl chlorides and bromides in water using stable benzothiazole-based palladium (II) precatalysts. Tetrahedron 63:9642–9651

60. Arvela RK, Leadbeater NE (2005) Microwave-promoted Heck coupling using ultralow metal catalyst concentrations. J Org Chem 70:1786–1790

61. Arvela RK, Pasquini S, Larhed M (2007) Highly regioselective internal Heck arylation of hydroxyalkyl vinyl ethers by aryl halides in neat water. J Org Chem 72:6390–6396

62. Appukkuttan P, Dehaen W, der Eycken EV (2003) Transition-metal-free Sonogashira-type coupling reactions in water. Eur J Org Chem 2003:4713–4716

63. Alcida E, Najera C (2006) The first fluoride-free Hiyama reaction of vinylsiloxanes promoted by sodium hydroxide in water. Adv Synth Catal 348:2085–2091

64. Kaval N, Bisztray K, Dehaen W, Kappe CO, der Eycken EV (2003) Microwave-enhanced transition metal-catalyzed decoration of 2(1H)-pyrazinone scaffolds. Mol Divers 7:125–133

65. Miyazawa A, Tanaka K, Sakakura T, Tashiro M, Tashiro H, Surya Prakash GK, Olah GA (2005) Microwave-assisted direct transformation of amines to ketones using water as an oxygen source. Chem Commun 2104–2106

66. Kumar V, Sharma A, Sharma A, Sinha AK (2007) Remarkable synergism in methylimidazole-promoted decarboxylation of substituted cinnamic acid derivatives in basic water medium under microwave irradiation: a clean synthesis of hydroxylated (E)-stilbenes. Tetrahedron 63:7640–7646

67. Dallinger D, Kappe CO (2007) Microwave-assisted synthesis in water as solvent. Chem Rev 107:2563–2591

68. Garuti L, Roberti M, Pizzirani D (2007) Nitrogen-containing heterocyclic quinones: a class of potential selective antitumor agents. Mini Rev Med Chem 7:481–489

69. Sperry JB, Wright DL Furans (2005) Thiophenes and related heterocycles in drug discovery. Curr Opin Drug Discov Devel 8:723–740

70. Kappe CO (2002) High-speed combinatorial synthetics utilizing microwave irradiation. Curr Opin Chem Biol 6:314–320

71. Polshettiwar V, Varma RS (2008) Greener and expeditious synthesis of bio-active heterocycles using microwave irradiation. Pure Appl Chem 80:777–790

72. Roberts BA, Strauss CR (2005) Toward rapid, "green", predictable microwave-assisted synthesis. Acc Chem Res 38:653–661

73. Kappe CO (2004) Controlled microwave heating in modern organic synthesis. Angew Chem Int Ed 43:6250–6284

74. Ju Y, Varma RS (2004) Aqueous N-alkylation of amines using alkyl halides: direct generation of tertiary amines under microwave irradiation. Green Chem 6:219–221

75. Ju Y, Varma RS (2005) An efficient and simple aqueous N-heterocyclization of aniline derivatives: microwave-assisted synthesis of N-aryl azacycloalkanes. Org Lett 7:2409–2411

76. Ju Y, Varma RS (2005) Microwave-assisted cyclocondensation of hydrazine derivatives with alkyl dihalides or ditosylates in aqueous media: syntheses of pyrazole, pyrazolidine and phthalazine derivatives. Tetrahedron Lett 46:6011–6014

77. Ju Y, Varma RS (2006) Aqueous *N*-heterocyclization of primary amines and hydrazines with dihalides: microwave-assisted syntheses of *N*-azacycloalkanes, isoindole, pyrazole, pyrazolidine, and phthalazine derivatives. J Org Chem 71:135–141
78. Ju Y, Li C-J, Varma RS (2004) Microwave-assisted Cu (I) catalyzed solvent-free three component coupling of aldehyde, alkyne and amine. QSAR Comb Sci 23:891–894
79. Kim YJ, Varma RS (2004) Microwave-assisted preparation of cyclic ureas from diamines in the presence of ZnO. Tetrahedron Lett 45:7205–7208
80. Varma RS, Kumar D (1999) Microwave-accelerated three-component condensation reaction on clay: solvent-free synthesis of imidazo [1, 2-a] annulated pyridines, pyrazines and pyrimidones. Tetrahedron Lett 40:7665–7669
81. Kappe CO, Kumar D, Varma RS (1999) Microwave-assisted high-speed parallel synthesis of 4-aryl-3, 4-dihydropyrimidin-2(1H)-ones using a solventless Biginelli condensation protocol. Synthesis 10:1799–1803
82. Polshettiwar V, Varma RS (2007) Biginelli reaction in aqueous medium: a greener and sustainable approach to substituted 3, 4-dihydropyrimidin-2(1H)-ones. Tetrahedron Lett 48:7343–7346
83. Varma RS, Dahiya R (1998) An expeditious and solvent-free synthesis of 2-amino-substituted isoflav-3-enes using microwave irradiation. J Org Chem 63:8038–8041
84. Varma RS, Kumar D, Liesen PJ (1998) Solid state synthesis of 2-aroylbenzo[b]furans, 1,3-thiazoles and 3-aryl-5,6-dihydroimidazo [2,1-b][1,3] thiazoles from α-tosyloxyketones using microwave irradiation. J Chem Soc Perkin Trans −1 4093–4096
85. Polshettiwar V, Varma RS (2007) Tandem bis-aldol reaction of ketones: a facile one pot synthesis of 1, 3-dioxanes in aqueous medium. J Org Chem 72:7420–7422
86. Jeselnik M, Varma RS, Polanc S, Kocevar M (2001) Catalyst-free reactions under solvent-free conditions: microwave-assisted synthesis of heterocyclic hydrazones below the melting points of neat reactants. Chem Commun 1716–1717
87. Polshettiwar V, Varma RS (2007) Polystyrene sulfonic acid catalyzed greener synthesis of hydrazones in aqueous medium using microwaves. Tetrahedron Lett 48:5649–5652
88. Polshettiwar V, Varma RS (2008) Rapid access to bio-active heterocycles: one-pot solvent-free synthesis of 1, 3, 4-oxadiazoles and 1, 3, 4-thiadiazoles. Tetrahedron Lett 49:879–883
89. Nadagouda MN, Varma RS (2007) Microwave-assisted shape-controlled bulk synthesis of noble nanocrystals and their catalytic properties. Cryst Growth Des 7:686–690
90. Baruwati B, Varma RS (2009) High value products from waste: grape pomace extract – a three-in-one package for the synthesis of metal nanoparticles. ChemSusChem 2:1041–1044
91. Baruwati B, Polshettiwar V, Varma RS (2009) Glutathione promoted expeditious green synthesis of silver nanoparticles in water using microwaves. Green Chem 11:926–930
92. Nadagouda MN, Polshettiwar V, Varma RS (2009) Self-assembly of palladium nanoparticles: synthesis of nanobelts, nanoplates and nanotrees using vitamin B_1 and their application in carbon-carbon coupling reactions. J Mater Chem 19:2026–2031
93. Nadagouda MN, Varma RS (2006) Green and controlled synthesis of gold and platinum nanomaterials using vitamin B_2: density-assisted self-assembly of nanospheres, wires and rods. Green Chem 8:516–518
94. Nadagouda MN, Varma RS (2007) A greener synthesis of core (Fe, Cu)-shell (Au, Pt, Pd and Ag) nanocrystals using aqueous vitamin C. Cryst Growth Des 7:2582–2587
95. Nadagouda MN, Varma RS (2008) Green synthesis of silver and palladium nanoparticles at room temperature using coffee and tea extract. Green Chem 10:859–862
96. Nadagouda MN, Varma RS (2008) Microwave-assisted shape controlled bulk synthesis of Ag and Fe nanorods in poly (ethylene glycol) solutions. Cryst Growth Des 8:291–295
97. Nadagouda MN, Varma RS (2007) Synthesis of thermally stable carboxymethyl cellulose/metal biodegradable nanocomposite films for potential biological applications. Biomacromolecules 8:2762–2767
98. Nadagouda MN, Varma RS (2007) Microwave-assisted synthesis of cross-Linked poly (vinyl alcohol) nanocomposites comprising single-wall carbon nanotubes (SWNT), multi-wall carbon nanotubes (MWNT) and buckminsterfullerene (C-60). Macromol Rapid Commun 28:842–847

99. Polshettiwar V, Nadagouda MN, Varma RS (2007) Synthesis and applications of micro-pine structured nano-catalyst. Chem Commun 6318–6320
100. Polshettiwar V, Baruwati B, Varma RS (2009) Self-assembly of metal oxides into three-dimensional nanostructures: synthesis and application in catalysis. ACS Nano 3:728–736
101. Mamedov AA, Belov A, Giersig M, Mamedova NN, Kotov NA (2001) Nanorainbows: graded semiconductor films from quantum dots. J Am Chem Soc 123:7738–7739
102. Alivisatos P (1996) Semiconductor clusters, nanocrystals, and quantum dots. Science 271:933–937
103. Klimov VI, Mikhailovsky AA, Xu S, Malko A, Hollingsworth JA, Leatherdale CA, Eisler H-J, Bawendi MG (2000) Optical gain and stimulated emission in nanocrystal quantum dots. Science 290:314–317
104. Sundar VC, Eisler H-J, Bawendi MG (2002) Room-temperature, tunable gain media from novel II-VI nanocrystal-titania composite matrices. Adv Mater 14:739–743
105. Bruchez M, Moronne M, Gin P, Weiss S, Alivisatos AP (1998) Semiconductor nanocrystals as fluorescent biological labels. Science 281:2013–2016
106. Chan WCW, Nie S (1998) Quantum dot bioconjugates for ultrasensitive nonisotopic detection. Science 281:2016–2018
107. Michalet X, Pinaud FF, Bentolila LA, Tsay JM, Doose S, Li JJ, Sundaresan G, Wu AM, Gambhir SS, Weiss S (2005) Quantum dots for live cells, in vivo imaging, and diagnostics. Science 307:538–544
108. Murray CB, Norris DJ, Bawendi MG (1993) Synthesis and characterization of nearly monodisperse CdE (E = sulfur, selenium, tellurium) semiconductor nanocrystallites. J Am Chem Soc 115:8706–8715
109. Qian H, Qiu X, Li L, Ren J (2006) Microwave-assisted aqueous synthesis: a rapid approach to prepare highly luminescent ZnSe(S) alloyed quantum dots. J Phys Chem B 110:9034–9040
110. Zhao Y, Hong J-M, Zhu J-J (2004) Microwave-assisted self-assembled ZnS nanoballs. J Cryst Growth 270:438–445
111. Zhu J, Palchik O, Chen S, Gedanken A (2000) Microwave assisted preparation of CdSe, PbSe, and $Cu_{2-x}Se$ nanoparticles. J Phys Chem B 104:7344–7347
112. Schumacher W, Nagy A, Waldman WJ, Dutta PK (2009) Direct synthesis of aqueous CdSe/ZnS-based quantum dots using microwave irradiation. J Phys Chem C 113:12132–12139
113. Qian H, Li L, Ren J (2005) One-step and rapid synthesis of high quality alloyed quantum dots (CdSe–CdS) in aqueous phase by microwave irradiation with controllable temperature. Mater Res Bull 40:1726–1736
114. Lu A-H, Salabas EL, Schuth F (2007) Magnetic nanoparticles: synthesis, protection, functionalization, and application. Angew Chem Int Ed 46:1222–1244
115. Polshettiwar V, Varma RS (2009) Nanoparticle-supported and magnetically recoverable palladium (Pd) catalyst: a selective and sustainable oxidation protocol with high turnover number. Org Biomol Chem 7:37–40
116. Polshettiwar V, Baruwati B, Varma RS (2009) Nanoparticle-supported and magnetically recoverable nickel catalyst: a robust and economic hydrogenation and transfer hydrogenation protocol. Green Chem 11:127–131
117. Polshettiwar V, Nadagouda MN, Varma RS (2008) Synthesis and applications of micro-pine structured nano-catalyst. Chem. Commun 6318–6320
118. Baruwati B, Guin D, Manorama SV (2007) Pd on surface-modified $NiFe_2O_4$ nanoparticles: a magnetically recoverable catalyst for Suzuki and Heck reactions. Org Lett 9:5377–5380
119. Guin D, Baruwati B, Manorama SV (2007) Pd on amine-terminated ferrite nanoparticles: a complete magnetically recoverable facile catalyst for hydrogenation reactions. Org Lett 9:1419–1421
120. Polshettiwar V, Varma RS (2010) Green chemistry by nano-catalysis. Green Chem 12:743–754
121. Loupy A, Varma RS (2006) Microwave effects in organic synthesis: mechanistic and reaction medium considerations. Chim Oggi 24:36–40

122. Strauss CR, Varma RS (2006) Microwaves in green and sustainable chemistry. Top Curr Chem 266:199–231
123. Kappe CO, Dallinger D (2009) Controlled microwave heating in modern organic synthesis: highlights from the 2004–2008 literature. Mol Divers 13:71–193
124. Polshettiwar V, Nadagouda MN, Varma RS (2009) Microwave-assisted chemistry: a rapid and sustainable route to synthesis of organics and nanomaterials. Aust J Chem 62:16–26
125. Gabriel C, Gabriel S, Grant EH, Halstead BSJ, Mingos DMP (1998) Dielectric parameters relevant to microwave dielectric heating. Chem Soc Rev 27:213–224
126. Poliokoff M, Licence P (2007) Sustainable technology: green chemistry. Nature 450:810–812
127. Polshettiwar V, Varma RS (2008) Olefin ring closing metathesis and hydrosilylation reaction in aqueous medium by Grubbs second generation ruthenium catalyst. J Org Chem 73:7417–7419
128. Hoz A, Diaz-Ortiz A, Moreno A (2005) Microwaves in organic synthesis. Thermal and non-thermal microwave effects. Chem Soc Rev 34:164–178
129. Nilsson P, Larhed M, Hallberg A (2001) Highly regioselective, sequential, and multiple palladium- catalyzed arylations of vinyl ethers carrying a coordinating auxiliary: an example of a Heck triarylation process. J Am Chem Soc 123:8217–8225
130. Kaiser NFK, Bremberg U, Larhed M, Moberg C, Hallberg A (2000) Fast, convenient, and efficient molybdenum-catalyzed asymmetric allylic alkylation under noninert conditions: an example of microwave-promoted fast chemistry. Angew Chem Int Ed 39:3596–3598
131. Bogdal D, Lukasiewicz M, Pielichowski J, Miciak A, Sz B (2003) Microwave-assisted oxidation of alcohols using Magtrieve™. Tetrahedron 59:649–653
132. Razzak T, Kremser JM, Kappe CO (2008) Investigating the existence of nonthermal/specific microwave affects using silicon carbide heating elements as power modulators. J Org Chem 73:6321–6329
133. Leadbeater NE, Torrenius HM (2002) A study of the ionic liquid mediated microwave heating of organic solvents. J Org Chem 67:3145–3148
134. Cadierno V, Francos J, Gimeno J (2008) Selective ruthenium-catalyzed hydration of nitriles to amides in pure aqueous medium under neutral conditions. Chem Eur J 14:6601–6605
135. Polshettiwar V, Varma RS (2009) Nanoparticle-supported and magnetically recoverable ruthenium hydroxide catalyst: effcient hydration of nitriles to amides in aqueous medium. Chem Eur J 15:1582–1586
136. Brogan AP, Dickerson TJ, Janda KD (2006) Enamine-based aldol organocatalysis in water: are they really all wet? Angew Chem Int Ed 45:8100–8102
137. Hayashi Y, Samanta S, Gotoh H, Ishikawa H (2008) Asymmetric Diels-Alder reactions of,-unsaturated aldehydes catalyzed by a diarylprolinol silyl ether salt in the presence of water. Angew Chem Int Ed 47:6634–6637
138. Huang J, Zhang X, Armstrong DW (2007) Highly efficient asymmetric direct stoichiometric aldol reactions on/in water. Angew Chem Int Ed 46:9073–9077
139. Blackmond DG, Armstrong A, Coombe V, Wells A (2007) Water in organocatalytic processes: debunking the myths. Angew Chem Int Ed 46:3798–3800
140. Polshettiwar V, Baruwati B, Varma RS (2009) Magnetic nanoparticle-supported glutathione: a conceptually sustainable organocatalyst. Chem Commun 1837–1839
141. Polshettiwar V, Varma RS (2010) Nano-organocatalyst: magnetically retrievable ferrite-anchored glutathione for microwave-assisted Paal-Knorr reaction, aza-Michael addition and pyrazole synthesis. Tetrahedron 66:1091–1097
142. Luque R, Baruwati B, Varma RS (2010) Magnetically separable nanoferrite-anchored glutathione: aqueous homocoupling of arylboronic acids under microwave irradiation. Green Chem 12:1540–1543. doi:10.1039/C0GC00083C
143. Polshettiwar V, Varma RS (2008) Ring-fused aminals: catalyst and solvent-free microwave-assisted α-amination of nitrogen heterocycles. Tetrahedron Lett 49:7165–7167

144. Varma RS, Naicker KP, Liesen PJ (1998) Microwave-accelerated crossed Cannizzaro reaction using barium hydroxide under solvent-free conditions. Tetrahedron Lett 3:8437–8440
145. Pillai UR, Sahle-Demessie E, Namboodiri VV, Varma RS (2002) An efficient and ecofriendly oxidation of alkenes using iron nitrate and molecular oxygen. Green Chem 4:495–497
146. Kumar D, Chandra Sekhar KVG, Dhillon H, Rao VS, Varma RS (2004) An expeditious synthesis of 1-aryl-4-methyl-1, 2, 4-triazolo [4, 3-a] quinoxalines under solvent-free conditions using iodobenzene diacetate. Green Chem 6:156–157
147. Kumar D, Sundaree MS, Patel G, Rao VS, Varma RS (2006) Solvent-free facile synthesis of novel α-tosyloxy β-keto sulfones using [hydroxy(tosyloxy)iodo] benzene. Tetrahedron Lett 47:8239–8241
148. Kumar D, Sundaree MS, Rao VS, Varma RS (2006) A facile one-pot synthesis of β-keto sulfones from ketones under solvent-free conditions. Tetrahedron Lett 47:4197–4199
149. Varma RS (2008) Chemical activation by mechanochemical mixing, microwave, and ultrasonic irradiation. Green Chem 10:1129–1130
150. Polshettiwar V, Varma RS (2007) Tandem bis-aza-Michael addition reaction of amines in aqueous medium promoted by polystyrenesulfonic acid. Tetrahedron Lett 48:8735–8738
151. Kumar D, Reddy VB, Mishra BG, Rana RK, Nadagouda MN, Varma RS (2007) Nanosized magnesium oxide as catalyst for the rapid and green synthesis of substituted 2-amino-2-chromenes. Tetrahedron 63:3093–3097
152. Skouta R, Varma RS, Li CJ (2005) Efficient Trost's γ-addition catalyzed by reusable polymer-supported triphenylphosphine in aqueous media. Green Chem 7:571–575
153. Ju Y, Kumar D, Varma RS (2006) Revisiting nucleophilic substitution reactions: microwave-assisted synthesis of azides, thiocyanates, and sulfones in an aqueous medium. J Org Chem 71:6697–6700
154. Namboodiri VV, Varma RS (2001) Microwave-accelerated Suzuki cross-coupling reaction in polyethylene glycol (PEG). Green Chem 3:146–148
155. Kumar D, Patel G, Mishra BG, Varma RS (2008) Ecofriendly polyethylene glycol (PEG)-promoted Michael addition reactions of α, β-unsaturated compounds. Tetrahedron Lett 49:6974–6976
156. Keh CCK, Namboodiri VV, Varma RS, Li C-J (2002) Direct formation of tetrahydropyranols via catalysis in ionic liquid. Tetrahedron Lett 43:4993–4996
157. Li Z, Wei C, Varma RS, Li C-J (2004) Three-component coupling of aldehyde, alkyne, and amine catalyzed by silver in ionic liquid. Tetrahedron Lett 45:2443–2446
158. Yang X-F, Wang M, Varma RS, Li C-J (2003) Aldol- and Mannich-type reactions via in situ olefin migration in ionic liquid. Org Lett 5:657–660
159. Yang X-F, Wang M, Varma RS, Li C-J (2004) Ruthenium-catalyzed tandem olefin migration aldol and Mannich-type reactions in ionic liquid. J Mol Catal A Chem 214:147–154
160. Yoo K, Namboodiri VV, Varma RS, Smirniotis PG (2004) Ionic liquid-catalyzed alkylation of isobutane with 2-butene. J Catal 222:511–519
161. Namboodiri VV, Varma RS, Sahle-Demessie E, Pillai UR (2002) Selective oxidation of styrene to acetophenone in the presence of ionic liquids. Green Chem 4:170–173
162. Nadagouda MN, Hoag GE, Collins JB, Varma RS (2009) Green synthesis of Au nanostructures at room temperature using biodegradable plant surfactants. Cryst Growth Des 9:4979–4983
163. Nadagouda MN, Castle A, Murdock RC, Hussain SM, Varma RS (2010) In vitro biocompatibility of nanoscale zerovalent iron particles (nZVI) synthesized using tea polyphenols. Green Chem 12:114–122
164. Moulton MC, Braydich-Stolle LK, Nadagouda MN, Kunzelman S, Hussain SM, Varma RS (2010) Synthesis, characterization and biocompatibility of "green" synthesized silver nanoparticles using tea polyphenols. Nanoscale 2:763–770
165. Hoag GE, Collins JB, Holcomb JL, Hoag JR, Nadagouda MN, Varma RS (2009) Degradation of bromothymol blue by 'greener' nano-scale zerovalent iron synthesized using tea polyphenols. J Mater Chem 19:8671–8677

166. Virkutyte J, Varma RS (2010) Fabrication and visible-light photocatalytic activity of novel Ag/TiO$_{2-x}$N$_x$ photocatalyst. New J Chem 34:1094–1096
167. Virkutyte J, Baruwati B, Varma RS (2010) Visible light induced photobleaching of methylene blue over melamine doped TiO$_2$ nanocatalyst. Nanoscale 2(7):1109–1111

Books and Reviews

Ahluwalia VK, Varma RS (2008) Alternative energy processes in chemical synthesis microwave, ultrasound and photo activation. Narosa Publishing House, New Delhi. ISBN 978-81-7319-848-9

Ahluwalia VK, Varma RS (2009) Green solvents for organic synthesis. Narosa Publishing House, New Delhi. ISBN 978-81-7319-964-6

Clark JH, Macquarrie D (2002) Handbook of green chemistry and technology. Blackwell Science, Oxford

Kappe CO, Stadler A (2005) Microwaves in organic and medicinal chemistry. Wiley-VCH, Weinheim, p 410

Kappe CO, Dallinger D, Murphree SS (2009) Practical microwave synthesis for organic chemists – strategies, instruments, and protocols. Wiley-VCH, Weinheim, p 296

Matlack AS (2001) Introduction to green chemistry. Marcel Deckers, New York

Nadagouda MN, Varma RS (2009) Risk reduction via greener synthesis of noble metal nanostructures and nanocomposites. In: Linkov I, Steevens J (eds) Nanomaterials: risks and benefits-proceedings of the NATO advanced workshop. Springer, Faro, pp 209–218

Polshettiwar V, Varma RS (2009) Environmentally benign chemical synthesis via mechanochemical mixing and microwave irradiation. In: Ballini R (ed) Eco-friendly synthesis of fine chemicals, RSC green chemistry book series. RSC, Cambridge, England, pp 275–292

Polshettiwar V, Varma RS (2009) Non-conventional energy sources for green synthesis in water (microwave, ultrasound, and photo). In: Li C-J, Anastas PT (eds) Handbook series, Handbook of green chemistry, Vol. 5: reactions in water. Wiley-VCH, Weinheim. ISBN 978-3-527-31574-1

Polshettiwar V, Varma RS (eds) (2010) Aqueous microwave chemistry: synthesis and applications, vol 7, RSC green chemistry series. Royal Society Chemistry, Cambridge, UK

Strauss CR, Varma RS (2006) Microwaves in green and sustainable chemistry. In: Larhed M, Olofsson K (eds) Microwave methods in organic synthesis, vol 266, Series in topics in current chemistry. Springer, Heidelberg, pp 199–231

Varma RS (2000) Environmentally benign organic transformations using microwave irradiation under solvent-free conditions. In: Anastas PT, Tundo P (eds) Green chemistry: challenging perspectives. Oxford University Press, Oxford, pp 221–244

Varma RS (2000) Expeditious solvent-free organic syntheses using microwave irradiation. In: Anastas PT, Heine L, Williamson T (eds) Green chemical syntheses and processes, Chapter 23, vol 767, ACS symposium series. American Chemical Society, Washington, DC, pp 292–312

Varma RS (2001) Microwave organic synthesis. In: Geller E (ed) McGraw-Hill Yearbook of Science and Technology 2002. McGraw-Hill, New York, pp 223–225

Varma RS (2006) Microwave technology: chemical synthesis applications. In: Seidel A (ed) Kirk-Othmer on-line encyclopedia of chemical technology, vol 16, 5th edn. Wiley, Hoboken, pp 538–594

Varma RS, Ju Y (2005) Microwaves in organic synthesis. In: Afonso CAM, Crespo JG (eds) Solventless reactions (SLR), Chapter 2.2. Wiley-VCH, Weinheim, pp 53–87

Varma RS, Ju Y (2006) Organic synthesis using microwaves and supported reagents. In: Loupy A (ed) Microwaves in organic sSynthesis, Chapter 8, 2nd edn. Wiley-VCH, Weinheim, pp 362–415

Chapter 6
Nanotoxicology in Green Nanoscience

Leah Wehmas and Robert L. Tanguay

Glossary

Embryonic development	The molecular signaling, cell divisions, cell rearrangements, and cell differentiation that lead to tissues, organs, and structures of an organism.
High throughput	A method of stream-lining, often through automation, testing procedures to rapidly conduct thousands of experiments.
Nanotechnology	As defined by the National Nanotechnology Initiative, "nanotechnology is the understanding and control of matter at the nanoscale, at dimensions between approximately 1 and 100 nm, where unique phenomena enable novel applications."
Structure activity relationships (SARS)	A method of relating how structural and physiochemical properties of a compound influence biological activity.
Tiered approach	An approach that optimizes identification of potentially hazardous compounds through testing and systematic interpretation of results [1].
Toxicology testing	Testing to examine and understand the adverse effects of physical, biological, or chemical compounds on organisms and the environment with the objective of mitigation or prevention [2].

This chapter was originally published as part of the Encyclopedia of Sustainability Science and Technology edited by Robert A. Meyers. DOI:10.1007/978-1-4419-0851-3

L. Wehmas (✉) • R.L. Tanguay
Department of Environmental and Molecular Toxicology, Environmental Health Sciences Center, Oregon State University, Corvallis, OR 97333, USA
e-mail: wehmasl@onid.orst.edu; Robert.Tanguay@oregonstate.edu

P.T. Anastas and J.B. Zimmerman (eds.), *Innovations in Green Chemistry and Green Engineering*, DOI 10.1007/978-1-4614-5817-3_6,
© Springer Science+Business Media New York 2013

Definition of the Subject

Nanotechnology holds great promise for future economical and technological advances, yet health and safety concerns regarding nanomaterials persist. As an emerging technology, nanotechnology is in the unique position to proactively address health and safety concerns throughout the product life cycle. Green chemistry aims to create benign compounds in a way that prevents pollution and reduces waste throughout every stage of production. Through green nanoscience, the principles of green chemistry can be applied toward making high performance, yet inherently safe nanomaterials. Successful application of green chemistry principles to assess nanomaterial health and safety requires efficient, predictive, high-throughput nanotoxicity testing. With these approaches, designers and manufacturers of nanomaterials can assess nanotoxicity early in production to redesign or replace hazardous nanomaterials.

Introduction

Nanotechnology is a field of technology, engineering, and science that involves shaping and rearranging matter near the atomic level to make materials, particles, or structures with at least one dimension on the scale of 1–100 nm [3]. At a size that is approximately one billionth that of a meter, distinctive mechanical, optical, and physiochemical properties emerge [3]. Nanotechnology exploits these novel nanoscale properties for unique applications in electronics, targeted drug delivery, personal care products, textiles, etc. [4–9].

The numerous applications and continued innovation in nanotechnology have contributed to the exponential growth in the industry. Within 5 years nanotechnology is projected to expand from a $3.8 billion industry to a $2.5 trillion industry by 2015 [10] making it a significant contributor to the global economy. This rapid growth indicates that nanotechnology is transitioning from a discovery to production industry [11, 12]. This scale-up in nanomaterial production, while contributing significantly to economic growth and industrial improvement, will also significantly increase the potential for environmental and human exposure to nanomaterials [13]. Therefore, the swift assessment of nano-environmental and human safety (nanoEHS) risks is necessary. While numerous nanomaterial toxicity studies have been conducted, conflicting results make it difficult to adequately assess nanoEHS. The unique properties that make nanomaterials an attractive technology (surface chemistry, surface area, size, shape, core material functionalization, aggregation, etc.) may also contribute to novel biological effects as a result of nanomaterial exposure [12]. Toxicology will play an important role in elucidating the mechanisms of those interactions [14] and will be referred to as nanotoxicology when applied to nanomaterials.

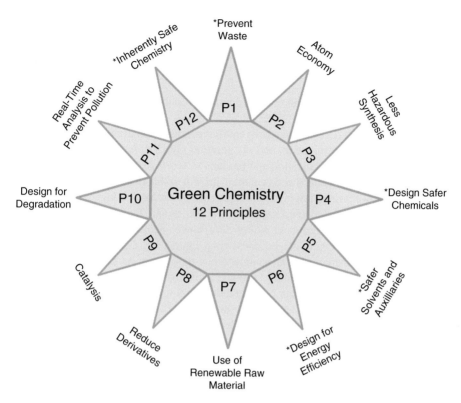

Fig. 6.1 The 12 principles of green chemistry. *Asterisks* designate principles that can be applied to nanotoxicology

The emergence of nanotechnology occurred in tandem with the introduction of the concepts of green chemistry [15] making it a technology ideally suited for incorporation of the principles of green chemistry (Fig. 6.1) early in the industry [16]. When nanomaterials are considered for commercialization, collaborative work between the producers of nanomaterials and toxicologists is necessary to prioritize [17] and design comprehensive, efficient, high-throughput nanotoxicity tests to address nanoEHS. Faster and predictive nanotoxicity tests will require an in vivo approach at several levels of biological organization including molecular → cellular → tissue → organ → whole organism effects. Toxicity testing implemented early in the nanomaterial design process will allow for refinement or replacement of unsafe nanomaterials earlier in the development pipeline. This will enable the production of inherently safer nanomaterials through the incorporation of the 12 principles of green chemistry [15], a major part of the green nanoscience research approach [16].

This entry will explore the role of toxicology in implementing green nanoscience, the methodology of incorporating nanotoxicology in order to directly and indirectly address the principles of green chemistry in green nanoscience, and the importance of utilizing robust models for nanotoxicity testing.

Nanotoxicology's Role in Green Nanoscience

Application of Green Chemistry

When one thinks of nanotechnology and nanomaterials consumer products such as sunscreens, cosmetics, or antimicrobial socks come to mind, yet there are several innovative nanomaterials that may have environmental remediation applications in oil spill cleanup, water purification, and air filtration [18–20]. As many nanomaterials move closer to commercialization and production levels expand, the potential environmental health applications and economic growth in this field could greatly benefit the society. However, the full benefit and promise of nanotechnology may be greatly impeded, or never realized, until it gains acceptance as safe by the general public and lawmakers. To gain widespread safety acceptance, nanoEHS concerns must be addressed. This will require green nanoscience approaches. Green nanoscience [16] aims to produce inherently safer nanomaterials by combining the 12 principles of green chemistry with nanoscience to provide guidelines to design effective materials with minimized waste and hazard throughout the life cycle [15]. Nanotechnology is a new and evolving industry, and is in the unique position to proactively apply green chemistry principles early in the design and synthesis of nanomaterials.

Material scientists and chemists will not be able to apply all the principles of green chemistry to nanoscience efficiently without collaborations with other scientific disciplines. According to Anastas and Eghbali, there has been a general lack of concern toward addressing principles 4 (P4) of green chemistry which involves designing compounds, such as nanomaterials, in a way to minimize overall hazard [21]. Green nanoscience can avoid the folly of other adopters of green chemistry and alleviate nanoEHS concerns through incorporation of nanotoxicology. The old paradigm of conducting toxicity testing at the end of a development process will be neither sustainable nor suitable to meet the goals of green nanotoxicology. Instead, nanotoxicology must be implemented early and often in green nanomaterial design and synthesis. Nanotoxicology will provide data that can be used in the refinement or replacement of unsafe nanomaterials through direct incorporation of P4 of green chemistry. In addition, it can be used to evaluate the safety of feed stock materials required for the synthesis of nanomaterials such as solvents or other synthesis by-products (P5). This will aid green nanoscience in designing benign nanomaterials (P12). Using nanotoxicology to address P4, P5, and P12 in green nanoscience may be fairly obvious (Fig. 6.1). What may not be so obvious is how nanotoxicology can be applied to P1 and P6 (Fig. 6.1) depending on the testing platform and models used which will be covered later in this entry.

Toxicology Models

Traditionally, mammalian toxicity tests are favored and thought to be the most predictive for assessing the human health and safety of compounds. Mammalian models share a high degree of genetic similarity with humans making resulting toxicity data easier to extrapolate to humans [14], thereby addressing P4 of green chemistry. However, concerns about animal welfare have generated a need to restrict mammalian use to only essential testing and to find alternative testing approaches [22]. Maintaining, housing, and testing mammals require significant infrastructure, materials, and time. Cumulatively, these requirements make these tests low throughput, expensive, and impractical for the broad-based nanoEHS research needs. Additionally, a large amount of test compound is necessary for mammalian toxicity assays making testing nanomaterials particularly challenging as they are often synthesized in small batches during the discovery phase [11]. Larger synthetic needs also create more waste which conflicts with green chemistry P1. Therefore, in vitro assays have become a popular alternative to whole animal tests as infrastructure requirements and maintenance needs are greatly reduced. These assays are also high-throughput and utilize far less test material. While in vitro tests can provide valuable data on a specific pathway or processes, the hazard and risk predictability of these assays for whole animals is quite limited. In vitro assays are inherently less complex as they most often consist of single cell types in cultured dishes. In addition, specific cell types express unique gene products. Since the critical gene products that play roles in nanotoxicological responses are not known in advance, it is a challenge to select the most appropriate in vitro system for nanoEHS research. To increase the sensitivity of in vitro toxicity screens, multiple in vitro assays are assembled to account for routes of exposure, organ effects, chemical deposition, etc. [23]. This reduces the economy of the in vitro approach to toxicity testing as it becomes necessary to conceptually rebuild an organism through numerous assays to obtain the required complexity for predictive results. These caveats will make it challenging to obtain reliable nanoEHS results in the near future. As eluded to in the introduction, the challenges of green nanoscience in conducting nanotoxicity testing for rapid, comprehensive assessment of nanoEHS necessitates the use of a robust and predictive in vivo model that can be adapted to high-throughput screening. Here the embryonic zebrafish (*Danio rerio*) is proposed as the ideal model for nanotoxicity tests.

The Embryonic Zebrafish Model

Background

There are several basic attributes of the embryonic zebrafish model that make it ideal for nanotoxicology and green nanoscience. The husbandry requirements of the zebrafish are minimal. As a small vertebrate model with high fecundity (200–300 eggs per day), short intervals between female spawns (2–3 days), and quick maturation, many zebrafish can be housed in a small area reducing infrastructure costs [24]. Thousands of eggs can be produced daily for experimentation. Aside from ease of care and maintenance, additional benefits of the zebrafish model include transparent eggs and external embryogenesis which facilitates examination of early developmental processes in real time [24].

The transparency of embryos during development make the zebrafish model exceptionally well suited to high-throughput testing platforms which will be necessary to alleviate backlog in demand for quick nanotoxicity evaluations. During early development complex molecular signaling events and organogenesis make the embryo highly susceptible to environmental perturbations [14]. Exposure to nanomaterials at this critical life stage can disrupt organ function, cellular pathfinding, and/or tissue morphology which can be easily tracked in the transparent zebrafish embryos throughout the duration of a 5-day exposure, noninvasively [25].

Molecular Applications

What began as a vertebrate model of developmental genetics is now extensively used a model for infectious diseases and immune system function [26–28], to understand hematopoiesis and related diseases [28, 29], for aging [30, 31], oncology [32–34], cardiovascular development [35, 36], kidney development and disease [37–39], for identifying the etiology of diseases of the eye [40, 41] and ear [42, 43], for understanding mechanisms of drug abuse [44–46], and for regenerative medicine [47–49]. Importantly, the zebrafish genome has been sequenced [50] and this information along with the conserved anatomy and physiology between fish and humans is being exploited by the biomedical research community (reviewed in [51–55]). Clearly, there are exciting possibilities to using zebrafish as a model for mechanistic-based nanotoxicity studies.

With the remarkable conservation between the zebrafish and human genome, it is anticipated that nearly all the toxicologically relevant pathways also function in zebrafish. The most well-studied toxicologically relevant pathway in zebrafish is the aryl hydrocarbon receptor (AHR) signal transduction pathway. Over the past few years, the zebrafish AHR pathway has been identified and completely characterized (reviewed in [56–58]). This has served as proof of concept and demonstrates that toxicology pathways can be efficiently examined in zebrafish.

A first step in elucidating the role of a gene in a biological process or in mediating responses to external stimuli (i.e., toxic responses) is to manipulate the expression level of the gene product. Producing knockout mice using embryonic stem cells has revolutionized biomedical research [59] and has been useful for toxicological studies in rodents [60–62]. Despite extensive efforts, embryonic stem cells for the zebrafish have yet to be developed [63], but a number of approaches have been effectively used to define the role of individual genes in zebrafish. The most common approach is the use of morpholinos which are chemically modified oligonucleotides that are resistant to chemical and enzymatic degradation yet maintain base pairing properties [64]. Morpholinos have been shown to bind to and block translation of mRNA in embryonic zebrafish [65–67]. Morpholinos can also be designed to bind to complement sequences at intron-exon splice junctions to inhibit proper mRNA processing. This approach is especially effective in zebrafish embryos [68–71].

Splice blocking morpholinos are useful to target genes that have alternative translation start sites, and to generate mispliced transcripts that result in dominant-negative proteins [68]. The successful application of morpholinos in zebrafish is evident by the large number of published studies using this approach (http://www.gene-tools.com/Publications).

With the availability of the entire zebrafish genome, conceivably any zebrafish gene can be targeted individually, or in combinations, using morpholinos and other antisense techniques. Thus, the function of mammalian orthologs can be rapidly evaluated in zebrafish. There are, however, limitations to these approaches. The gene repression effects are transient, which are most suitable for evaluating gene functions during development, (up to approximately 3–5 days post fertilization). Efforts to effectively deliver morpholinos to zebrafish at older life stages or to specific cellular targets are not routine. To overcome this limitation, heritable knockout approaches have been developed.

Despite the lack of a technology for gene knockout by homologous recombination, researchers have generated a large number of mutant fish lines. The majority of these were created using the chemical mutagen, ethylnitrosourea (ENU), followed by phenotypic screening for developmental defects [72, 73]. Such screens facilitate the identification of genes essential for biological processes with no prior functional knowledge of the genes. To date, large-scale zebrafish mutant searches have led to the identification of thousands of mutants with unique embryonic and larval development defects [74, 75]. Toxicologists can use available mutants that phenocopy the effects of toxicant exposure to help identify molecular targets. Furthermore, since genome-wide genetic screens are routine in zebrafish, there is a significant untapped potential to use forward genetics to identify modifiers of susceptibility to xenobiotic insult. Since chemical mutagenesis is non-biased, there is a high probability that novel targets can be identified in zebrafish. For instance, a genetic screen for increased susceptibility to nanomaterial exposure could be used to identify suppressor pathways. Screening for resistance to chemical exposure has the potential to identify genes that modulate the toxic response. With the power of comparative genomics, the identification of novel genes can be rapidly evaluated using integrated approaches in mammals. An advantage of ENU-mediated mutagenesis is that small mutations are created; however, the identification of the mutated gene remains a challenge. The once tedious and expensive approach of positional cloning has successfully been used to identify hundreds of the mutated genes [76, 77], and has evolved into a more efficient "targeted" methodology. For example, the Targeting Induced Local Lesions in Genomes (TILLING) method can now be used in zebrafish to identify and/or select for loss-of-function alleles (knockouts) [78]. TILLING is conceptually simple; random mutations are created using traditional chemical mutagenesis followed by high-throughput PCR-based screening to identify point mutations in the gene of interest. Therefore, TILLING is a targeted approach and, in theory, every zebrafish gene can now be knocked out. Insertional mutagenesis is an alternative approach to generate zebrafish mutants and has the benefit that the interrupted gene can be immediately identified [79–81]. An emerging method utilizing zinc finger nucleases (ZFNs) holds promise in bypassing the inability to generate knockout zebrafish by homologous recombination. Zinc finger nucleases are engineered to bind and cleave specific sections of DNA. Application of ZFNs has been successful in generating targeted germline mutations in the zebrafish [82].

Numerous techniques have been developed to introduce foreign DNA into zebrafish including microinjection [83], retroviral infection [79, 80], biolistic delivery [84], and electroporation [85]. The applications for transgenic zebrafish to unravel mechanisms of toxicity appear limitless. Conditional promoters such as the hsp70 [86] and tetracycline responsive [87] can be used to control transgene expression. Functional analysis in zebrafish can also be achieved using ectopic expression vectors under the control of tissue-specific promoters reviewed in Refs. [88, 89]. The number of transgenic zebrafish lines is rapidly growing with numerous lines useful for toxicological research. For example, transgenic fish that express fluorescent reporter genes in response to stimuli have been used to evaluate estrogenic chemical exposure [90], to visualize calcium signaling in vivo [91, 92], to respond to AHR ligands [93], etc. In addition to assessing the transcriptional responses to xenobiotics, fluorescent transgenes have proven valuable to monitor the dynamic movements of cells. Since embryonic zebrafish are transparent, toxicity targets can be noninvasively monitored, in real time. Examples include the use of transgenic lines with neuronal-specific expression of GFP [94] to assess the impact of nicotine on motor neuron development [45], the use of sonic hedgehog responsive fish to evaluate the impact of TCDD on CNS development [95], and the use of heat shock responsive fish to localize the targets of cadmium toxicity [96]. Finally, transgenic zebrafish lines have been developed to detect mutations in vivo [97].

An emerging area is the use of zebrafish as a model to elucidate the impact of xenobiotics on CNS development and function. Zebrafish possess all typical senses including vision, olfaction, taste, touch, balance, and hearing; and their sensory pathways share an overall homology with humans [98]. Cognitive behavioral tests suggest that anatomic substrates of cognitive behavior are also conserved between fish and other vertebrates. Thus, similar to observations of hippocampal lesions in mammals, lesions of the structural homolog of the hippocampus in fish selectively impair spatial memory [99]. Zebrafish behavior is altered by exposure to ethanol [46, 100, 101], nicotine [45], cocaine [44], antipsychotics [102], pesticides [103, 104], and metals such as mercury [105]. Recent studies indicate that broad-based chemical genetic screens can be used to identify novel, clinically relevant neuroactive compounds [106]. These initial behavioral studies suggest that the zebrafish model may be ideal to investigate the long-term consequences on CNS function following transient exposures to xenobiotics.

Similar to other toxicological models, "omic" based approaches are now routine using zebrafish. With the availability of the genome sequence, custom and commercial cDNA and oligonucleotide microarrays are currently available. These microarrays have been effectively used to define developmental stage and tissue-specific transcriptomes [107–112] and to identify gene expression changes in response to xenobiotic exposure. Microarray technology has been successfully applied to toxicology. For instance, the gene expression changes in response to

TCDD [49, 58, 113], hypoxia [114], 4-nonylphenol [115], and chlorpromazine [110] have been reported. Further understanding of genome-wide changes in transcript levels will be possible through application of RNA sequencing technology [116] to examine how the entire zebrafish transcriptome changes with exposure to toxicants. The amenability of the zebrafish to all the molecular biology techniques recently described will be very important in determining the mechanisms by which nanomaterials can cause adverse reactions in humans and other organisms.

High-Throughput Platform

The extensive growth in the field of nanotechnology has created a pipeline problem for efficient nanoEHS assessment. High-throughput nanotoxicity testing approaches can alleviate this problem. The embryonic zebrafish model is already a moderate-throughput platform, but readily amenable to high-throughput nanotoxicity testing with the incorporation of robotics and automation.

Zebrafish embryos develop in a chorion, which is an acellular envelope that acts as a protective barrier. Normally, when reared at 28°C, zebrafish hatch at between 48 and 72 hour post fertilization (hpf). To increase the sensitivity of the embryonic assay, it is critical to initiate exposures as early as possible. The earliest feasible stage to remove the chorions from the zebrafish embryos is at 6 hpf via enzymatic digestion. The methods for removing the chorion from large groups of embryos have been optimized, eliminating a potential barrier to uptake from the surrounding water [25].

A potential bottleneck to high-throughput embryonic zebrafish assays is the placement of dechorionated zebrafish embryos into 96- or 384-well assay plates. For toxicity exposures, it is necessary to sort embryos by identifying and selecting healthy eggs at similar developmental stage. This is a delicate process because the embryos can easily rupture. Incorporation of precision robotic plate loaders with visual systems to sort embryos could greatly increase the productivity of the embryonic zebrafish assay. The development of such a system has recently been achieved [117]. Microinjection of embryos is a molecular technique that introduces exogenous chemicals, nanomaterials, DNA, RNA, protein, or synthetic oligonucleotides into zebrafish embryos. These techniques are powerful investigative tools that are time intensive. Precision robotics with computer vision and control could hasten this arduous task and increase injection success rates [118].

Perhaps the greatest hurdle in developing high-throughput embryonic zebrafish nanotoxicity tests involves screening for sometimes subtle phenotypic changes. Zebrafish embryos can display a wide range of malformations that must be evaluated by laboratory personnel [25]. Experienced toxicologists may require up to 30 min per plate depending on the number of endpoints investigated. Introduction of automated, image-based screening methods would greatly enhance throughput. After devising automated plate preparation and embryo handling, the next step to developing a high-throughput assay is to set up imaging systems that can automatically focus and capture images [119]. There must be sufficient data storage capabilities to save thousands of high content images. In addition, software algorithms are necessary to process the image data [119], and the software must be capable of efficiently evaluating phenotypic changes. Commercial software analyzing in vitro cell cultures is currently available [119]; however, the embryonic zebrafish presents an additional challenge of orientation of the embryo which can influence image content. Finally, data mining and statistical analyses tools are needed to distinguish and report significant results which can then be confirmed and investigated further if necessary [119]. This will require a high amount of computing power. At this point in automation, researchers must pay special attention to the quality of the data analysis such as minimizing the generation of false positives and negatives [120]. Addressing the steps necessary to high-throughput assay development will enhance the efficiency of the zebrafish assay while resulting in more objective data. This will help address the nanoEHS concerns [119]. Currently, there are laboratories that are independently developing such automated systems.

The Zebrafish as a Green Model

Since the development of the 12 principles of green chemistry in the early 1990s [15], the movement toward adopting these principles in research has increased [21]. Many of the aforementioned benefits of the embryonic zebrafish model make it ideal for use in green toxicology and nanotoxicology. This section will discuss how the zebrafish model applies to the principles of green chemistry.

While the principles of green chemistry generally refer to synthesis and design of chemical compounds, they can be adapted to non-chemistry applications. Of the 12 principles of green chemistry (Fig. 6.1), use of the zebrafish model can help green nanoscience address P1 and P6. Principle 1 of green chemistry pertains to reduction in waste (Fig. 6.1) and use of the zebrafish as an in vivo, vertebrate model reduces the space, use of materials and equipment, and generation of waste required for maintaining mammals. For example, in our laboratory a standard nanotoxicity assay involves performing a concentration-response exposure in three 96-well plates with 100 μL of solution per well. Each plate contains an n = 16 per treatment and controls that contain no test material. The chemical treatments include a fivefold dilution series; therefore, an exposure with three plates of five treatments uses low milligram quantities of exposure material reducing waste generation (P1). While the zebrafish cannot completely replace the use of mammalian testing, it can be used as a high-throughput screen to prioritize nanomaterials for mammalian tests which will reduce their use.

With the great demand for toxicity testing on many new and existing compounds, mammalian testing cannot keep pace. Therefore, introduction of the zebrafish assay can increase the energy efficiency of toxicity testing in accord with

P6 of green chemistry (Fig. 6.1). Maintaining and conducting toxicity tests on mammals requires much more energy input than zebrafish toxicity testing. Even with in vitro systems, energy is required in maintaining the cells and conducting the tests. If numerous in vitro tests are required to obtain human-relevant toxicity data, then the energy efficiency of zebrafish testing is superior.

In one assay, the zebrafish model effectively combines the predictability of mammalian toxicity tests with the speed and economy of in vitro assays. Zebrafish embryonic assays are increasingly becoming high-throughput through automation and development. As the molecular signaling cascades, physiology, and anatomy are well conserved between humans and zebrafish, nanotoxicity testing in zebrafish should be highly relevant. To date, many phenotypic endpoints of toxicity in zebrafish can be applied to understanding conditions in humans. For instance, the induction of bradycardia in zebrafish embryos has been applied to identify pre-clinical drugs with risks for heart arrhythmias in humans [121]. Berghmans et al. demonstrated a correlation between the effects of pharmacologically active compounds in humans or mammals with the effects in embryonic zebrafish through cardiac, intestinal, and visual assays [122]. These results support the zebrafish model's predictive capabilities and utility to help green nanoscience achieve the aim of designing inherently safer nanomaterials (P12).

Nanotoxicology and the Zebrafish Model

The zebrafish model has been used to design a high-throughput, three-tiered testing approach that evaluates adverse effects at several levels of organization (Fig. 6.2). While this tiered testing scheme evaluates the safety of nanomaterials, the major objective of this assay is to understand the physiochemical properties of

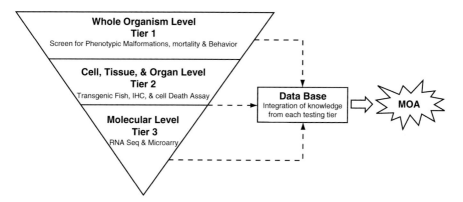

Fig. 6.2 Diagram outlining the tiers of nanotoxicity testing (1–3) and the types the laboratory techniques used to understand effects at the different levels of organization. Data from all three tiers are integrated to determine the mode of action (MOA) of a nanomaterial

nanomaterials that result in toxicity. A critical first step is the availability of well-characterized nanomaterials, where the size, number of atoms, agglomeration state, shape, surface chemistry, charge, and purity are precisely known [11, 123]. Tier one involves a waterborne exposure of dechorionated zebrafish embryos to the well-characterized nanomaterials to evaluate effects at the whole organism level. The exposure is continuous for 5 days beginning at 6 hpf and finishing at 120 hpf. During the assay, zebrafish embryos are screened for phenotypic effects at two time points (24 and 120 hpf) which manifest as mortality or phenotypic changes including increased spontaneous movement, developmental delays, or notochord malformations at 24 hpf. At 120 hpf, viable larval fish are screened for pericardial edema; yolk sac edema; or eye, nose, snout, jaw, axis, fin, brain, otic, notochord, or circulatory malformations using multi-well microscope imaging (Fig. 6.3). To evaluate impacts on the CNS development and behavior, two assays have been developed for screening at 120 hpf. The effects on motility and tactile response are evaluated by determining if fish display normal predator avoidance behavior by swimming away from a light touch to the head or tail. The second behavioral assay exploits the well-studied phenomenon of increased larval zebrafish motor activity in the dark [124]. The locomotor activity of nanomaterial exposed zebrafish larvae is recorded using Videotrack V3 software (Viewpoint Life Sciences, Inc., France) for 10 min in the dark following a 10 min rest period in the light. The results reveal whether nanomaterial exposure impacts this complex light-induced behavior.

To understand zebrafish nanomaterial absorption, it is necessary to experimentally measure nanomaterial uptake so that the relationship between in vivo nanoparticle dosimetry and response can be modeled. Determination of the internal dose of nanoparticles in embryonic zebrafish requires the use of Instrumental Neutron Activation Analysis (INAA) or Inductively Coupled Plasma-Optical Emission Spectrometry/Mass Spectrometry (ICP-OES/ICP-MS respectively). Nanomaterials that elicit phenotypic or behavioral responses in the zebrafish advance to tier two of testing to elucidate effects occurring at the cellular, tissue, or organ level. Use of transgenic zebrafish with fluorescent reporters and/or immunohistochemistry can help pinpoint the transport and deposition of a nanomaterial by tracking where adverse effects occur. Examples of two transgenic lines are illustrated in Figs. 6.4 and 6.5. Figure 6.4 illustrates a live 4-day-old zebrafish under non-excitatory light (top panel), and the same live fish images under excitatory light (bottom panel). The green fluorescent protein specifically labels the secondary motor neurons in these fish. Figure 6.5 illustrates a high magnification view of the spinal cord of a live zebrafish illustrating the detailed structures of the spinal neurons as they innervate into the adjacent muscle. These and other transgenic fish lines are useful tools when nanomaterial exposures are suspected to disrupt either the development or migration of spinal neurons. Another in vivo tier-2 assay is to determine if nanomaterial exposure results in increased cell death (Fig. 6.6). In this example, exposure to fullerenes leads to an increase in cellular death throughout the live zebrafish embryo [125].

Tier three testing focuses on the molecular effects of nanomaterial exposure. RNA seq or microarrays can be used to evaluate changes in gene expression profiles. Figure 6.7 illustrates the power and complexity of this approach.

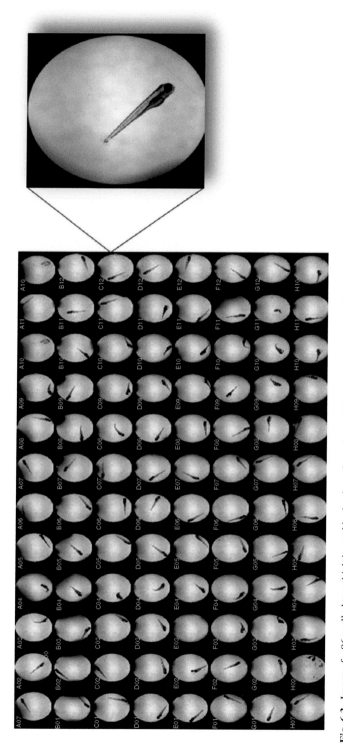

Fig. 6.3 Image of a 96-well plate which is used in the three tiers of zebrafish nanotoxicity assay. The plate contains zebrafish larva (120 hpf) after 5 days of exposure. *Right image* shows an enlargement of a single zebrafish in an exposure well

Fig. 6.4 Live 96-hour-old transgenic zebrafish (islet 1-gfp) that expresses the green fluorescent protein expressed specifically in spinal secondary motor neurons (*bottom image*). *Top image* is the transgenic zebrafish imaged under non-excitatory light

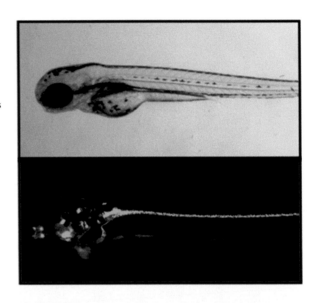

Fig. 6.5 High-resolution confocal image of live transgenic zebrafish (nbt-gfp) expressing the green fluorescent protein in primary motor neurons

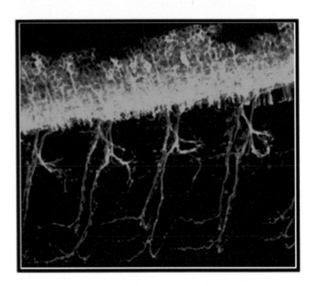

Following exposure to four different chemicals, the impact of the exposures on all genes in the genome was assessed, and provided distinct changes in gene expression. Known gene relationships (ontology) can be used to find particular pathways that may be impacted by nanomaterial exposure. The information from tiers two and three also enter the database. Data across all three tiers of nanotoxicity testing are integrated to define the mode of nanomaterial toxicity.

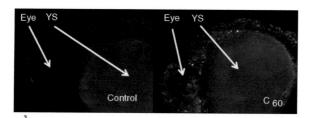

Fig. 6.6 Representative example of live 24-hour-old zebrafish embryos stained with acridine orange (AO). AO stains cells undergoing cellular death assay. AO binds to nucleic acids and fluoresces orange when cells enter lysosomes for degradation. Greater fluorescence in the nanoparticle (carbon 60 fullerene-C_{60}) exposed embryo indicates enhanced cellular death compared to the control. *YS* yolk sac

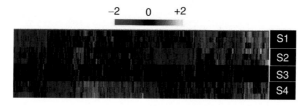

Fig. 6.7 Representative microarray heat map displaying changes in gene expression following exposure to four different chemicals. The *rows* across the heat map (labeled *S1*, *S2*, *S3*, and *S4*) represent three different samples (in triplicate), while each *column* indicates a different gene. *Blue* signifies a decrease in gene expression, *black* signifies no changes, and *yellow* signifies an increase in gene expression

Future Directions

Regardless of the many marketable nanomaterials and nanomaterial-based products poised for industrial production, few have actually been commercialized. A partial explanation for this lag can be assigned to difficulty in scaling up the nanosynthesis process from bench level to production level and still maintaining the quality, purity, and efficiency of the small batches [12]. However, a lack of nanoEHS data for confident assessment of the risks posed by exposure to nanomaterials remains a significant problem. The most critical need is to identify the design rules that define specific physiochemical properties of nanomaterials that predict biologic effects (biotic-nano models based on structure activity relationships) [126].

A major bottleneck to ultra high-throughput screening in the zebrafish assay will be caused by time constraints in data processing resulting from the large volume of data generated from high content imaging and large sample sizes. To avoid increases in total handling time of the screening process, increased computing power and reduction in the time required for data analyses and interpretation of the results are necessary [120]. Improvements in data analyses and interpretation will necessitate improvements and innovations in computer software. Improvements in computer software and whole organism-based imaging technology will also be necessary in augmenting the sensitivity of the zebrafish assay. For example, improvements in and increased dissemination of Cognition Network Technology (CNT) image analysis, a method that attempts to imitate human thought processes, over pixel-based image analysis methods could potentially allow for capturing subtle phenotypes that cannot be seen by the human eye irrespective of the orientation of the embryo [19]. Another method of increasing the sensitivity of the zebrafish assay may be through the addition of a high-throughput adaptation of micro-CT imaging. Micro-CT imaging, which is already used in mammalian research, is able to capture three-dimensional images of an organism noninvasively [127]. This technology could be used to capture slight but perhaps significant morphological responses that are presently not practical to assess on large numbers of animals.

Analysis of the changes in gene expression that occur with toxicant exposure has become a powerful tool in toxicology for determining toxicity mechanisms. With the development of RNA-Seq and organism-wide changes in gene transcripts will allow faster identification of molecular targets of compounds [116]. RNA-Seq also generates large amounts of data which must be processed and analyzed. Future work must be directed at determining the best way to analyze RNA-Seq data and in automating the processing and analysis procedures. This will assist in timely gene expression analysis of effects resulting from nanomaterial exposure in the zebrafish assay.

Adoption of high-throughput embryonic zebrafish assay will alleviate the backlog of nanotoxicity tests necessary for nanomaterial risk assessment, which will not be feasible to evaluate the toxicity of all nanomaterials [128]. Therefore, it is imperative that nanotoxicity screens be conducted in a way to accurately link the physiochemical properties of nanomaterials to toxicological endpoints, effectively creating structure activity relationships (SARS) [126]. By developing nano-SARS, the toxicity of new nanomaterials becomes increasingly predictive of human safety. It is realistic to say that extensive development of nano-SARS will occur over the next decade. There are collaborative efforts ongoing for establishing the foundation for such models [126]. For example, a few nanotoxicity studies have demonstrated that surface chemistry [129, 130] and size [129, 131, 132] may be the most important predictors of nanomaterial activity.

A framework must be also created to organize and curate the data for nano-SARS. To be successful, the framework must include three key guidelines. First, the nanomaterials tested in the toxicity assays must be well characterized [11, 123]. Accurate characterization of nanomaterials remains a major immediate objective in green nanoscience. Second, libraries of well-characterized nanomaterials with systematic variations in structural or physiochemical properties must be assembled. By screening these libraries in the embryonic zebrafish assay and focusing on one property at a time, one can begin to elucidate how nanomaterial core size, core composition, and functionalization influence biological activity [128]. Third, the data gathered from these tests must be assembled into an open database where others can access the information. The success of this framework requires close collaborative efforts between toxicologists, nanomaterial scientists, regulatory agencies, and industry. With the use of efficient vertebrate toxicological models like zebrafish and the development of accurate SARS, benign by design nanomaterials will be closer to realization.

Bibliography

1. Hushon J, Clerman R, Wagner B (1979) Tiered testing for chemical hazard assessment. Environ Sci Technol 13:1202–1207
2. Society of Toxicology (2005) How do you define toxicology? Soc Toxicol Commun. http://www.toxicology.org/ai/pub/si05/SI05_Define.asp
3. NSET/NEHI (2011) NNI environmental, health, and safety research strategy fact sheet. http://nano.gov/sites/default/files/pub_resource/2011_ehs_strategy_fact_sheet_locked.pdf
4. Forrest DR (2001) Molecular nanotechnology. IEEE Instrum Meas Mag 4(3):11–20
5. Lecoanet H, Wiesner MR (2004) Assessment of the mobility of nanomaterials in groundwater acouifers. Abs Pap Am Chem Soc 227:U1275–U1275
6. Lecoanet HF, Bottero JY, Wiesner MR (2004) Laboratory assessment of the mobility of nanomaterials in porous media. Environ Sci Technol 38:5164–5169
7. Lecoanet HF, Wiesner MR (2004) Velocity effects on fullerene and oxide nanoparticle deposition in porous media. Environ Sci Technol 38:4377–4382
8. Okamoto Y (2001) Ab initio investigation of hydrogenation of C-60. J Phys Chem A 105:7634–7637
9. Sun O, Wang Q, Jena P, Kawazoe Y (2005) Clustering of Ti on a C-60 surface and its effect on hydrogen storage. J Am Chem Soc 127:14582–14583
10. Lux Research (2009) The recession's ripple effect on nanotech. Lux Research Inc., New York
11. Dahl J, Maddux BLS, Hutchison JE (2007) Green nanosynthesis. Chem Rev 107:2228–2269
12. Hutchison JE (2008) Greener nanoscience: a proactive approach to advancing applications and reducing implications of nanotechnology. ACS Nano 2:395–402
13. Thomas K, Sayre P (2005) Research strategies for safety evaluation of nanomaterials, Part I: evaluating the human health implications of exposure to nanoscale materials. Toxicol Sci 87:316–321
14. NRC (2000) Scientific Frontiers in developmental toxicology and risk assessment. National Academy Press, Washington, DC, pp 1–327
15. Anastas PT, Warner JC (1998) Green chemistry: theory and practice. Oxford University Press, Oxford/New York, xi, 135 p
16. McKenzie LC, Hutchinson J (2004) Green nanoscience: an integrated approach to greener products, processes and applications. Chem Today 22:30–33
17. Abbott BD, Perdew GH, Buckalew AR, Birnbaum LS (1994) Interactive regulation of Ah and glucocorticoid receptors in the synergistic induction of cleft palate by 2,3,7,8-tetrachlorodibenzo-*p*-dioxin and hydrocortisone. Toxicol Appl Pharmacol 128:138–150
18. Podgórski A, Balazy A, Gradon L (2006) Application of nanofibers to improve the filtration efficiency of the most penetrating aerosol particles in fibrous filters. Chem Eng Sci 61:6804–6815
19. Savage N, Diallo MS (2005) Nanomaterials and water purification: opportunities and challenges. J Nanoparticle Res 7:331–342
20. Yuan J, Liu X, Akbulut O, Hu J, Suib SL et al (2008) Superwetting nanowire membranes for selective absorption. Nat Nano 3:332–336
21. Anastas P, Eghbali N (2010) Green chemistry: principles and practice. Chem Soc Rev 39:301–312
22. Lein P, Silbergeld E, Locke P, Goldberg AM (2005) In vitro and other alternative approaches to developmental neurotoxicity testing (DNT). Environ Toxicol Pharmacol 19:735–744
23. Oberdorster G, Maynard A, Donaldson K, Castranova V, Fitzpatrick J et al (2005) Principles for characterizing the potential human health effects from exposure to nanomaterials: elements of a screening strategy. Part Fibre Toxicol 2:8
24. Detrich HW, Westerfield M, Zon LI (eds) (1999) The zebrafish biology. Academic, San Diego, 391 p
25. Truong L, Harper SL, Tanguay RL (2011) Evaluation of embryotoxicity using the zebrafish model. Methods Mol Biol 691:271–279

26. van der Sar AM, Appelmelk BJ, Vandenbroucke-Grauls CM, Bitter W (2004) A star with stripes: zebrafish as an infection model. Trends Microbiol 12:451–457
27. Trede NS, Langenau DM, Traver D, Look AT, Zon LI (2004) The use of zebrafish to understand immunity. Immunity 20:367–379
28. Traver D, Herbomel P, Patton EE, Murphey RD, Yoder JA et al (2003) The zebrafish as a model organism to study development of the immune system. Adv Immunol 81:253–330
29. de Jong JL, Zon LI (2005) Use of the zebrafish to study primitive and definitive hematopoiesis. Annu Rev Genet 39:481–501
30. Gerhard GS (2003) Comparative aspects of zebrafish (*Danio rerio*) as a model for aging research. Exp Gerontol 38:1333–1341
31. Keller ET, Murtha JM (2004) The use of mature zebrafish (*Danio rerio*) as a model for human aging and disease. Comp Biochem Physiol C Toxicol Pharmacol 138:335–341
32. Spitsbergen J, Kent M (2003) The state of the art of the zebrafish model for toxicology and toxicologic pathology research – advantages and current limitations. Toxicol Pathol 31:62–87
33. Amatruda JF, Shepard JL, Stern HM, Zon LI (2002) Zebrafish as a cancer model system. Cancer Cell 1:229–231
34. Moore JL, Gestl EE, Cheng KC (2004) Mosaic eyes, genomic instability mutants, and cancer susceptibility. Methods Cell Biol 76:555–568
35. Chen JN, Fishman MC (2000) Genetic dissection of heart development. Ernst Schering Res Found Workshop 29:107–122
36. Beis D, Bartman T, Jin SW, Scott IC, D'Amico LA et al (2005) Genetic and cellular analyses of zebrafish atrioventricular cushion and valve development. Development 132:4193–4204
37. Drummond IA (2004) Zebrafish kidney development. Methods Cell Biol 76:501–530
38. Hentschel DM, Park KM, Cilenti L, Zervos AS, Drummond I et al (2005) Acute renal failure in zebrafish: a novel system to study a complex disease. Am J Physiol Renal Physiol 288:F923–F929
39. Drummond IA (2005) Kidney development and disease in the zebrafish. J Am Soc Nephrol 16:299–304
40. Bahadori R, Huber M, Rinner O, Seeliger MW, Geiger-Rudolph S et al (2003) Retinal function and morphology in two zebrafish models of oculo-renal syndromes. Eur J Neurosci 18:1377–1386
41. McMahon C, Semina EV, Link BA (2004) Using zebrafish to study the complex genetics of glaucoma. Comp Biochem Physiol C Toxicol Pharmacol 138:343–350
42. Whitfield TT (2002) Zebrafish as a model for hearing and deafness. J Neurobiol 53:157–171
43. Nicolson T (2005) The genetics of hearing and balance in zebrafish. Annu Rev Genet 39:9–22
44. Darland T, Dowling JE (2001) Behavioral screening for cocaine sensitivity in mutagenized zebrafish. Proc Natl Acad Sci USA 98:11691–11696
45. Svoboda KR, Vijayaraghavan S, Tanguay RL (2002) Nicotinic receptors mediate changes in spinal motoneuron development and axonal pathfinding in embryonic zebrafish exposed to nicotine. J Neurosci 22:10731–10741
46. Gerlai R, Lahav M, Guo S, Rosenthal A (2000) Drinks like a fish: zebra fish (*Danio rerio*) as a behavior genetic model to study alcohol effects. Pharmacol Biochem Behav 67:773–782
47. Poss KD, Keating MT, Nechiporuk A (2003) Tales of regeneration in zebrafish. Dev Dyn 226:202–210
48. Akimenko MA, Mari-Beffa M, Becerra J, Geraudie J (2003) Old questions, new tools, and some answers to the mystery of fin regeneration. Dev Dyn 226:190–201
49. Andreasen EA, Mathew LK, Tanguay RL (2006) Regenerative growth is impacted by TCDD: gene expression analysis reveals extracellular matrix modulation. Toxicol Sci 92:254–269
50. Vogel G (2000) Genomics. Sanger will sequence zebrafish genome. Science 290:1671
51. Zon LI (1999) Zebrafish: a new model for human disease. Genome Res 9:99–100

52. Ackermann GE, Paw BH (2003) Zebrafish: a genetic model for vertebrate organogenesis and human disorders. Front Biosci 8:d1227–d1253
53. Rubinstein AL (2003) Zebrafish: from disease modeling to drug discovery. Curr Opin Drug Discov Devel 6:218–223
54. Wixon J (2000) Featured organism: *Danio rerio*, the zebrafish. Yeast 17:225–231
55. Dodd A, Curtis PM, Williams LC, Love DR (2000) Zebrafish: bridging the gap between development and disease. Hum Mol Genet 9:2443–2449
56. Hahn M (2002) Aryl hydrocarbon receptors: diversity and evolution(1). Chem Biol Interact 141:131
57. Tanguay RL, Andreasen EA, Walker MK, Peterson RE (2003) Dioxin toxicity and aryl hydrocarbon receptor signaling in fish. In: Schecter A (ed) Dioxins and health. Plenum Press, New York, pp 603–628
58. Carney SA, Chen J, Burns CG, Xiong KM, Peterson RE et al (2006) AHR activation produces heart-specific transcriptional and toxic responses in developing zebrafish. Mol Pharmacol 70:549–561
59. Muller U (1999) Ten years of gene targeting: targeted mouse mutants, from vector design to phenotype analysis. Mech Dev 82:3–21
60. Ryffel B (1997) Impact of knockout mice in toxicology. Crit Rev Toxicol 27:135–154
61. Rudolph U, Mohler H (1999) Genetically modified animals in pharmacological research: future trends. Eur J Pharmacol 375:327–337
62. Gonzalez FJ (2002) Transgenic models in xenobiotic metabolism and toxicology. Toxicology 181–182:237–239
63. Fan L, Collodi P (2002) Progress towards cell-mediated gene transfer in zebrafish. Brief Funct Genomic Proteomic 1:131–138
64. Summerton J, Weller D (1997) Morpholino antisense oligomers: design, preparation, and properties. Antisense Nucleic Acid Drug Dev 7:187–195
65. Nasevicius A, Ekker SC (2000) Effective targeted gene 'knockdown' in zebrafish. Nat Genet 26:216–220
66. Nasevicius A, Ekker SC (2001) The zebrafish as a novel system for functional genomics and therapeutic development applications. Curr Opin Mol Ther 3:224–228
67. Nasevicius A, Larson J, Ekker SC (2000) Distinct requirements for zebrafish angiogenesis revealed by a VEGF-A morphant. Yeast 17:294–301
68. Draper BW, Morcos PA, Kimmel CB (2001) Inhibition of zebrafish fgf8 pre-mRNA splicing with morpholino oligos: a quantifiable method for gene knockdown. Genesis 30:154–156
69. Yan YL, Miller CT, Nissen RM, Singer A, Liu D et al (2002) A zebrafish sox9 gene required for cartilage morphogenesis. Development 129:5065–5079
70. Knight RD, Nair S, Nelson SS, Afshar A, Javidan Y et al (2003) Lockjaw encodes a zebrafish tfap2a required for early neural crest development. Development 130:5755–5768
71. Imamura S, Kishi S (2005) Molecular cloning and functional characterization of zebrafish ATM. Int J Biochem Cell Biol 37:1105–1116
72. Haffter P, Granato M, Brand M, Mullins MC, Hammerschmidt M et al (1996) The identification of genes with unique and essential functions in the development of the zebrafish, *Danio rerio*. Development 123:1–36
73. Driever W, Solnica-Krezel L, Schier AF, Neuhauss SC, Malicki J et al (1996) A genetic screen for mutations affecting embryogenesis in zebrafish. Development 123:37–46
74. Abdelilah S, Solnica-Krezel L, Stainier DY, Driever W (1994) Implications for dorsoventral axis determination from the zebrafish mutation janus. Nature 370:468–471
75. Stainier DY, Fouquet B, Chen JN, Warren KS, Weinstein BM et al (1996) Mutations affecting the formation and function of the cardiovascular system in the zebrafish embryo. Development 123:285–292
76. Talbot WS, Schier AF (1999) Positional cloning of mutated zebrafish genes. Methods Cell Biol 60:259–286

77. Brownlie A, Donovan A, Pratt SJ, Paw BH, Oates AC et al (1998) Positional cloning of the zebrafish sauternes gene: a model for congenital sideroblastic anaemia. Nat Genet 20:244–250

78. Henikoff S, Till BJ, Comai L (2004) TILLING. Traditional mutagenesis meets functional genomics. Plant Physiol 135:630–636

79. Amsterdam A, Burgess S, Golling G, Chen W, Sun Z et al (1999) A large-scale insertional mutagenesis screen in zebrafish. Genes Dev 13:2713–2724

80. Chen W, Burgess S, Golling G, Amsterdam A, Hopkins N (2002) High-throughput selection of retrovirus producer cell lines leads to markedly improved efficiency of germ line-transmissible insertions in zebra fish. J Virol 76:2192–2198

81. Golling G, Amsterdam A, Sun Z, Antonelli M, Maldonado E et al (2002) Insertional mutagenesis in zebrafish rapidly identifies genes essential for early vertebrate development. Nat Genet 31:135–140

82. Meng X, Noyes MB, Zhu LJ, Lawson ND, Wolfe SA (2008) Targeted gene inactivation in zebrafish using engineered zinc-finger nucleases. Nat Biotechnol 26:695–701

83. Meng A, Tang H, Yuan B, Ong BA, Long Q et al (1999) Positive and negative cis-acting elements are required for hematopoietic expression of zebrafish GATA-1. Blood 93:500–508

84. Torgersen J, Collas P, Alestrom P (2000) Gene-gun-mediated transfer of reporter genes to somatic zebrafish (*Danio rerio*) tissues. Mar Biotechnol (NY) 2:293–300

85. Powers DA, Hereford L, Cole T, Chen TT, Lin CM et al (1992) Electroporation: a method for transferring genes into the gametes of zebrafish (*Brachydanio rerio*), channel catfish (*Ictalurus punctatus*), and common carp (*Cyprinus carpio*). Mol Mar Biol Biotechnol 1:301–308

86. Halloran MC, Sato-Maeda M, Warren JT, Su F, Lele Z et al (2000) Laser-induced gene expression in specific cells of transgenic zebrafish. Development 127:1953–1960

87. Huang CJ, Jou TS, Ho YL, Lee WH, Jeng YT et al (2005) Conditional expression of a myocardium-specific transgene in zebrafish transgenic lines. Dev Dyn 233:1294–1303

88. Linney E, Hardison NL, Lonze BE, Lyons S, DiNapoli L (1999) Transgene expression in zebrafish: a comparison of retroviral-vector and DNA-injection approaches. Dev Biol 213:207–216

89. Linney E, Udvadia AJ (2004) Construction and detection of fluorescent, germline transgenic zebrafish. Methods Mol Biol 254:271–288

90. Bogers R, Mutsaerds E, Druke J, De Roode DF, Murk AJ et al (2006) Estrogenic endpoints in fish early life-stage tests: luciferase and vitellogenin induction in estrogen-responsive transgenic zebrafish. Environ Toxicol Chem 25:241–247

91. Ashworth R, Brennan C (2005) Use of transgenic zebrafish reporter lines to study calcium signalling in development. Brief Funct Genomic Proteomic 4:186–193

92. Higashijima S, Masino MA, Mandel G, Fetcho JR (2003) Imaging neuronal activity during zebrafish behavior with a genetically encoded calcium indicator. J Neurophysiol 90:3986–3997

93. Mattingly CJ, McLachlan JA, Toscano WA Jr (2001) Green fluorescent protein (GFP) as a marker of aryl hydrocarbon receptor (AhR) function in developing zebrafish (*Danio rerio*). Environ Health Perspect 109:845–849

94. Higashijima S, Hotta Y, Okamoto H (2000) Visualization of cranial motor neurons in live transgenic zebrafish expressing green fluorescent protein under the control of the islet-1 promoter/enhancer. J Neurosci 20:206–218

95. Hill A, Howard CV, Strahle U, Cossins A (2003) Neurodevelopmental defects in zebrafish (*Danio rerio*) at environmentally relevant dioxin (TCDD) concentrations. Toxicol Sci 76:392–399

96. Blechinger SR, Warren JT Jr, Kuwada JY, Krone PH (2002) Developmental toxicology of cadmium in living embryos of a stable transgenic zebrafish line. Environ Health Perspect 110:1041–1046

97. Amanuma K, Takeda H, Amanuma H, Aoki Y (2000) Transgenic zebrafish for detecting mutations caused by compounds in aquatic environments. Nat Biotechnol 18:62–65
98. Scalzo FM, Levin ED (2004) The use of zebrafish (*Danio rerio*) as a model system in neurobehavioral toxicology. Neurotoxicol Teratol 26:707–708
99. Rodriguez F, Lopez JC, Vargas JP, Broglio C, Gomez Y et al (2002) Spatial memory and hippocampal pallium through vertebrate evolution: insights from reptiles and teleost fish. Brain Res Bull 57:499–503
100. Gerlai R (2003) Zebra fish: an uncharted behavior genetic model. Behav Genet 33:461–468
101. Carvan MJ 3rd, Loucks E, Weber DN, Williams FE (2004) Ethanol effects on the developing zebrafish: neurobehavior and skeletal morphogenesis. Neurotoxicol Teratol 26:757–768
102. Giacomini NJ, Rose B, Kobayashi K, Guo S (2006) Antipsychotics produce locomotor impairment in larval zebrafish. Neurotoxicol Teratol 28:245–250
103. Levin ED, Swain HA, Donerly S, Linney E (2004) Developmental chlorpyrifos effects on hatchling zebrafish swimming behavior. Neurotoxicol Teratol 26:719–723
104. Bretaud S, Lee S, Guo S (2004) Sensitivity of zebrafish to environmental toxins implicated in Parkinson's disease. Neurotoxicol Teratol 26:857–864
105. Samson JC, Goodridge R, Olobatuyi F, Weis JS (2001) Delayed effects of embryonic exposure of zebrafish (*Danio rerio*) to methylmercury (MeHg). Aquat Toxicol 51:369–376
106. Kokel D, Bryan J, Laggner C, White R, Cheung CY et al (2010) Rapid behavior-based identification of neuroactive small molecules in the zebrafish. Nat Chem Biol 6:231–237
107. Corredor-Adamez M, Welten MC, Spaink HP, Jeffery JE, Schoon RT et al (2005) Genomic annotation and transcriptome analysis of the zebrafish (*Danio rerio*) hox complex with description of a novel member, hox b 13a. Evol Dev 7:362–375
108. Lo J, Lee S, Xu M, Liu F, Ruan H et al (2003) 15000 unique zebrafish EST clusters and their future use in microarray for profiling gene expression patterns during embryogenesis. Genome Res 13:455–466
109. Linney E, Dobbs-McAuliffe B, Sajadi H, Malek RL (2004) Microarray gene expression profiling during the segmentation phase of zebrafish development. Comp Biochem Physiol C Toxicol Pharmacol 138:351–362
110. van der Ven K, De Wit M, Keil D, Moens L, Van Leemput K et al (2005) Development and application of a brain-specific cDNA microarray for effect evaluation of neuro-active pharmaceuticals in zebrafish (*Danio rerio*). Comp Biochem Physiol B Biochem Mol Biol 141:408–417
111. Mathavan S, Lee SG, Mak A, Miller LD, Murthy KR et al (2005) Transcriptome analysis of zebrafish embryogenesis using microarrays. PLoS Genet 1:260–276
112. Clark MD, Hennig S, Herwig R, Clifton SW, Marra MA et al (2001) An oligonucleotide fingerprint normalized and expressed sequence tag characterized zebrafish cDNA library. Genome Res 11:1594–1602
113. Handley-Goldstone HM, Grow MW, Stegeman JJ (2005) Cardiovascular gene expression profiles of dioxin exposure in zebrafish embryos. Toxicol Sci 85:683–693
114. Ton C, Stamatiou D, Liew CC (2003) Gene expression profile of zebrafish exposed to hypoxia during development. Physiol Genomics 13:97–106
115. Hoyt PR, Doktycz MJ, Beattie KL, Greeley MS Jr (2003) DNA microarrays detect 4-nonylphenol-induced alterations in gene expression during zebrafish early development. Ecotoxicology 12:469–474
116. Wang Z, Gerstein M, Snyder M (2009) RNA-Seq: a revolutionary tool for transcriptomics. Nat Rev Genet 10:57–63
117. Mandrell D, Moore A, Jephson C, Sarker M, Lang C et al (2011) Automated zebrafish chorion removal and single embryo transfer: optimizing throughput of zebrafish developmental toxicity screens. J Lab Autom (submitted)
118. Wang W, Liu X, Gelinas D, Ciruna B, Sun Y (2007) A fully automated robotic system for microinjection of zebrafish embryos. PLoS One 2:e862

119. Carpenter AE (2007) Image-based chemical screening. Nat Chem Biol 3:461–465
120. Mayr LM, Fuerst P (2008) The future of high-throughput screening. J Biomol Screen 13:443–448
121. Milan DJ, Peterson TA, Ruskin JN, Peterson RT, MacRae CA (2003) Drugs that induce repolarization abnormalities cause bradycardia in zebrafish. Circulation 107:1355–1358
122. Berghmans S, Butler P, Goldsmith P, Waldron G, Gardner I et al (2008) Zebrafish based assays for the assessment of cardiac, visual and gut function–potential safety screens for early drug discovery. J Pharmacol Toxicol Methods 58:59–68
123. Grassian VH (2008) When size really matters: size-dependent properties and surface chemistry of metal and metal oxide nanoparticles in gas and liquid phase environments†. J Phys Chem C 112:18303–18313
124. MacPhail RC, Brooks J, Hunter DL, Padnos B, Irons TD et al (2009) Locomotion in larval zebrafish: influence of time of day, lighting and ethanol. Neurotoxicology 30:52–58
125. Usenko CY, Harper SL, Tanguay RL (2007) *In vivo* evaluation of carbon fullerene toxicity using embryonic zebrafish. Carbon 45:1891–1898
126. Harper SL, Dahl JL, Maddux BLS, Tanguay RL, Hutchison JE (2008) Proactively designing nanomaterials to enhance performance and minimize hazard. I J Nanotechnol 5:124–142
127. Holdsworth DW, Thornton MM (2002) Micro-CT in small animal and specimen imaging. Trends Biotechnol 20:S34–S39
128. Harper SL, Usenko C, Hutchinson JE, Maddux BLS, Tanguay RL (2008) *In vivo* biodistribution and toxicity depends on nanomaterial composition, size, surface functionalization and route of exposure. J Exp Nanosci 3:195–206
129. Magrez A, Kasas S, Salicio V, Pasquier N, Seo JW et al (2006) Cellular toxicity of carbon-based nanomaterials. Nano Lett 6:1121–1125
130. Nel A, Xia T, Madler L, Li N (2006) Toxic potential of materials at the nanolevel. Science 311:622–627
131. Colvin V (2003) The potential environmental impact of engineered nanomaterials. Nat Biotechnol 21:1166–1170
132. Jiang W, KimBetty YS, Rutka JT, ChanWarren CW (2008) Nanoparticle-mediated cellular response is size-dependent. Nat Nano 3:145–150

Chapter 7
New Polymers, Renewables as Raw Materials

Richard P. Wool

Glossary

Composite materials	Are engineered or naturally occurring materials made from two or more constituent materials with significantly different physical or chemical properties which remain separate and distinct at the microscopic scale.
Lignin	Is a complex chemical compound most commonly derived from wood and an integral part of the secondary cell walls of plants and some algae.
Triglyceride	Is an ester derived from glycerol and three fatty acids. It is the main constituent of vegetable oil and animal fats.

Definition of the Subject

This entry describes how to make high volume materials out of thin-air using CO_2 and sunlight.

This chapter was originally published as part of the Encyclopedia of Sustainability Science and Technology edited by Robert A. Meyers. DOI:10.1007/978-1-4419-0851-3

R.P. Wool (✉)
Department of Chemical Engineering, Center for Composite Materials, Affordable Composites from Renewable Sources (ACRES), University of Delaware, Newark, DE 19716-3144, USA
e-mail: wool@UDel.Edu

P.T. Anastas and J.B. Zimmerman (eds.), *Innovations in Green Chemistry and Green Engineering*, DOI 10.1007/978-1-4614-5817-3_7,
© Springer Science+Business Media New York 2013

Introduction

Recent advances in genetic engineering, composite science, and natural fiber development offer significant opportunities for developing new, improved materials from renewable resources that can biodegrade or be recycled, enhancing global sustainability. A wide range of high-performance, low-cost materials can be made using plant oils, natural fibers, and lignin. By selecting the fatty acid distribution function of plant oils via computer simulation and the molecular connectivity, chemical functionalization and molecular architecture can be controlled to produce linear, branched, or cross-linked polymers. These materials can be used as pressure-sensitive adhesives, elastomers, rubbers, foams, and composite resins. This entry describes the chemical pathways that were used to modify plant oils and allow them to react with each other and various comonomers to form materials with useful properties.

Polymers and polymeric composite materials have extensive applications in the aerospace, automotive, marine, infrastructure, military, sports, and industrial fields. These lightweight materials exhibit excellent mechanical properties, high corrosion resistance, dimensional stability, and low assembly costs. Traditionally, polymers and polymeric composites have been derived from petroleum; however, as the applications for polymeric materials increase, finding alternative sources of these materials has become critical. In recent years, the Affordable Composites from Renewable Sources (ACRES) program at the University of Delaware has developed a broad range of chemical routes to use natural triglyceride oils to make polymers and composite materials [1, 2]. These materials have economic and environmental advantages that make them attractive alternatives to petroleum-based materials.

Natural oils, which can be derived from both plant and animal sources, are abundant in most parts of the world, making them an ideal alternative to chemical feedstocks. These oils are predominantly made up of triglyceride molecules, which have the structure shown in Fig. 7.1. Triglycerides are composed of three fatty acids joined at a glycerol juncture. The most common oils contain fatty acids that vary from 14 to 22 carbons in length with 0–3 double bonds per fatty acid. Table 7.1 shows the fatty acid distributions of several common oils [3]. Exotic oils are composed of fatty acids with other types of functionalities, such as epoxies, hydroxyls, cyclic groups, and furanoid groups [4]. Because of the many different fatty acids present, at a molecular level these oils are composed of many different types of triglycerides with numerous levels of unsaturation. With newly developed genetic engineering techniques, the variation in unsaturation can be controlled in plants such as soybean, flax, and corn; however, some oils are better suited to polymer resin development.

Besides applications in the foods industry, triglyceride oils have been used extensively to produce coatings, inks, plasticizers, lubricants, and agrochemicals [5–11]. In the polymers field, the use of these oils as toughening agents was investigated. Barrett [12] has reviewed an extensive amount of work on using these oils to produce interpenetrating networks (IPNs). It was found that IPNs formed by triglycerides could increase the toughness and fracture resistance in conventional thermoset polymers [13–16]; see also [17–25]. In these works, the

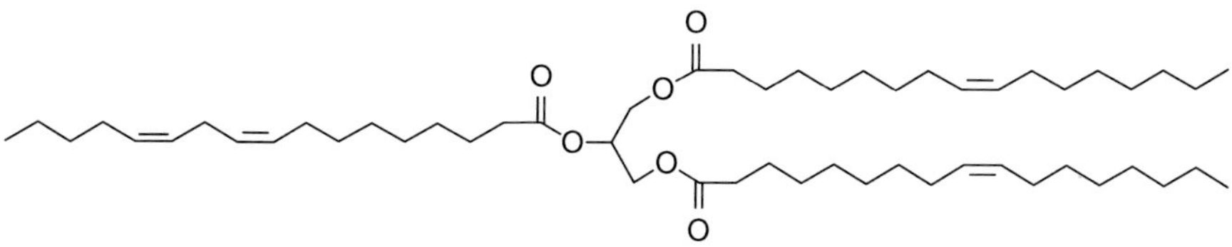

Fig. 7.1 Triglyceride molecule, the major component of natural oils

Table 7.1 Fatty acid distribution in various plant oils

Fatty acid	# C: # DB	Canola	Corn	Cotto- nseed	Linseed	Olive	Palm	Rapeseed	Soybean	High oleic[a]
Myristic	14:0	0.1	0.1	0.7	0.0	0.0	1.0	0.1	0.1	0.0
Myristoleic	14:1	0.0	0.0	0.0	0.0	0.0	0.0	0.0	0.0	0.0
Palmitic	16:0	1.1	10.9	21.6	5.5	13.7	41.4	3.0	11.0	6.4
Palmitoleic	16:1	0.3	0.2	0.6	0.0	1.2	0.2	0.2	0.1	0.1
Margaric	17:0	0.1	0.1	0.1	0.0	0.0	0.1	0.0	0.0	0.0
Margaroleic	17:1	0.0	0.0	0.1	0.0	0.0	0.0	0.0	0.0	0.0
Stearic	18:0	1.8	2.0	2.6	3.5	2.5	1.1	1.0	1.0	3.1
Oleic	18:1	60.9	25.4	18.6	19.1	71.1	39.3	13.2	23.4	82.6
Linoleic	18:2	21.0	59.6	51.4	15.3	10.0	10.0	13.2	53.2	2.3
Linolenic	18:3	8.8	1.2	0.7	56.6	0.6	0.4	9.0	7.8	3.7
Arachidic	20:0	0.7	0.4	0.3	0.0	0.9	0.3	0.5	0.3	0.2
Gadoleic	20:1	1.0	0.0	0.0	0.0	0.0	0.0	9.0	0.0	0.4
Eicosadienoic	20:2	0.0	0.0	0.0	0.0	0.0	0.0	0.7	0.0	0.0
Behenic	22:0	0.3	0.1	0.2	0.0	0.0	0.1	0.5	0.1	0.3
Erucic	22:1	0.7	0.0	0.0	0.0	0.0	0.0	49.2	0.0	0.1
Lignoceric	24:0	0.2	0.0	0.0	0.0	0.0	0.0	1.2	0.0	0.0
Average # DB/ triglyceride		3.9	1.5	3.9	6.6	2.8	1.8	3.8	1.6	3.0

[a]Genetically engineered high oleic acid content soybean oil (DuPont)

functional triglyceride was a minor component in the polymer matrix acting solely as a modifier to improve the physical properties of the main matrix. Consequently, the triglyceride-based materials were low-molecular-weight (M_W), lightly cross-linked materials incapable of displaying the necessary rigidity and strength required for structural applications by themselves.

Synthetic Pathways for Triglyceride-Based Monomers

Triglycerides contain active sites amenable to chemical reaction: the double bond, the allylic carbons, the ester group, and the carbons alpha to the ester group. These active sites can be used to introduce polymerizable groups on the triglyceride using the same techniques applied in the synthesis of petrochemical-based polymers. The key step is to reach a higher level of M_W and cross-link density, as well as to incorporate chemical functionalities known to impart stiffness in a polymer network (e.g., aromatic or cyclic structures). Figure 7.2 illustrates several synthetic pathways that accomplish this [1].

In structures **5, 6, 7, 8**, and **11**, the double bonds of the triglyceride are used to functionalize the triglyceride with polymerizable chemical groups. From the natural triglyceride, it is possible to attach maleates (**5**) [6–11] or to convert the unsaturation to epoxy (**7**) [26–28] or hydroxyl functionalities (**8**) [29, 30]. Such transformations make the triglyceride capable of reaction via ring-opening or polycondensation polymerization. These particular chemical pathways are also accessible via natural epoxy and hydroxyl functional triglycerides [12, 14–16] It is also possible to attach vinyl functionalities to the epoxy and hydroxyl functional triglycerides. Reaction of the epoxy functional triglyceride with acrylic acid incorporates acrylates onto the triglyceride (**6**), while reaction of the hydroxylated triglyceride with maleic anhydride incorporates maleate half-esters and esters onto the triglyceride (**11**). These monomers can then be blended with a reactive diluent, similar to most conventional vinyl ester resins and cured by free-radical polymerization.

The second method for synthesizing monomers from triglycerides is to convert the triglyceride to monoglycerides through a glycerolysis (**3A**) reaction or an amidation reaction (**2, 3B**) [31–36]. Monoglycerides are used in surface coatings, commonly referred to as alkyd resins, because of their low cost and versatility [32]. In those applications, the double bonds of the monoglyceride are reacted to form the coating. However, monoglycerides also can react through the alcohol groups via polycondensation reactions with a comonomer, such as a diacid, epoxy, or anhydride. Alternatively, maleate half-esters can be attached to these monoglycerides (**9**) allowing them to undergo free-radical polymerization.

The third method is to functionalize the unsaturation sites as well as reduce the triglyceride into monoglycerides. This can be accomplished by glycerolysis of an unsaturated triglyceride, followed by hydroxylation or by glycerolysis of a hydroxy functional triglyceride. The resulting monomer can then be reacted with maleic anhydride, forming a monomer capable of polymerization by the free-radical mechanism [1].

Although the structure of triglycerides is complex in nature, it is possible to characterize some aspects of it using proton nuclear magnetic spectroscopy (^{1}H-NMR) and Fourier transform infrared (FTIR) spectroscopy. A typical ^{1}H NMR spectrum of soybean oil is shown in Fig. 7.3, with peak assignments. The two sets of peaks at 1.0–1.4 ppm are produced by the four glycerol methylene protons per triglyceride [0]. The triplet set of peaks at 2.3 ppm is produced by the six protons in the alpha position relative to the carbonyl groups. The peak at 0.9 ppm is produced by the nine methyl protons per triglyceride at the end of each fatty acid chain. These three groups of peaks provide a standard by which other peaks can be used to quantitatively characterize functional groups in the triglyceride.

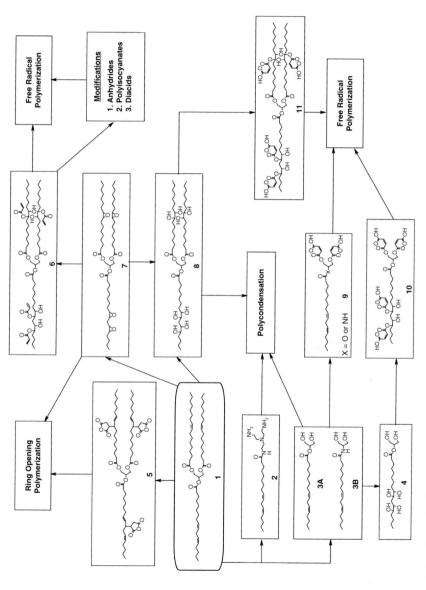

Fig. 7.2 Chemical pathways leading to polymers from triglyceride molecules [1]

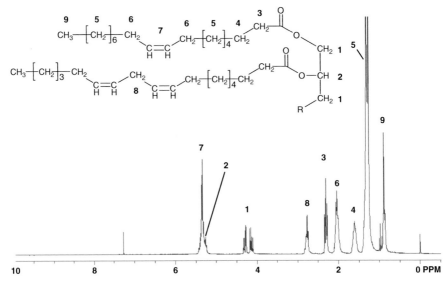

Fig. 7.3 ^{1}H-NMR spectrum of soybean oil (CDCl$_3$). R represents a third fatty acid

In this entry, the focus is on three triglyceride monomers, shown in Fig. 7.4, which have been found to be promising candidates for use in the composites and engineering plastics fields. They are acrylated epoxidized soybean oil (AESO), the maleinized soybean oil monoglyceride (SOMG/MA), and maleinized hydroxylated soybean oil (HSO/MA). These monomers, when used as a major component of a molding resin, have shown properties comparable to conventional polymers and composites, and these properties will be presented. In addition, they can be used as a matrix in synthetic and natural fiber-reinforced composites.

Acrylated Epoxidized Soybean Oil

Acrylated epoxidized oils (Fig. 7.4) are synthesized from the reaction of acrylic acid with epoxidized triglycerides. Epoxidized triglycerides can be found in natural oils, such as vernonia plant oil, or can be synthesized from more common unsaturated oils, such as soybean oil or linseed oil, by a standard epoxidation reaction [37]. The natural epoxy oil, vernonia oil, has a functionality of 2.8 epoxy rings per triglyceride [4]. Epoxidized soybean oil is commercially available and is generally sold with a functionality of 1.1–1.6 epoxy rings per triglyceride, which can be identified via ^{1}H NMR [20, 38]. Epoxidized linseed oil is also commercially available when higher epoxy content is required. Predominantly, these oils are used as alternative plasticizers in polyvinyl chloride in place of phthalates [39–41], but their use as a toughening agent also was explored [20, 23–25, 42]. With the

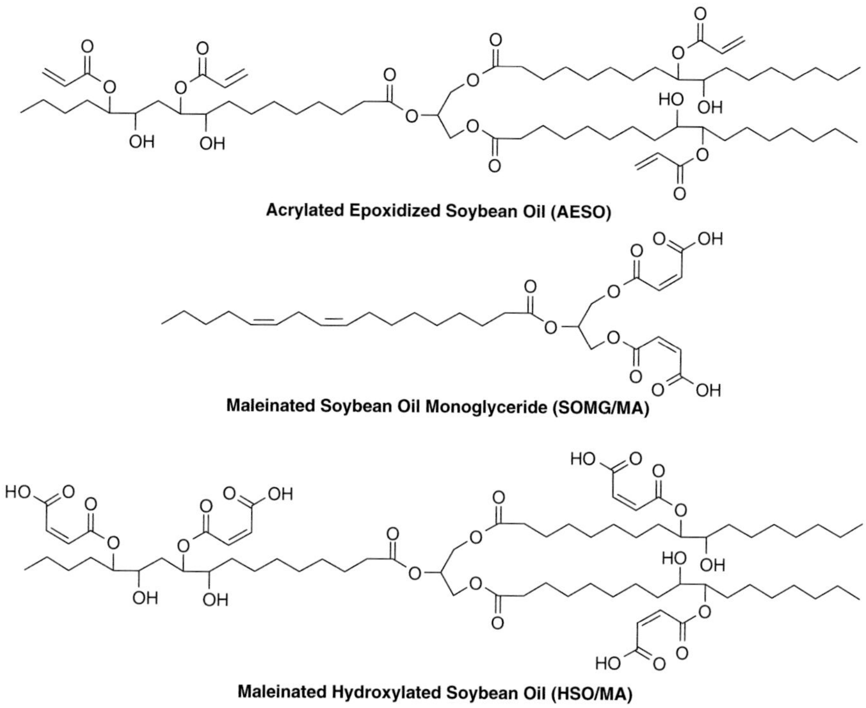

Acrylated Epoxidized Soybean Oil (AESO)

Maleinated Soybean Oil Monoglyceride (SOMG/MA)

Maleinated Hydroxylated Soybean Oil (HSO/MA)

Fig. 7.4 Triglyceride based monomers

addition of acrylates, the triglyceride can be reacted via addition polymerization. AESO was used extensively in surface coatings and is commercially manufactured in forms such as Ebecryl 860 [7, 43, 44]. Urethane and amine derivatives of AESO have also been developed for coating and ink applications [8, 9, 45].

The reaction of acrylic acid with epoxidized soybean oil occurs through a standard substitution reaction and was found to have first-order dependence with respect to epoxy concentration and second-order dependence with respect to acrylic acid concentration [46]. However, epoxidized oleic methyl ester was found to display second-order dependence on both epoxy and acrylic acid concentrations [47]. Although the reaction of epoxidized soybean oil with acrylic acid is partially catalyzed by the acrylic acid, the use of additional catalysts is common. Tertiary amines, such as N, N-dimethylaniline, triethylamine, and 1,4-diazobicyclo[2.2.2]octane, are commonly used [38, 48]. Additionally, more selective organometallic catalysts have been developed that reduce the amount of epoxy homopolymerization [49, 50].

AESO can be blended with a reactive diluent, such as styrene, to improve its processability and to control the polymer properties to reach a range acceptable for structural applications. By varying the amount of styrene, it is possible to produce polymers with different moduli and glass transition temperatures. Polymer properties can also be controlled by changing the M_W of the monomer or the functionality of the acrylated triglyceride. Consequently, a range of properties, and therefore applications, can be found. Subsequent to the acrylation reaction, the triglyceride contains both residual amounts of unreacted epoxy rings as well as newly formed hydroxyl groups, both of which can be used to further modify the triglyceride by reaction with a number of chemical species, such as diacids, diamines, anhydrides, and isocyanates. The approach presented here is to oligomerize the triglycerides with reagents that have chemical structures conducive to stiffening the polymer, such as cyclic or aromatic groups. Reacting AESO with cyclohexanedicarboxylic anhydride (Fig. 7.5a) forms oligomers, increasing the entanglement density as well as introducing stiff cyclic rings to the structure. Reaction of the AESO with maleic acid (Fig. 7.5b) also forms oligomers and introduces more double bonds. Although it is desirable to maximize the conversion of hydroxyls or epoxies, the viscosity increases dramatically at high levels of conversion. Eventually, this can lead to gelation, so the reaction must be carefully monitored. After oligomerization, the modified AESO resin can be blended with styrene and cured in the same manner as the unmodified AESO resin.

Maleinized Soybean Oil Monoglyceride

Maleinized soybean oil monoglyceride (Fig. 7.4) is synthesized from the triglyceride oil in two steps [33]. The first is a standard glycerolysis reaction to convert the triglycerides into monoglycerides by reacting triglycerides with glycerol; see [31]. The product is generally a mixture of mono- and diglycerides, as illustrated in Fig. 7.6. Using excess glycerol can aid in conversion. Additionally, the reaction can be run in solvent or in the presence of an emulsifier catalyst [34]. Once the reaction is completed, it is possible to separate a portion of the unreacted glycerol by cooling the product rapidly [33]. The presence of glycerol is not detrimental to the end polymer, because it can be reacted with maleic anhydride in the same manner as the monoglycerides and incorporated into the end polymer network.

The maleinization of the soybean oil monoglyceride (SOMG) mixture at temperatures below 100°C produces monoglyceride, diglyceride, and glycerol maleate half-esters. This reaction makes no attempt to produce a polyester, and the half-ester formation is expected to proceed at low temperatures in the presence of either acid or base catalysts without any by-products. A good indication of the success of this reaction is to follow the signal intensity ratio of maleate vinyl protons to fatty acid vinyl protons ($N_M \backslash N_{FA}$) in the ^{1}H NMR spectrum. The use of 2-methylimidazole and triphenyl antimony as catalysts was shown to be successful when conducting the reaction at temperatures of 80–100°C with a 3:2 weight ratio of glycerides to maleic anhydride ($N_M \backslash N_{FA} = 0.85$) [33, 51]. Once these maleates have been added, the monoglycerides can react via addition polymerization. Because maleates are relatively unreactive with each other, the addition of styrene increases the polymerization conversion and imparts rigidity to the matrix.

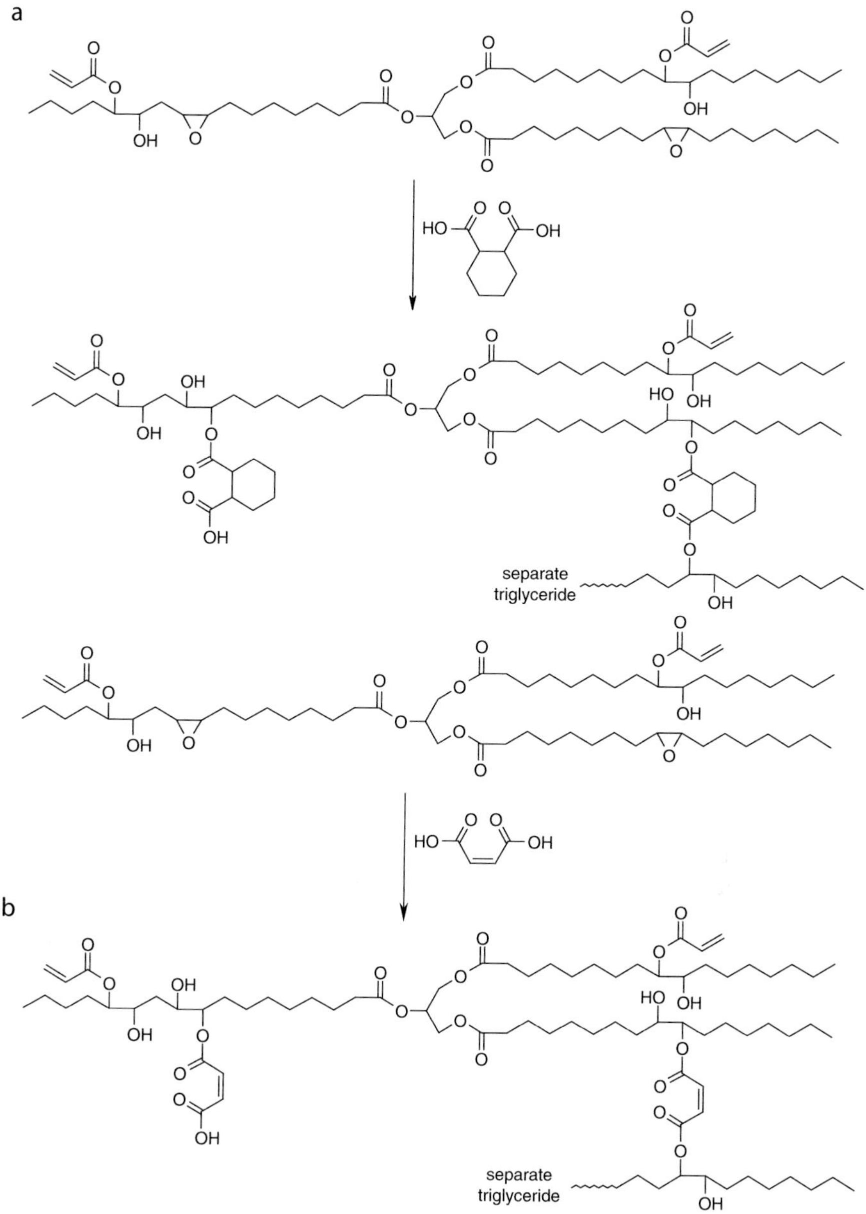

a

b

Fig. 7.5 (**a**) Modification of AESO by reaction with cyclohexane dicarboxylic anhydride. (**b**) Modification of AESO by reaction with maleic acid

To increase the glass transition temperature (T_g) and modulus of the SOMG/MA polymer, more rigid diols can be added during the maleinization reaction. Two such diols are neopentyl glycol (NPG) and bisphenol A (BPA), which may increase the rigidity of the end polymer network. Although their addition to the maleinization

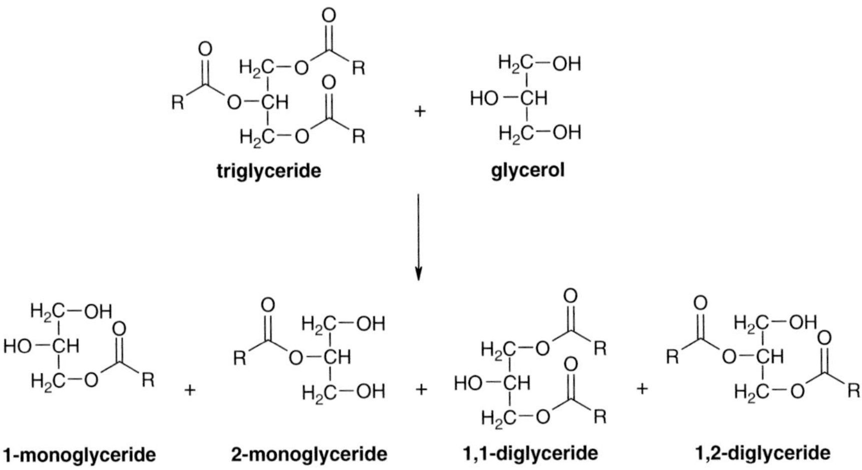

Fig. 7.6 Glycerolysis of triglycerides to form mixtures of mono- and diglycerides

mixture will reduce the renewable resource content of the final resin, they should result in higher T_g values for the end polymer. The synthesis of maleate half-esters of organic polyols, including NPG and BPA and the cross-linking of the resulting maleate half-esters with a vinyl monomer such as styrene, has been reported [52, 53]. The literature also abounds with examples of unsaturated polyesters prepared from NPG and maleic anhydride with other polyols and diacids [54–57]. However, the copolymers of NPG and BPA bis maleate half-esters with SOMG maleate half-esters are new.

The properties of the SOMG/MA polymer as well as the effect of adding NPG and BPA on the mechanical properties of the final polymers are presented here. For this purpose, mixtures of SOMG/NPG and SOMG/BPA, prepared at the same weight ratio, were maleinized, and the copolymers of the resulting maleates with styrene were analyzed for their mechanical properties and compared to that of SOMG maleates.

Maleinized Hydroxylated Oil

Maleinized hydroxylated oil (HO/MA) is synthesized in a manner similar to both the AESO and the SOMG/MA monomers. The double bonds of an unsaturated oil are used to attach the polymerizable groups by converting the double bonds of the triglyceride to hydroxyl groups. The hydroxyls can then be used to attach maleates. As shown in Fig. 7.2, there are two routes to synthesize the hydroxylated triglyceride. The first is through an epoxidized intermediate. By reacting the epoxidized triglyceride with an acid, the epoxies can be easily converted to hydroxyl groups [29, 58].

Alternatively, the hydroxylated oil can be synthesized directly from the unsaturated oil, as described in [1]. After hydroxylation, the oil can be reacted with maleic anhydride to functionalize the triglyceride with maleate half-esters. A molar ratio of 4:1 anhydride to triglyceride was used in all cases, and the reaction was catalyzed with N,N-dimethylbenzylamine. Once the maleinization reaction is finished, the monomer resin can be blended with styrene similar to the other resins presented here.

SOPERMA: Soybean Oil Pentaerythritol Glyceride Maleates

In the preceding sections, the preparation of soybean oil monoglyceride maleates (SOMG/MA) by E. Can [65] was reported. Soybean oil was reduced to monoglycerides through a glycerolysis reaction, and the glycerolysis product, which was a mixture of mono- and diglycerides as well as unreacted triglycerides and free glycerol, was reacted with maleic anhydride to convert the free hydroxyls to maleate half-esters, thus allowing them to free radically polymerize. The use of pentaerythritol instead of glycerol in the same synthetic route, as shown in Fig. 7.7, offers certain advantages, such as the presence of more hydroxyl groups and, therefore, more reactive sites for malination, which should result in a higher cross-link density for the resulting polymers. There are no standard conditions for the alcoholysis of the triglyceride oils with pentaerythritol; the reactant molar ratios and reaction conditions change according to the end use of the product. Soybean pentaerythritol molar ratios of SO:PER = 1:2 and SO:PER = 1:3 were used, which should give mixtures of monoglycerides and pentaerythritol monoesters as the main products. Higher amounts of pentaerythritol were avoided because this would further decrease the triglyceride content of the formulation. The idealized reaction scheme for the soybean oil pentaerythritol alcoholysis reaction is shown in Fig. 7.7.

The soybean oil pentaerythritol alcoholysis reactions were carried out for the first time by E. Can at 230–240°C for 0.5, 2, and 5.5 h. A reaction temperature of 180–190°C was also employed. $Ca(OH)_2$ was used as a catalyst at a concentration of 1% of the total weight of the oil and the polyol. $Ca(OH)_2$ forms soap with the free fatty acids in the oil and promotes the reaction at least in part by increasing the solubility of pentaerythritol in the oil. $Ca(OH)_2$ was reported to be an effective catalyst for the glycerolysis of soybean oil; it increases the monoglyceride yield and reduces the triglyceride content of the glycerolysis product. The amount of reactants and catalysts used for the soybean oil pentaerythritol alcoholysis reactions and the resulting malination reactions, at different mole ratios, are shown in Table 7.2.

For the malination of the alcoholysis products, maleic anhydride was in a 1:1 molar ratio with the number of hydroxyls on pentaerythritol used in the alcoholysis reaction, thus molar ratios of SO:PER:MA = 1:2:8 and 1:3:12 were employed. A molar ratio of SO:PER:MA = 1:3:7.62 was also used to reduce the unreacted maleic anhydride content of the latter reaction. A reaction temperature of 95–100°C was used, because when lower reaction temperatures were used, for example, 60°C,

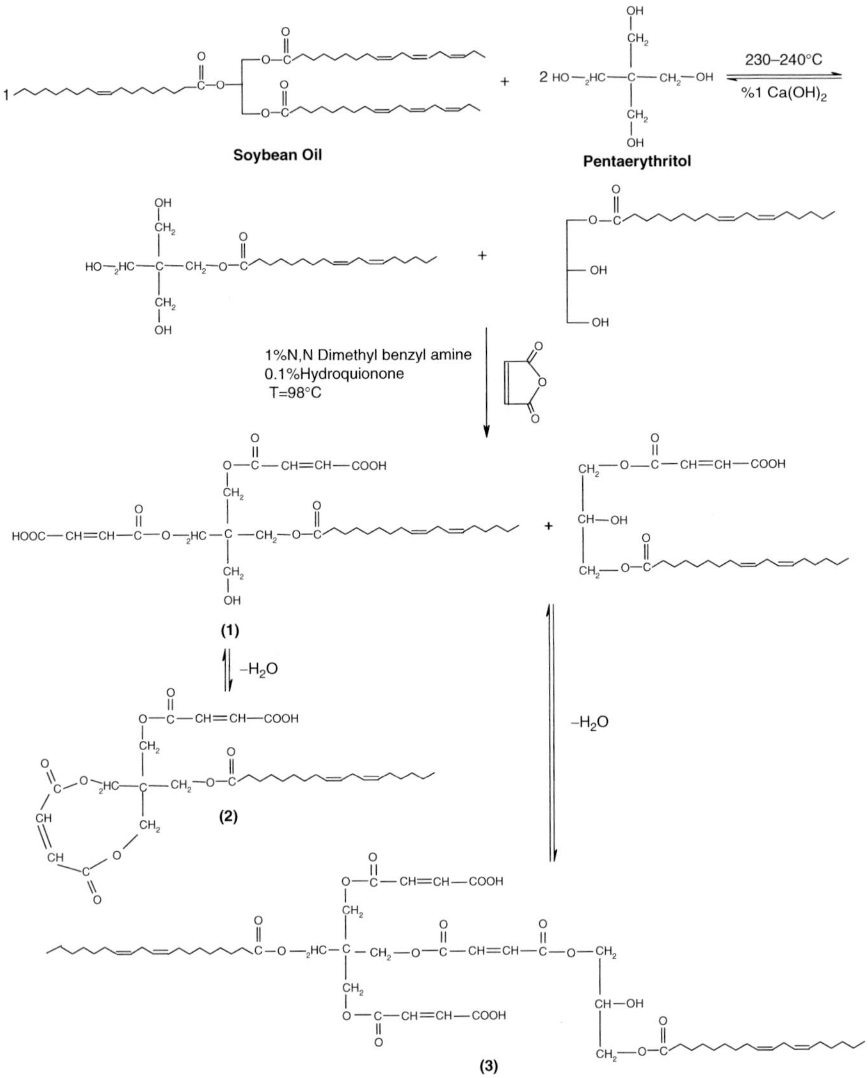

Fig. 7.7 The reaction scheme for soybean oil pentaerythritol alcoholysis and malination reaction (SOPERMA)

the reaction rate was significantly lower and therefore not preferred. Temperatures above 100°C led to gelation of the product due to polyesterification of both the maleate half-esters and maleic anhydride with the free hydroxyls and were therefore avoided. Table 7.2 shows the amount of the reactants used for the malination reactions done at different SO:PER:MA molar ratios. The product at room temperature was a light brown solid. The SOPERMA(1:3:12) and SOPERMA(1:2:8) products were prepared in a similar manner by changing the molar ratios of the reactants in the formulation.

Table 7.2 The amount of ingredients used in the alcoholysis of soybean oil with pentaerythritol and the following malination reactions

Reactants	SO:PER:MA (1:2:8)		SO:PER:MA (1:3:12)		SO:PER:MA (1:3:7.62)	
	Weight of reactant (g)	Moles of reactant	Weight of reactant (g)	Moles of reactant	Weight of reactant (g)	Moles of reactant
Soybean oil	5	0.0057	5	0.0057	5	0.0057
Pentaerythritol	1.554	0.01143	2.331	0.01714	2.331	0.01714
Ca(OH)$_2$	0.0655	–	0.0733	–	0.0733	–
Maleic anhydride	1.479	0.04571	6.720	0.06857	1.267	0.0435
N,N dimethyl benzyl amine	0.1103	–	0.1405	–	0.116	–
Hydroquinone	0.011	–	0.014	–	0.0116	–

SOGLYME: Soybean Oil Monoglyceride Methacrylates

Soybean oil monoglyceride methacrylates (SOGLYME) were synthesized by E. Can [66]. Soybean oil monoglyceride methacrylates were prepared in a two-step process. First, soybean oil was glycerolized in the presence of Ca(OH)$_2$ as catalyst at 230–240°C for 5 h. The glycerolysis of soybean oil under these conditions gives a product with an equilibrium mixture containing the monoglycerides, diglycerides, and the two starting materials [63]. The glycerolysis product was then reacted with methacrylic anhydride at 55°C to form the methacrylate esters of the glycerides and methacrylic acid. Pyridine, which is an effective catalyst in the reaction of methacrylic anhydride with alcohols, was used as the catalyst. Hydroquinone was used to inhibit the radical polymerization of the reactive methacrylate esters. The idealized reaction schemes for the glycerolysis and methacrylation are shown in Fig. 7.8.

The glycerolysis of soybean oil was carried out in a 1:2.4 molar ratio of soybean oil to glycerol. For the following methacrylation reaction, methacrylic anhydride was in a 1:1 molar ratio with the number of hydroxyls on glycerol used in the glycerolysis reaction, thus the molar ratio of soybean oil, glycerol, and methacrylic anhydride was SO:GLYC:ME = 1:2.4:7.2. The amount of reactants and catalysts used for the soybean oil glycerolysis and methacrylation reactions are shown in Table 7.3. The resulting SOGLYME product was a light yellow liquid.

COPERMA: Castor Oil Pentaerythritol Glyceride Maleates

Castor oil is not commonly used in alkyd resin formulations and there are few reports on the alcoholysis of castor oil triglycerides. For the preparation of castor oil-based monomers, castor oil was first alcoholized with glycerol, pentaerythritol, and an aromatic diol; BPA propoxylate and the alcoholysis products were then malinated, as shown in Fig. 7.9 [66]. Bisphenol A propoxylate was used specifically

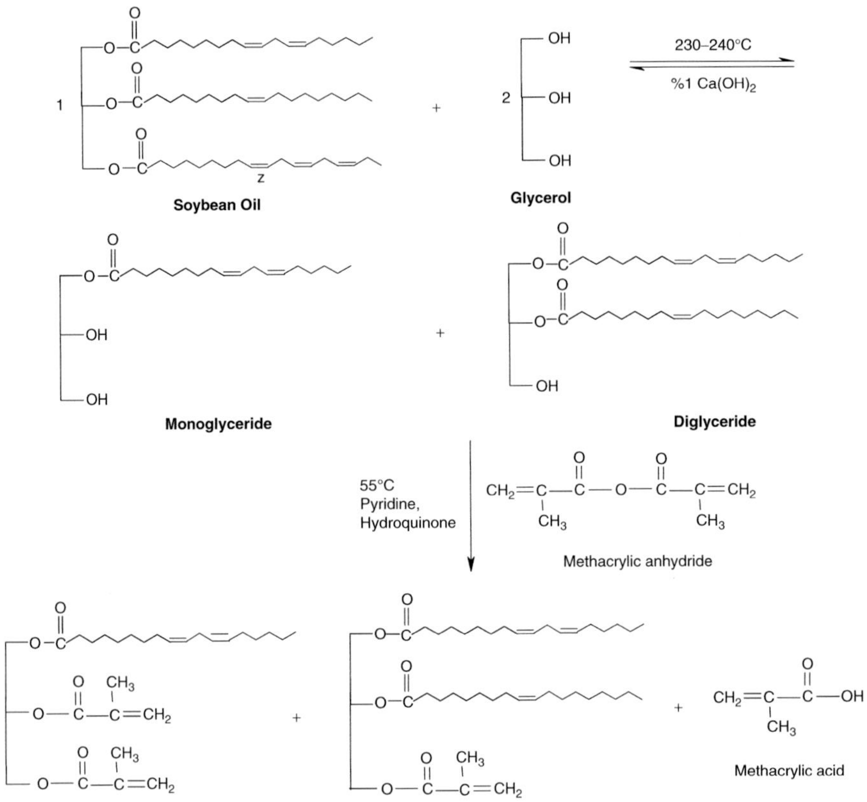

Fig. 7.8 The reaction scheme for soybean oil glycerolysis and methacrylation reactions (SOGLYCMA)

Table 7.3 Reactant amounts used in soybean oil glycerolysis and methacrylation reactions (SO: GLYC:ME = 1:2.4:7.2)

Component	Weight(g)	Moles
Soybean oil	119.93	0.1371
Glycerol	30.30	0.329
Calcium hydroxide	0.7514	–
Methacrylic anhydride	152.14	0.987
Pyridine	3.024	–
Hydroquinone	0.302	–

to introduce the rigid aromatic rings onto the triglyceride structure. The maleate esters of castor oil alcoholysis products have never been synthesized before, thus the castor oil-based monomers presented here are totally new resins. The alcoholysis reactions of castor oil were carried out for 2 h at 230–240°C in the presence of $Ca(OH)_2$ as catalyst, similar to the soybean oil alcoholysis reactions.

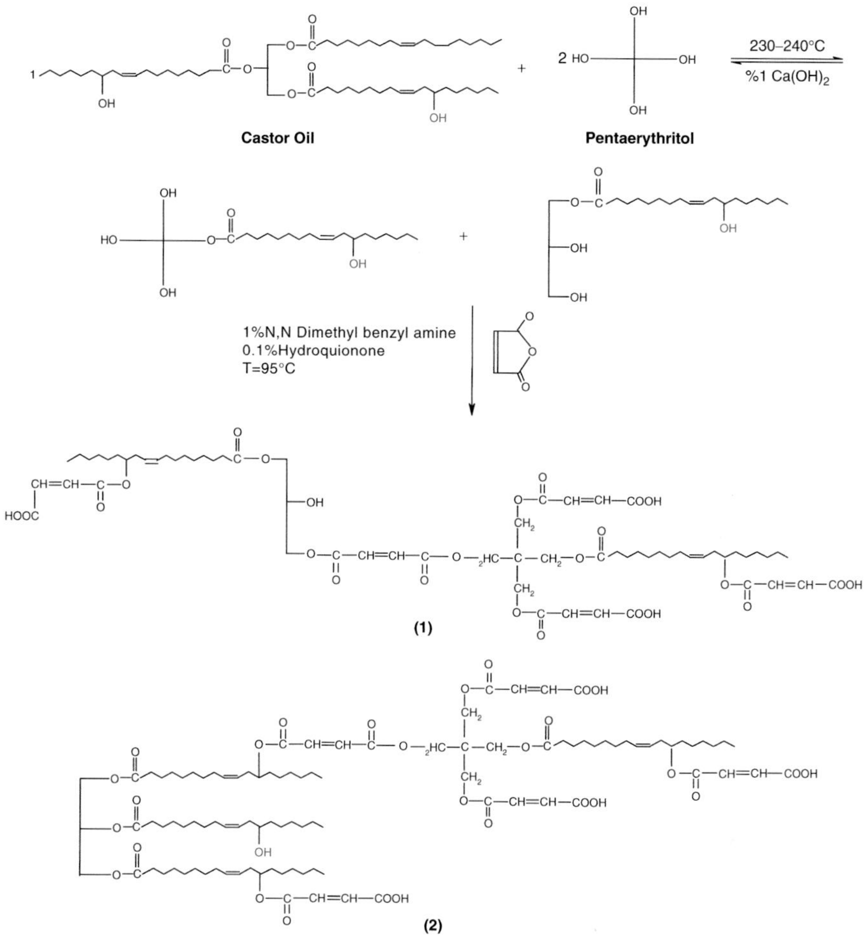

Fig. 7.9 The reaction scheme for castor oil pentaerythritol alcoholysis and malination reactions (COPERMA)

The malination reactions were carried out for 5 h at 98°C to ensure the completeness of the malination of the secondary hydroxyls of castor oil. *N,N*-Dimethylbenzylamine, which is reported to be an effective catalyst for the malination of hydroxylated oils, was used as a catalyst. Castor oil was also directly malinated to see the effect of the alcoholysis step on the mechanical properties of the resulting polymers.

The molar ratio of castor oil to maleic anhydride was 1:3 for malination of castor oil; therefore, the reaction was carried out in an excess of maleic anhydride assuming that 1 mol castor oil contains 2.7 mol of hydroxyls. The reactants used in this reaction as well as their mole numbers and masses are given in Table 7.1. The COPERMA product was a light brown solid (Table 7.4).

Table 7.4 Reactant amounts used in castor oil pentaerythritol alcoholysis and malination reactions (CO:PER:MA = 1:2:10.7)

Component	Weight(g)	Moles
Castor oil	120	0.13
Pentaerythritol	35.33	0.26
Calcium hydroxide	0.778	
Maleic anhydride	178.99	1.826
N,N dimethyl benzyl amine	3.35	
Hydroquinone	0.335	

Table 7.5 Reactant amounts used in castor oil glycerolysis and malination reactions (CO:GLY: MA = 1:2.2:9.3)

Component	Weight(g)	Moles
Castor oil	100	0.108
Glycerol	21.9	0.238
Calcium hydroxide	1.22	–
Maleic anhydride	98.08	1.001
N,N dimethyl benzyl amine	2.20	–
Hydroquinone	0.220	–

COGLYMA: Castor Oil Monoglyceride Maleates

The castor oil glycerolysis reaction was carried out in a molar ratio of CO:GLY = 1:2.2 using reactants shown in Table 7.5. The reaction mixture was heated to 230–240°C and agitated under N_2 atmosphere for 2 h at this temperature. The reaction product at room temperature was a light brown liquid. The idealized structures of both the reactants and products for the glycerolysis reaction are shown in Fig. 7.10. Both unreacted castor oil and excess glycerol exist as by-products in the reaction.

For the malination of the castor oil glycerolysis product, maleic anhydride was used in a molar ratio sufficient to malinate the hydroxyls of both castor oil and the glycerol used in the glycerolysis reaction. Thus, the molar ratio of castor oil (CO), glycerol (GLY), and maleic anhydride (MA) was CO:GLY:MA = 1:2.2:9.3. The castor oil glycerolysis product as prepared above was heated to 90°C with mechanical stirring, then the specified amounts of maleic anhydride and hydroquinone were added. The mixture was stirred at this temperature until the maleic anhydride melted and mixed with the castor oil glycerolysis product. *N,N*-Dimethyl-benzylamine was added and the reaction mixture was heated to 98°C. The mixture was agitated at this temperature for 5 h. The reaction product at room temperature was a light brown solid.

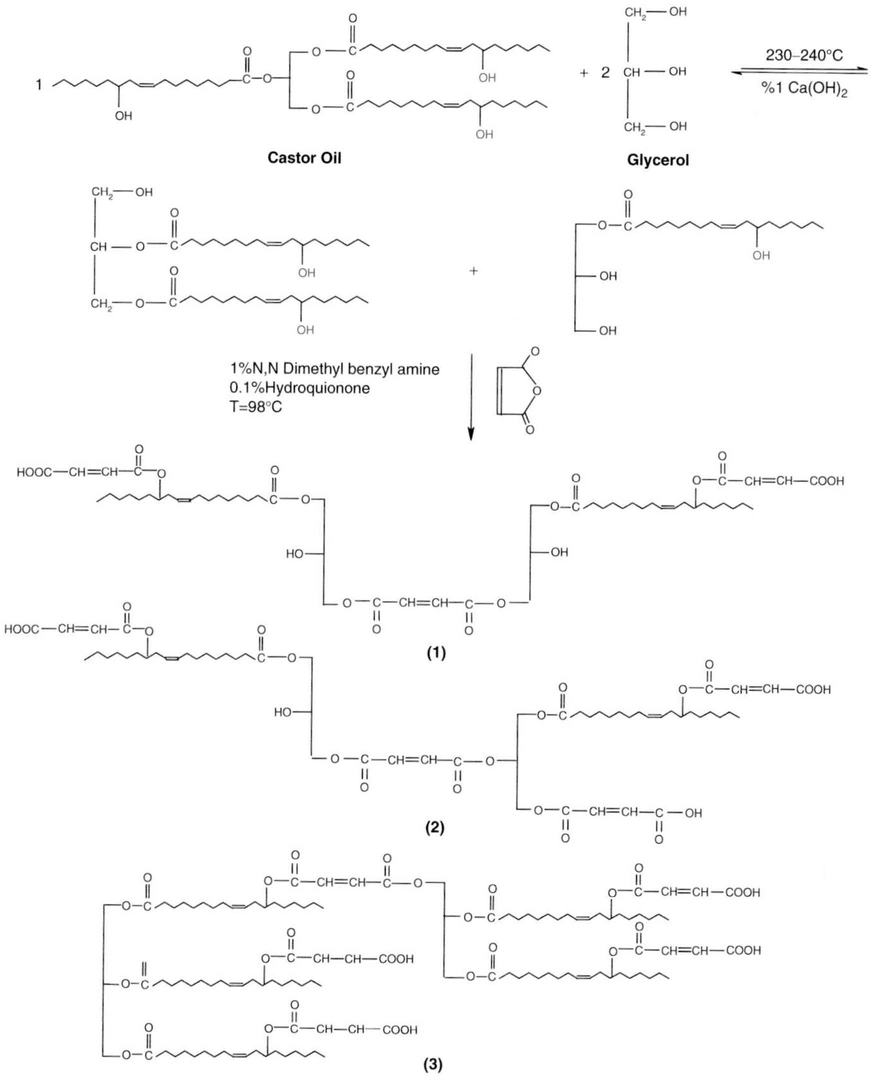

Fig. 7.10 The reaction scheme for castor oil glycerolysis and malination reactions (COGLYMA)

COBPPRMA: Castor Oil Bisphenol a Propoxylate Glyceride Maleates

The castor oil (CO), pentaerythritol (PER) alcoholysis reaction was carried out in a molar ratio of CO:PER = 1:2 [66]. The reaction product at room temperature was a light brown liquid. The idealized structures of both the reactants and products for the alcoholysis reaction are shown in Fig. 7.11. Both unreacted castor oil and excess glycerol exist as by-products in the reaction. For the malination of the castor oil

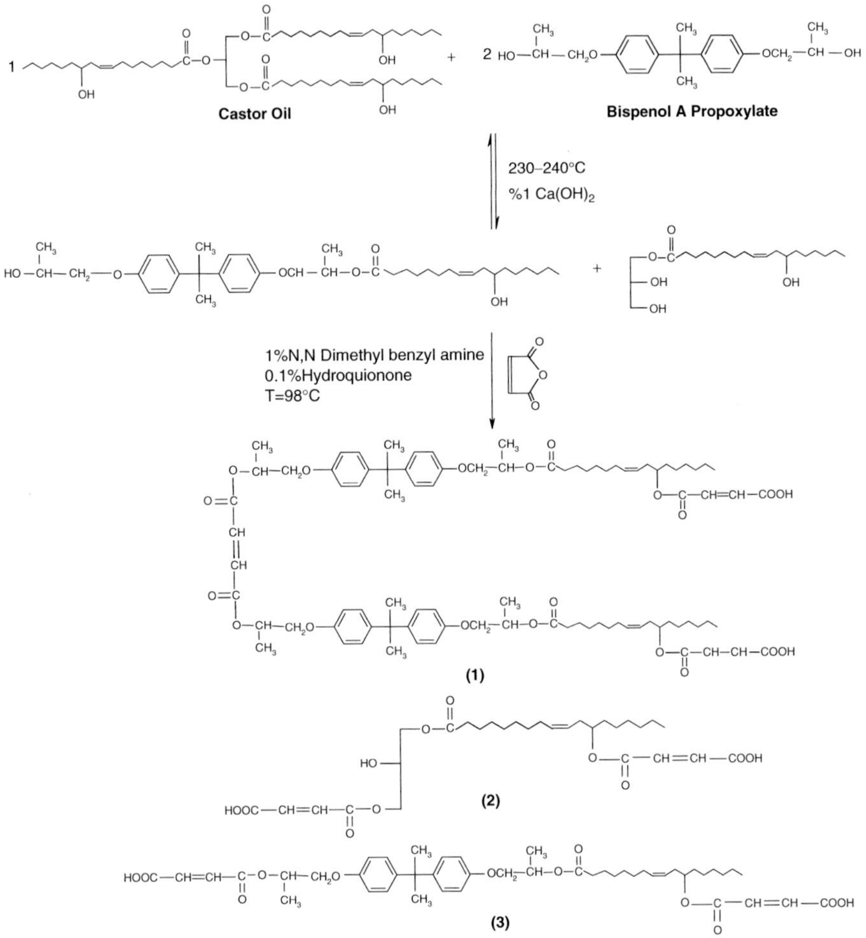

Fig. 7.11 The reaction scheme for castor oil Bisphenol A propoxylate alcoholysis and malination reactions (COBPAPRMA)

pentaerythritol alcoholysis product, maleic anhydride was used in a molar ratio to malinate the hydroxyls of both castor oil and the pentaerythritol used in the alcoholysis reaction. Thus, the molar ratio of castor oil (CO), pentaerythritol (PER), and maleic anhydride (MA) was CO:PER:MA = 1:2:10.7. The malinated product at room temperature was a light brown solid.

COMA: Castor Oil Maleates

COMA consists of the castor oil (CO) which was directly maleated on the three hydroxyl groups using maleic anhydride (MA). The ratios of reactants used are shown in Table 7.6 [66].

Table 7.6 Reactant amounts used in malination of castor oil (CO:MA = 1:3)

Component	Weight(g)	Moles
Castor oil	100	0.108
Maleic anhydride	31.82	0.325
N,N dimethyl benzyl amine	1.32	–
Hydroquinone	0.132	–

Polymers from Plant Oils

Acrylated Epoxidized Soybean Oil Polymers

Acrylated epoxidized soybean oil (AESO), shown in Fig. 7.4, was examined for its ability to produce high T_g and high modulus polymers. A commercial form of AESO, Ebecryl 860, was blended with various amounts of styrene to determine the effect of blending on mechanical and dynamic mechanical properties. The AESO used had an average functionality of approximately three acrylates per triglyceride as determined by ^{1}H NMR [38]. The optimal number of acrylates per triglyceride to obtain maximum stiffness and strength is about five acrylic acid groups per triglyceride. An example ^{1}H NMR spectrum of AESO is shown in Fig. 7.12. Similar to soybean oil, the triplet peak at 2.3 ppm can be used as a basis for the protons present alpha to the carbonyls in the triglyceride. The three peaks in the range of 5.8–6.5 ppm represent the three protons of the acrylate group.

Styrene monomer was blended with the AESO along with a free-radical initiator, 2,5-dimethyl-2,5-di(2-ethylhexanoylperoxy) hexane. The addition of styrene to any type of unsaturated polyester is common practice in the composite liquid molding resin field. Its low cost and low viscosity improve the price and processability of the resin. For triglyceride-based polymers, the styrene also imparts a rigidity that the triglyceride does not naturally possess. The amount of initiator used was 1.5 wt% of the total resin weight (AESO plus styrene). For tensile testing of the polymers, samples were prepared in accordance with ASTM D 638. The resin was cured at 60°C for 12 h, followed by 125°C for 1.5 h. Samples for dynamic mechanical analysis (DMA) testing were prepared by pouring resin into a rubber gasket between two metal plates covered with aluminum foil. Samples were cured at 65°C for 1.5 h and postcured at 125°C for 1.5 h.

Synthesis of Modified Acrylated Epoxidized Soybean Oil Polymers

To improve the properties of the AESO-based resins, modified forms of the AESO were synthesized. These modifications involved partially reacting epoxidized soybean oil with acrylic acid and reacting the remaining epoxies with anhydrides or diacids. A more detailed explanation of the synthesis of partially acrylated

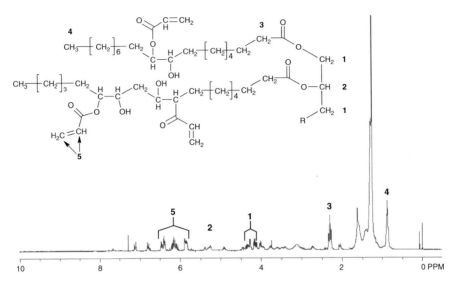

Fig. 7.12 ¹H-NMR spectrum for acrylated epoxidized soybean oil (Ebecryl 860, UCB Chemicals Co.)

epoxidized soybean oil can be found in other sources [38]. In summary, a mixture of epoxidized soybean oil was mixed with a stoichiometric amount of acrylic acid (about 1,500 g ESO to 460 g acrylic acid). Hydroquinone was added as a free-radical inhibitor in the amount of 0.07 wt% of the total reactants' weight, as well as 1,4-diazobicyclo[2.2.2]octane to act as a catalyst in the amount of 0.1 wt% of the total reactants' weight. This was reacted at 95°C for about 11 h, after which it was allowed to cool to room temperature. The resulting product had approximately 1.7 acrylates/triglyceride and 0.4 residual epoxy/triglyceride according to ¹H NMR. The remaining 2.3 epoxies were lost to epoxy homopolymerization [38].

The first modification was the reaction of AESO with cyclohexanedicarboxylic anhydride (CDCA), as illustrated earlier in Fig. 7.5a. In a typical reaction, the synthesized AESO was reacted with 7.4% of its weight in CDCA and 0.1% of its weight in 2-methyl imidazole, which catalyzes the reaction [38]. After reacting at 110°C for about 3 h, the majority of the anhydride and epoxy groups was consumed, as indicated by FTIR spectroscopy. The second modification was the reaction of AESO with maleic acid (Fig. 7.5b). This was accomplished by reacting the synthesized AESO with 11% of its weight in maleic acid [38]. The reaction was held at approximately 80°C for 4 h, during which consumption of the epoxies was again confirmed by FTIR spectroscopy.

The modified resins were then blended with styrene and initiator in the amounts of 66 wt% modified AESO, 33 wt% styrene, and 1 wt% 2,5-dimethyl-2,5-di(2-ethylhexanoylperoxy) hexane initiator. After curing at 65°C for 1.5 h and postcuring at 125°C for 1.5 h, the polymers' dynamic mechanical properties were analyzed and compared to the unmodified AESO resin.

Maleinized Soybean Oil Monoglyceride Resin Synthesis

The maleinized soybean oil monoglyceride (Fig. 7.4) was synthesized by breaking the triglycerides into monoglyceride and then functionalizing the alcohol groups with maleic anhydride. The glycerolysis reaction was done by heating the triglycerides in the presence of glycerol and a catalyst. In a typical reaction, glycerol was heated at 220–230°C for 2 h under an N_2 atmosphere to distill off any water present [33]. The amount of soybean oil reacted with the glycerol was 4 g soybean oil to 1 g glycerol, a molar ratio of 1.75 mol glycerol to 1 mol triglyceride. The soybean oil was added in five portions to the glycerol, each portion 1 h apart. With the first portion, commercial soap was added in the amount of 1% of the total oil amount to act as an emulsifier and catalyst. The solution was heated at 230°C under N_2 while being stirred. After 5.5 h, the reaction was immediately cooled to room temperature with an ice bath, causing glycerol to separate from the mixture. On removal of this layer, approximately 90% of the reaction solution, consisting of glycerides and glycerol, was recovered.

Maleinization of the mixture was accomplished by heating 60 g of glyceride/glycerol mixture to about 80°C while being stirred. Maleic anhydride was then added in the amount of 40 g. As the anhydride melted, 0.6 g triphenyl antimony was added as a catalyst along with 0.01 g hydroquinone. The reaction was complete after 5.5 h, according to FTIR and ^1H NMR, resulting in a mixture of maleinized glycerides and glycerol (SOMG/MA) [33].

Maleinized Soybean Oil Monoglyceride/Neopentyl Glycol Resin Synthesis

Modifying the procedure given in [53], SOMG/NPG/MA resin was synthesized as follows [51]. Forty-five grams of SOMG was placed into a 250-mL round-bottom flask equipped with a temperature controller and a magnetic stirrer and then heated to 125°C. Fifteen grams of NPG (0.144 mol) was then added to SOMG, and as the NPG melted, 58.3 g maleic anhydride was added. As the three compounds formed a homogenous solution, 0.06 g triphenyl antimony catalyst and 0.015 g hydroquinone were added. The solution was stirred for 6.5 h at 120°C. ^1H NMR analysis of the product showed the formation of both the SOMG and NPG maleate and later fumarate vinyl groups. The product was a light yellow viscous liquid at room temperature.

Maleinized Soybean Oil Monoglyceride/Bisphenol a Resin Synthesis

The preparation of maleates of BPA and ethylene and propylene oxide adducts of BPA was reported in [52]. For this work, SOMG and BPA were maleinized as a mixture [51]. Forty-five grams of SOMG was placed into a 250-mL round-bottom flask equipped with a temperature controller and a magnetic stirrer and heated to 125°C. Fifteen grams of BPA (0.0657 mol) was added to the SOMG, and as BPA dissolved, 42.88 g maleic anhydride (0.4375 mol) was added. As the three compounds formed a homogenous solution, 0.6 g triphenyl antimony and 0.01 g hydroquionone were also added. The solution was then stirred for 9 h at 125°C until maleic anhydride consumption was completed. The ^{1}H NMR analysis of the product showed the formation of both the SOMG and BPA maleate and later fumarate vinyl groups. The reaction product was an orange-colored viscous liquid (98 g) at room temperature.

Copolymerization of the Maleates with Styrene

The copolymerization of SOMG/MA, SOMG/NPG/MA, and SOMG/BPA/MA with styrene were all run under the same conditions for comparison of the mechanical properties of the resulting polymers. For this purpose, a certain weight ratio of the maleate mixture was mixed with 35% of its own weight of styrene in a closed vial. All of the maleate products were found to be soluble in styrene. tert-Butyl peroxybenzoate radical initiator, 2% by weight of the total mixture, was added. Nitrogen gas sparging and vacuum degassing were carried out for 5 min. The solution was then transferred to a rectangular rubber gasket mold sandwiched between two steel plates. The resin-filled mold was heated to 120°C at a rate of 5°C/min and was cured at this temperature for 3.5 h. It was then postcured at 150°C for 1 h. Samples were clear, homogeneous, and free of voids or gas bubbles. The polymer samples were polished and prepared for DMA, which was conducted in a three-point bending geometry on a Rheometrics Solids Analyzer II. The temperature was ramped from 30°C to 200°C at a rate of 5°C/min, with a frequency of 1 Hz and strain of 0.01%.

Malinated Hydroxylated Oil Polymer Synthesis

The HO/MA shown in Fig. 7.4 uses the unsaturation of the triglyceride to incorporate polymerizable groups. This monomer was used in a series of experiments to understand how triglyceride structure can affect the synthesis and dynamic mechanical properties of the end polymer [59]. Olive oil, cottonseed oil, soybean oil, safflower oil, linseed oil, triolein, and a genetically engineered high oleic soybean oil were converted into HO/MA resins. The levels of unsaturation for these oils are shown in Table 7.1. The fatty acid chain lengths for all of these oils are between 17.5 and 18 carbons, making the unsaturation level essentially the only difference among oils.

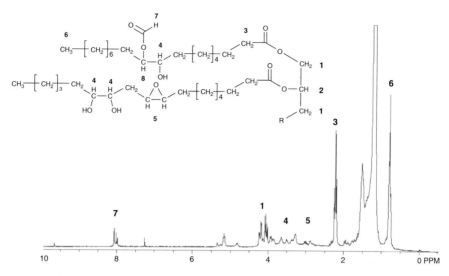

Fig. 7.13 ¹H-NMR of hydroxylated soybean oil. Treating the oil with formic acid and hydrogen peroxide results in conversion of the double bonds to hydroxy groups

Hydroxylation was done by stirring the oil (~100 g) vigorously in the presence of formic acid (150 mL) and 30% (aq) hydrogen peroxide (55 mL) at 25°C [30, 59]. The reaction time was 18 h to reach a maximum conversion of double bonds. Formic acid, peroxide, and water were then removed from the hydroxylated oil by dissolving the reaction mixture in diethyl ether and washing multiple times with water and then saturated (aq) sodium bicarbonate until a neutral pH was reached. The solution was then washed with saturated (aq) sodium chloride and dried over sodium sulfate. Finally, the ether was evaporated off under vacuum. The extent of hydroxylation can be characterized by ¹H NMR. An example ¹H NMR spectrum is presented in Fig. 7.13 with corresponding peak assignments [59]. The extent of hydroxylation has a linear dependence on the level of unsaturation. Generally, for every double bond present on the triglyceride, an average of 1.6 hydroxyls can be added [59].

The purified hydroxylated oil was reacted with maleic anhydride in a ratio of 1 mol triglyceride to 4 mol anhydride. The hydroxylated oil was heated to a temperature of about 80°C, and finely ground maleic anhydride was then added. Upon dissolving of the anhydride, *N,N*-dimethylbenzylamine was added to catalyze the reaction. The reaction was continued for 3 h, and the extent of maleinization was determined by ¹H NMR. An illustrative ¹H NMR is shown in Fig. 7.14 [59]. Under these reaction conditions, the extent of functionalization plateaus in the range of 2.1–2.8 maleates/triglyceride for all oils. Approximately 20–25% of the maleates attached to the triglycerides isomerize to form fumarate groups (*trans* confirmation). Unreacted maleic anhydride remained in the resin and was polymerized during the cure reaction. The HO/MA resins were then dissolved in styrene in a molar ratio of 7:1 styrene to HO/MA. Resins were cured using 2,5-dimethyl-

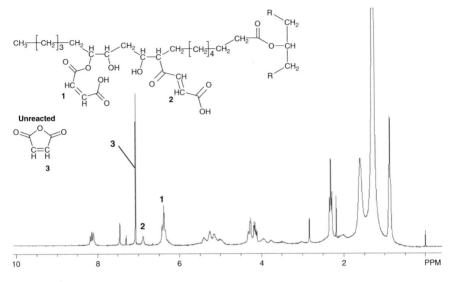

Fig. 7.14 ¹H-NMR of maleinized hydroxylated soybean oil. Peaks 1 and 2 represent the maleate half esters and fumarate half esters, respectively. Peak 3 represents unreacted maleic anhydride

2,5-di-(2-ethylhexanoylperoxy)hexane at 65°C for 1.5 h and postcured at 120°C for 1 h. DMA was conducted in a three-point bending geometry on a Rheometrics Solids Analyzer II. Temperature was ramped from 30°C to 175°C at a rate of 5°C/min, with a frequency of 1 Hz and strain of 0.01%.

SOPERMA and COPERMA Polymer Synthesis

The general-purpose unsaturated polyester (UP) resin is a linear polymer with the number average molecular weights in the range of 1,200–3,000 g/mol. Depending on the chemical composition and molecular weight, they can be viscous liquids or solids. The plant oil-based resins prepared are not linear polymers but similar mixtures of monomers or oligomers with different molecular weights. The number and weight average molecular weights of different species in each of these plant oil-based resins were in the range of ~300–2,000 g/mol as presented in section "Synthetic Pathways for Triglyceride-Based Monomers." With the exception of the COMA and SOGLYME resins, which were liquid at room temperature, all the malinated glyceride-based resins prepared were paste-like solids at room temperature. The melting points of these resins were in the range of 60–70°C, similar to the general-purpose UP resins whose melting points are in the range of 60–77°C.

To prepare the styrenated plant oil-based resins, which were solid at room temperature, the resin was first heated in an oil bath above its melting point (~70°C), then the necessary amount of styrene was added and the mixture was

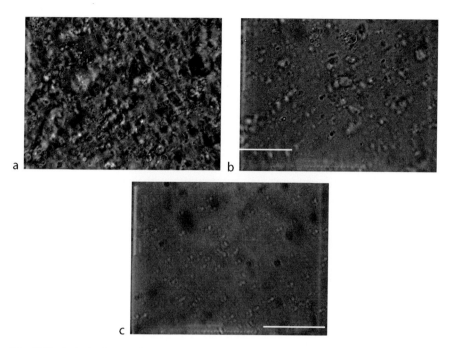

Fig. 7.15 Optical microscopic pictures of the SOPERMA-styrene mixtures (500×) (**a**) 80 wt% (**b**) 60 wt% (**c**) 20 wt% SOPERMA

agitated in the oil bath at a temperature of 80°C until styrene and the resin became totally mixed. The resin was then processed at room temperature for polymerization. For the resins that were liquid at room temperature, styrene was added, mixed, and processed at room temperature. Apart from the SOGLYME resin (methacrylated soybean oil monoglycerides), all the styrenated resins showed phase separation in the microscale. Figure 7.15 shows the optical microscopic pictures of the SOPERMA-styrene mixtures at decreasing SOPERMA concentrations at 500× magnification. The SOPERMA forms nonuniform droplets in the continuous styrene matrix in the 1–5-μm size range. As can be seen in Fig. 7.15, these droplets become less dense and more uniform in size as the concentration of SOPERMA decreases from 80 to 20 wt%. Similarly, COPERMA, COGLYMA, and COBPAPRMA resins containing 33 wt% styrene were found to be incompatible with styrene and exhibited phase separation.

The general-purpose UP resins based on propylene glycol, phthalic, and maleic anhydride are miscible with styrene. The molecular weights of the general-purpose UP resins have a number average molecular weight of 900 g/mol and a weight average molecular weight of 2,400 g with a polydispersity of 2.7. The acid number of these polyesters is around 50 mg KOH/g. The incompatibility of the malinated plant oil-based resins in styrene can be attributed to both the abundance of the acid groups of the maleate half-esters and the presence of high-molecular-weight species in these resins. The acid number of all the malinated glyceride-based resins was found to be above 200 mg KOH/g. This value is much higher than that of the general-purpose UP resins. The molecular species present in the malinated glyceride-based resins are highly polar compared to styrene, and the strong interactions of these molecules via hydrogen bonding between the acid groups result in the insolubility of the malinated glycerides in styrene. Thus, the SOGLYME resin based on methacrylated glycerides, both with the acrylic acid by-product and without acrylic acid, showed no phase separation in styrene. The methacrylate half-esters as shown earlier in Fig. 7.8 do not carry acid functionality, and oligomer formation during the methacrylation reaction cannot occur. Additionally, these molecules are less polar than maleate half-esters and they cannot interact via hydrogen bonding, which favors their compatibility with styrene. The SOPERMA resin was also found to be insoluble in more polar monomers such as acrylic acid and acrylonitrile. The immiscibility of the SOPERMA resin in these solvents shows that the molecular weight and strong interactions between the malinated glyceride molecules play an important role in the immiscibility of these molecules in polar solvents.

A liquid molding resin should have properties within a certain operating range to be successfully used in molding processes. Possibly the most stringent requirement is the resin's viscosity, which must range between 200 and 1,000 cP. At viscosities lower than 200 cP, air pockets will remain in the mold after injection. At viscosities greater than 1,000 cP, voids may occur in the part, the time required for injection increases, and problems with fiber wetting can arise during composite preparation.

The surface free energy of a liquid, also referred to as *surface tension*, determines most of the surface and interfacial properties such as wetting, adhesion, and adsorption. Surface tension results from an imbalance of molecular forces in a liquid. At the surface of the liquid, the liquid molecules are attracted to each other and exert a net force, pulling themselves together. High surface tension values mean that the molecules tend to interact strongly; thus, polar materials show high surface energy values. The surface energy of a liquid molding resin may be especially important for reinforcement of the resins by fibers. The wetting of a fiber with a liquid resin can be judged by the difference between the surface energies of the fiber and the resin. For the most desirable condition, proper wetting and spreading in resin transfer molding processes, the surface energy of the fiber should be high, whereas the surface energy of the resin should be low. Table 7.7 shows the specific gravity, viscosity (η), and surface energy values for the plant oil-based resins and general-purpose UP resins containing 33 wt% styrene.

As can be seen from Table 7.7, apart from the COMA and the SOGLYME resins, these resins show viscosities that are in a range that is suitable for liquid molding processes. Although the COMA and SOGLYME resins show low viscosities, the viscosities of these resins can easily be increased by decreasing the amount of the styrene diluent. It was found by Can [66] that as the weight fraction of styrene was increased, the viscosity decreased in an exponential manner. This result is very desirable since it shows that a small amount of comonomer can be used to make these resin systems much easier to process.

Table 7.7 The specific gravity, viscosity, η and surface energy values for the plant oil based resins (33 wt% styrene) and general purpose UP resins

Resin	Specific gravity (g/ml)	Viscosity, η (cP)	Surface energy (mN/m)
SOPERMA	0.94	343	27.38
COPERMA	1.06	363	28.84
COGLYCMA	1.04	213	26.36
COBPAPRMA	0.98	183	27.2
COMA	0.75	92	26.02
SOGLYME	0.87	51	22.7
GP-UP	1.14	200–2,000	24–30

The surface energy values of these resins show values closer to those of UP resins, and are significantly lower than those of vinyl ester resins (32–34 mN/m), which may have difficulty in wetting fiber substrates. Among the malinated resins, the surface free energy value is highest for the COPERMA resin and lowest for the COMA resin in proportion to the maleate content of the resins. A higher maleate content should result in a higher polarity and a higher surface free energy. The SOGLYME resin shows the lowest surface free energy among all the resins. The methacrylated glycerides do not carry an acid functional group as malinated glycerides and, therefore, are less polar than the corresponding maleates.

The curing of UP resin is accomplished via free-radical cross-linking polymerization between the UP molecules and styrene. The UP molecules are the cross-linkers, while the styrene acts as an agent to cross-link the adjacent polyester molecules. Similarly, in the plant oil-based resin systems, the functionalized glycerides act as the cross-linker units, and styrene is the agent that links the adjacent glyceride molecules. Styrene is the most commonly used vinyl monomer in unsaturated polyester resins due to its low viscosity, low cost, and reactivity with the unsaturated sites of polyesters. The unsaturation present on the UP backbone is very sluggish in homopolymerization. The reactivity ratio of styrene and maleic/fumaric acid esters is about 0, indicating that this system has a tendency to form alternating copolymers. Because the malinated plant oil-based resins were found to be insoluble in styrene at room temperature, it was especially important for us to determine the conversion of polymerization for the maleate and styrene monomers.

For the determination of the conversion and kinetics of the polymerization, the styrenated SOPERMA, COMA, and COPERMA resins were prepared as described earlier. The styrene concentration was 33 wt% for each resin. *tert*-Butyl peroxybenzoate was used as the initiator. The initiator concentration was 2 wt% for the SOPERMA and COMA resins and 1.5 wt% for the COPERMA resin. The curing of all resins was carried out at 120°C for 3 h for comparison of conversion and rate of polymerization in different resins. The SOPERMA and COPERMA resins were also postcured at 150°C and 160°C, respectively, for 1 h after 2 h at 120°C. Although, all the resins showed similar conversion versus time profiles during the isothermal cure, the final conversion values were different for each resin.

Table 7.8 lists the total conversion (α) and the conversion of maleates (α_{ma}) and styrene (α_{st}) for the COMA, SOPERMA, and COPERMA resins (33 wt% styrene) at

Table 7.8 The total conversion(α), the conversion of maleates, (α_{ma}), and styrene (α_{st}), for the SOPERMA, COMA and COPERMA resins (33 wt% styrene) at the end of 3 hours at 120°C

Resin	α_{ma}	α_{st}	α
COMA	0.998	0.921	0.967
SOPERMA	0.979	0.828	0.886
COPERMA	0.952	0.835	0.885

the end of 3 h at 120°C. As can be seen, the final maleate conversion decreases as the maleate content of the resin increases from COMA to COPERMA. The styrene conversion and thus the total conversion is also considerably lower for the SOPERMA and COPERMA resins than for the COMA resin. It is expected that the molecular mobility of the resin decreases as the cross-link density increases, resulting in lower total conversion. As a result, styrene monomer as well as some resin can be trapped in the network and cannot participate in polymerization. However, although the COPERMA resin has a higher maleate content and, therefore, a higher cross-link density than the SOPERMA resin, the styrene and thus the final conversions for the SOPERMA and COPERMA resins do not show a significant difference. The maleate conversion was higher than styrene conversion for all of the resins.

The total ultimate conversion of unsaturated polyesters ranges from 0.75 to 0.9 and increases with increasing temperatures. Similarly, the final conversions for the isothermal cure of the plant oil-based resins were lower than the complete conversion. During an isothermal cure, when the increasing glass transition temperature of the resin reaches the reaction temperature and the material evolves from the rubbery state to the glassy state, the rate of propagation becomes diffusion controlled. This process, referred to as *vitrification*, may virtually terminate the polymerization, limiting the conversion that can be reached isothermally. This was true in particular for the SOPERMA and COPERMA resins, which possess glass transition temperatures (T_g) when fully cured [T_g(SOPERMA) = 139°C and T_g(COPERMA) = 146°C] that are higher than the cure temperature. Thus a postcure was necessary for these systems to increase the conversion.

Properties of Plant Oil Resins

Viscoelastic and Mechanical Properties of AESO-Styrene Polymers

The storage moduli, E', of the AESO-styrene neat polymers at various temperatures and compositions are shown in Fig. 7.16. At room temperature, the polymers display moduli proportional to the amount of styrene present, which is expected from the tensile properties presented earlier. Additionally, at room temperature all of the polymers are in the transition phase from the glassy region to the rubbery plateau. Even at temperatures as low as −130°C, it does not appear that these polymers have reached a characteristic glassy plateau. At extremely low

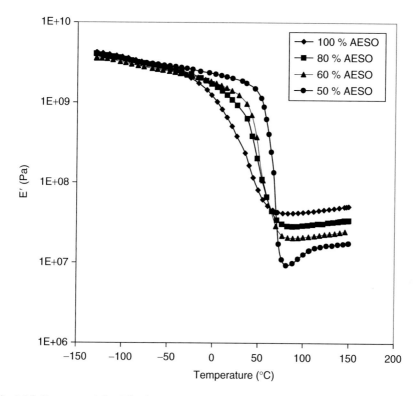

Fig. 7.16 Storage modulus (E′) of AESO-styrene copolymer as a function of temperature

temperatures, all compositions exhibit essentially equal moduli of about 4 GPa. At higher temperatures, the compositions show moduli inversely proportional to the amount of styrene present. According to rubber elasticity theory [60], the lower styrene content polymers have a higher cross-link density, as observed in Fig. 7.16.

The T_g is often designated by either the temperature at which the dynamic loss modulus E'' value is at a peak or the temperature at which the loss tangent tan δ exhibits a peak [61]. As shown in Fig. 7.17, all of the AESO-styrene copolymers exhibit two peaks in E''. A minor relaxation occurs in the range of −85°C to −95°C, showing little dependence on composition. The much larger relaxation, corresponding to the T_g, occurs in the range of −10°C to 60°C and also becomes sharper in nature with the addition of styrene. These peaks are shown in the tan δ graph in Fig. 7.18. The temperature at which these peaks occur exhibits a linear dependency on composition, increasing with the amount of styrene present in the system, as illustrated by Fig. 7.19.

The dynamic mechanical behavior just discussed is a combination of three factors: cross-link density, copolymer effects, and plasticization. As the amount of AESO increases, so does the number of multifunctional monomers. Therefore, the overall cross-link density will be greater with increasing amounts of AESO, as supported by the high-temperature moduli shown in Fig. 7.16. Increasing the cross-

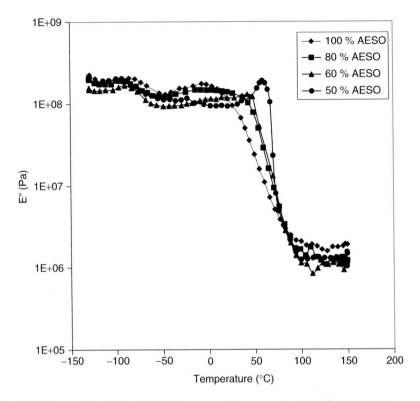

Fig. 7.17 Loss modulus (E″) of AESO-styrene copolymer as a function of temperature

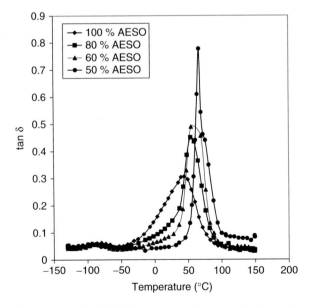

Fig. 7.18 Damping peak (tan δ) of AESO-styrene copolymer as a function of temperature

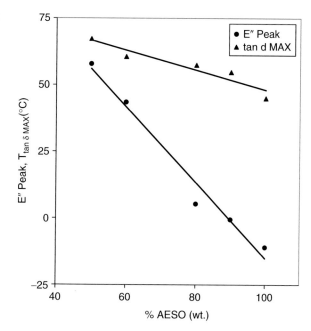

Fig. 7.19 E″ and tan δ peak temperatures of various compositions of AESO-styrene polymer

link density slows the transition in E' from glassy to rubbery behavior. Additionally, the tan δ peak broadens and decreases in height [61]. The copolymer effect occurs frequently when there are differences in the reactivity or structure of the different monomers. If one monomer is more reactive, it is depleted faster, causing polymer formed later in the reaction to be composed mostly of the slower reacting monomer. This causes heterogeneity in the composition of the total polymer. If these monomers differ from each other in their physical properties, such as very different T_g's, a general broadening of the glass–rubber transition is frequently observed, due to this gradient [61].

The other factor in the dynamic mechanical behavior, plasticization, is due to the molecular nature of the triglyceride. The starting soybean oil contains fatty acids that are completely saturated and cannot be functionalized with acrylates. Therefore, these fatty acids act in the same manner as a plasticizer, introducing free volume and enabling the network to deform more easily. The addition of even small amounts of plasticizer to polymers was known to drastically broaden the transition from glassy to rubbery behavior and reduce the overall modulus [61]. This plasticizer effect presents an issue that may be inherent to all natural triglyceride-based polymers that use the double bonds to add functional groups. However, with advances in genetic engineering, it may be possible to reduce this trend by reducing the amount of saturated fatty acids present, thus sharpening the glass–rubber transition. This issue is addressed later in the properties of HO/MA polymers produced from genetically engineered high oleic content oil and synthetic triolein oil. The existence of some saturated fatty acids, though, can contribute to improved toughness and ballistic impact resistance [62].

Fig. 7.20 Tensile modulus of AESO-styrene copolymers

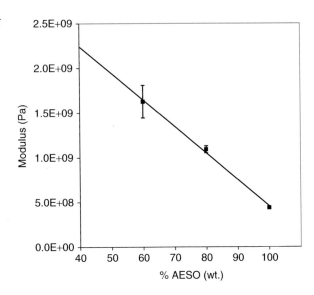

Tensile Properties of AESO-Styrene Polymers

The tensile moduli of three AESO-styrene copolymers at room temperature are shown in Fig. 7.20. The pure AESO polymer has a modulus of about 440 MPa. At a styrene content of 40 wt%, the modulus increases fourfold to 1.6 GPa. In this region, the dependency on composition appears to be fairly linear. The ultimate tensile strengths of these materials, as shown in Fig. 7.21, also show linear behavior. The pure AESO samples exhibited strengths of approximately 6 MPa, whereas the polymers with 40 wt% styrene show much higher strengths of approximately 21 MPa. Therefore, it is apparent that the addition of styrene drastically improves the properties of the end resin.

Dynamic Mechanical Behavior of Modified AESO Resins

The dynamic mechanical properties of the AESO polymers modified by cyclohexanedicarboxylic anhydride (CDCA) and maleic acid were better than the unmodified polymers. As shown in Fig. 7.22, the storage modulus at room temperature increases with both of these modifications. The storage modulus of the unmodified AESO resin at room temperature is 1.3 GPa, whereas the cyclohexanedicarboxylic anhydride modification increases the modulus to 1.6 GPa. The maleic acid modification provides the most improvement, raising the storage modulus to 1.9 GPa. The T_g, as indicated by the peak in tan δ, does not show any large increase from the anhydride modification, as shown in Fig. 7.23. However, the maleic acid modification shifts the tan δ peak by almost 40 °C, showing a peak at 105 °C. The increased broadness of the peak can be attributed to increased cross-link density.

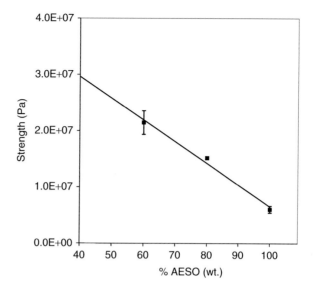

Fig. 7.21 Ultimate tensile strength of AESO-styrene copolymers

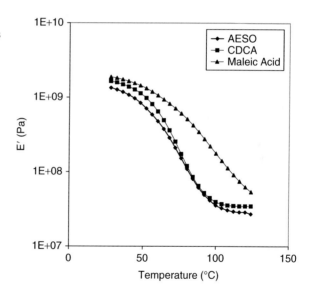

Fig. 7.22 Storage modulus (E′) of modified AESO resins as a function of temperature

SOMG/MA Polymer Properties

As seen in Fig. 7.24, the tan δ peak for the SOMG/MA polymer occurs at around 133°C, and the polymer has an E' value of approximately 0.92 GPa at room temperature. It is apparent that the glass transition is rather broad due to the broad molecular weight distribution of the SOMG maleates. The distribution of

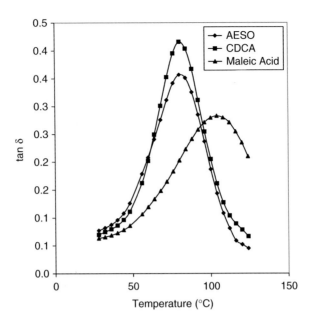

Fig. 7.23 Damping peak (tan δ) of modified AESO resins as a function of temperature. Peaks in tan δ were found at 81°C (CDCA modified) and 105°C (maleic acid modified) compared to 79°C for the synthesized AESO

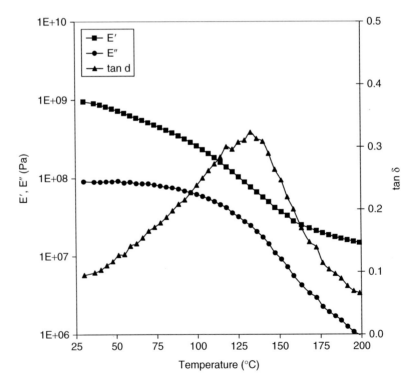

Fig. 7.24 Dynamic mechanical behavior for SOMG/MA polymer

soybean oil monoglyceride monomaleates, monoglyceride bismaleates, diglyceride monomaleates, and glycerol tris maleates was confirmed by mass spectral analysis, which was reported in a previous publication [63]. The tensile tests performed on the copolymers of SOMG maleates with styrene showed a tensile strength of 29.36 MPa and a tensile modulus of 0.84 GPa as calculated from the force displacement graph.

SOMG/NPG Maleates (SOMG/NPG/MA)

The DMA of SOMG/NPG/MA polymer showed a tan δ peak at approximately 145°C and an E' value of 2 GPa at room temperature. The 12°C increase in the T_g and the considerable increase in the modulus of the copolymers of SOMG/NPG maleates with styrene compared to that of the SOMG maleates can be attributed to the replacement of the flexible fatty acid chains by the rigid methyl groups of NPG. The overall dynamic mechanical behavior of the SOMG/NPG/MA polymer was very similar to that of the SOMG/MA shown in Fig. 7.21. Despite the higher T_g and modulus, there remained a broad glass transition. The tensile strength of the SOMG/NPG/MA polymer was found to be 15.65 MPa, whereas the tensile modulus was found to be 1.49 GPa.

Maleinized pure NPG polymerized with styrene (NPG/MA) was prepared to compare its properties with the SOMG/NPG/MA polymer [51]. DMA analysis of the NPG/MA showed a tan δ peak at approximately 103°C and an E' value of about 2.27 GPa at 35°C. The high T_g value observed for the SOMG/NPG/MA system (~145°C) is attributed to a synergistic effect of both the NPG and SOMG together since the T_g value observed for the NPG/MA system (~103°C) is much lower. This is probably due to the incorporation of the fatty acid unsaturation into the polymer in the SOMG/NPG/MA system. On the other hand, the comparatively higher E' value observed for the NPG maleates explains the increase in the E' observed for the SOMG/NPG/MA system compared to that of the SOMG/MA system. The decrease in tensile strength of the SOMG/NPG/MA system compared to that of SOMG/MA may be attributed to a broader molecular weight distribution of this system compared to that of the SOMG maleates.

SOMG/BPA Maleates (SOMG/BPA/MA)

The DMA of this polymer showed a tan δ peak at around 131°C and an E' value of 1.34 GPa at 35°C. The introduction of the rigid benzene ring on the polymer backbone considerably increased the modulus of the final polymer compared to that of the SOMG maleates. The T_g of this polymer, however, was not very different from that of the SOMG maleates (133°C). This was attributed to a lower yield in the maleinization of the BPA, as determined from ^{1}H NMR data [51]. Like the SOMG/NPG/MA polymer, the SOMG/BPA/MA displayed the characteristic gradual glass transition shown in Fig. 7.21.

HO/MA Dynamic Mechanical Polymer Properties

The dynamic mechanical properties of the HO/MA polymers were found to be better than those of the AESO polymers. Little variation was seen between the polymers made from the different oils. At room temperature, the E' for all of the oils existed between 1.45 and 1.55 GPa, showing no dependence on saturation level. The dynamic mechanical behavior was similar between the different oils, with the typical behavior being shown in Fig. 7.25. The temperatures at which a maximum in tan δ was exhibited ranged from 107°C to 116°C, which are all substantially higher than the AESO base resin. These properties are fairly close to those shown by conventional petroleum-based polymers. However, the distinctive triglyceride behavior still exists, in that the glass transitions are extremely broad and that, even at room temperature, the materials are not completely in a glassy state. Again this is probably due to the saturated fatty acids of the triglycerides that act as a plasticizer.

Although the extent of maleinization was approximately the same from oil to oil, it is possible to see how the slight differences affect the T_g. In Fig. 7.26, the tan δ peak temperature was plotted as a function of maleate functionality. Within this range, the behavior is linear, suggesting that if higher levels of functionalization are able to be reached, the properties should improve accordingly [64]. However, it is expected that past a certain extent of maleate functionality, the tan δ peak temperature dependence will plateau. Work is currently being pursued to test the limits of this behavior.

It was previously stated that the broadness in the glass transition may be inherent to all triglyceride-based polymers. However, work with genetically engineered oil and synthetic oil has shown that it is possible to reduce this characteristic. The genetically engineered high oleic soybean oil has an average functionality of three double bonds/triglyceride and has the fatty acid distribution shown in Table 7.1. The maleinized form of this oil had a functionality of two maleates/triglyceride. The properties of polymers from this material were compared to polymers from triolein oil, which is monodisperse, consisting only of oleic fatty acid esters (18 carbons long and one double bond). The maleinized triolein oil had a functionality of 2.1 maleates/triglyceride. Thus, the only difference between the two oils is the fatty acid distribution of the high oleic oil versus the monodisperse triolein oil.

The dynamic mechanical properties of polymers made from these oils are shown in Fig. 7.27. The T_g of these two polymers does not seem to differ much, judging from either their tan δ peak or the inflection in the E'. However, the broadness of the transitions does differ. It is apparent that the triolein polymer has a sharper E' transition from the glassy region to the rubbery region. This is evident also in the tan δ peak, which has a higher peak height. The transition is not yet as sharp as

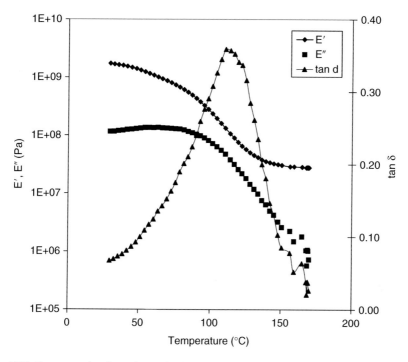

Fig. 7.25 Representative dynamic mechanical behavior for HO/MA polymers

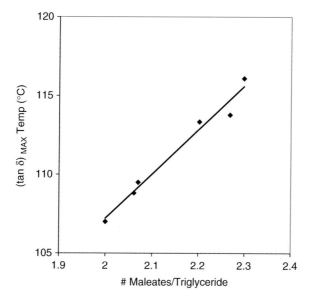

Fig. 7.26 Peak in tan δ as a function of maleate functionality

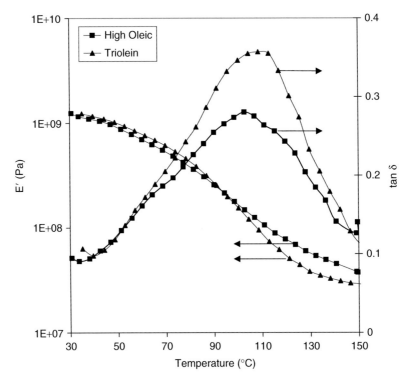

Fig. 7.27 Dynamic mechanical properties of polymers made from maleinized hydroxylated high oleic oil and triolein oil. The monodisperse triolein displays a sharper transition from the glassy region to the rubbery region

petroleum-based polymers. This is probably caused by the triolein monomer having a functionality of only two maleates/triglyceride. Consequently, there is still a plasticizer effect present, but this effect may be reduced by controlling the reaction conditions to reach higher conversions.

SOPERMA Polymer Properties

The typical DMA behavior of the SOPERMA-styrene polymer (40 wt% styrene) is shown in Fig. 7.28, where it can be seen that, at room temperature, these polymers are already in transition from the glassy region to the rubbery plateau. Most thermoset polymers show a distinct glassy region in which the modulus is independent of temperature. This is not observed for the SOPERMA-styrene polymers. The SOPERMA-styrene polymers show a very broad transition from the glassy to the rubbery state. Because of this broad transition, these polymers do not show a clear

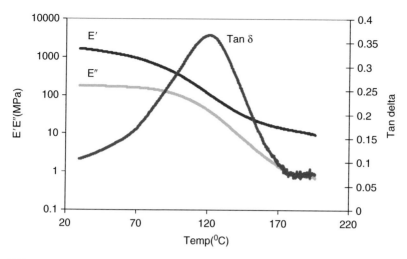

Fig. 7.28 Typical DMA behavior of the SOPERMA-styrene polymer (40 wt% styrene)

peak in the loss modulus E''. Thus, the tan δ curve is also very broad. The broad transition observed for the SOPERMA-styrene polymers is a result of two major effects. One major effect is the phase separation, which results in higher T_g SOPERMA-rich and lower T_g styrene-rich regions in the polymer matrix. Another effect that may result in a broad glass transition is the plasticizing effect of the fatty acids present in the SOPERMA monomer which are not functionalized. The transition from the glassy to the rubbery state broadens significantly with the addition of small amounts of plasticizers to polymers.

Figures 7.29a and b show the flexural modulus and flexural strength of the SOPERMA-styrene polymers as a function of styrene concentration. As can be seen both the flexural modulus and flexural strength of the polymers increase with increasing concentrations of styrene despite the decrease in cross-link density, v. Thus, the rigid aromatic structure of the styrene monomer as compared to the SOPERMA monomer with flexible fatty acid chains dominates the effect of cross-link density. The linear dependence of flexural modulus on styrene concentration above 30 wt% styrene follows Eq. 7.1:

$$E_f = 0.0278\left(W_{styrene}\right) + 0.2643, \tag{7.1}$$

where $W_{styrene}$ presents the weight percentage of styrene. This correlation predicts the flexural modulus of 100% polystyrene as 3.04 GPa. The flexural modulus values of standard polystyrenes of different grades are in the range of 2.9–3.8 GPa. In the same manner, the dependence of flexural strength on styrene concentration follows Eq. 7.2:

$$S_f = 0.6159\left(W_{styrene}\right) + 27.696 \tag{7.2}$$

Fig. 7.29 The change of (**a**) flexural modulus (**b**) flexural strength of SOPERMA-styrene polymers at increasing styrene weight percentages

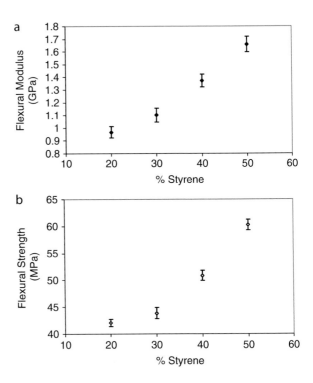

which predicts the flexural strength of the 100% polystyrene as 89.29 MPa. The strength values of standard polystyrene samples are in the range of 70–100 MPa.

The addition of butyrated kraft lignin to SOPERMA had a large effect on the polymer properties. The T_g's (as determined from the tan δ maximum) of the SOPERMA-styrene polymers as a function of lignin concentration are shown in Fig. 7.30a. As can be seen, there is a slow increase in T_g at low concentrations and then at 5 wt% lignin the T_g increases significantly. The T_g of the SOPERMA-lignin composite should be influenced by both the cross-link density of the system as well as the higher T_g of kraft lignin (167°C). The cross-link densities, as determined using the modulus in the rubbery region, are shown in Fig. 7.30b, where the cross-link density of the network increases with lignin until 5 wt% and then starts to decrease again. The increase in cross-link density with lignin addition may be attributed to specific interactions between the polymer matrix and the lignin molecule. The carboxylic acid groups of the SOPERMA monomer may interact with the available hydroxyl groups or thiol groups of the lignin molecule. Additionally, lignin may have effects on the kinetics of polymerization of both the SOPERMA and styrene monomer due to its inhibition effect on radical polymerization, which may affect the cross-link density. However, more work needs to be done to evaluate this effect. The cross-link density levels off at 5 wt% load and starts to decrease for higher concentrations. Thus, at this point the cross-link density must decrease due to the increase in the volume fraction of lignin, which cannot

Fig. 7.30 (a) Glass transition temperature, T_g, and (b) cross-link density (v) variation of the SOPERMA-styrene polymers as a function of butyrated lignin content

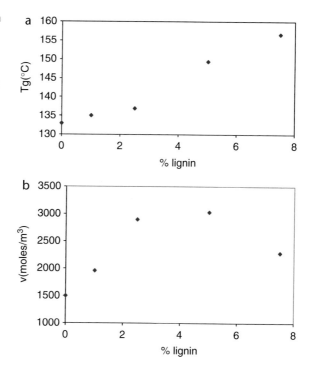

apparently interact with the matrix any more. The significant increase in T_g at 5 and 7.5 wt% lignin shows that the T_g of the system approaches the T_g of the kraft lignin (167°C) component at these high concentrations.

Figure 7.31 shows that both the flexural modulus and flexural strength of the SOPERMA-styrene polymers increase continuously with increasing lignin content of the resin. Because butyrated kraft lignin is dissolved in the polymer matrix, it should undergo the same strain as the polymer matrix. Thus, the modulus of the composite should increase with the introduction of the rigid aromatic structure of lignin to the system. Additionally, the cross-link density increase with lignin addition is also expected to increase the modulus. On the other hand, the increase in flexural strength with increasing lignin content may be attributed to both the increase in modulus with lignin addition and an increase in cross-link density up to 5 wt% lignin load.

SOGLYME Polymer Properties

As discussed in section "SOPERMA and COPERMA Polymer Synthesis," the soybean oil monoglyceride methacrylates (SOGLYME) were prepared by methacrylation of the soybean oil glycerolysis product by methacrylic anhydride,

Fig. 7.31 The change of (**a**) flexural modulus (**b**) flexural strength of SOPERMA-styrene polymers at increasing butyrated lignin content

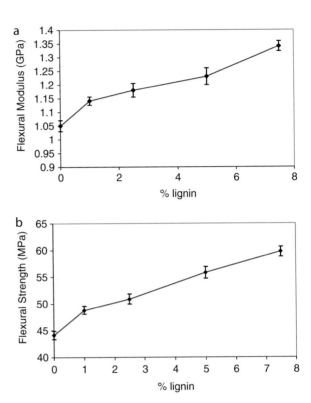

as shown earlier in Fig. 7.8. The methacrylated glycerides did not show phase separation in styrene. The crude methacrylated soybean oil monoglycerides contain methacrylic acid as a by-product. Methacrylic acid is itself a reactive diluent and acts as a comonomer in the system. Thus, the mechanical properties of the polymers prepared from this crude product (SOGLYME-MEA) as well as the polymers prepared using styrene as the third comonomer (SOGLYME-MEA-ST) will be examined. Properties of the polymers prepared from the acid extracted product with styrene (SOGLYCME-ST) will also be examined.

Figures 7.32a and b show the storage modulus (E'), loss modulus (E''), and tan δ values as a function of temperature for the polymers prepared from the crude methacrylated soybean oil monoglycerides that contain methacrylic acid as a by-product (SOGLYME-MEA) and the styrenated resin (SOGLYME-MEA-ST) (33 wt %) respectively. As can be seen, these polymers also show a broad transition from the glassy state to the rubbery state, similar to the SOPERMA-styrene polymers. The polymers prepared from these resins did not show phase separation, which means that the phase separation observed in the SOPERMA-styrene polymers is not the only factor responsible for the broad transition observed in these triglyceride-based polymers. The plasticization effect of the long flexible fatty acid chains present in the cross-linked monomer has a significant effect on the observed behavior. Additionally,

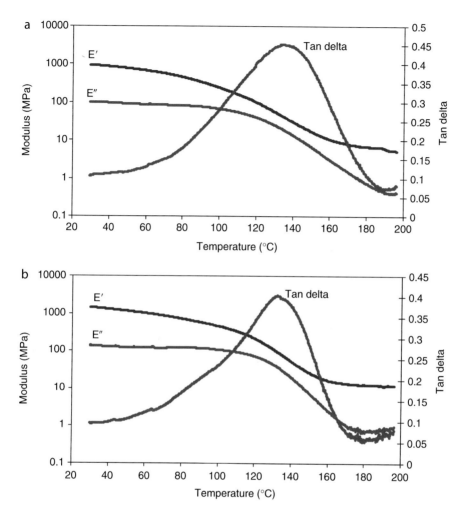

Fig. 7.32 Storage modulus (E′), loss modulus (E″) and tan δ values as a function of temperature for (**a**) SOGLYME-MEA (**b**) SOGLYME-MEA-STpolymers

as can be seen, the tan δ peak is broader for the SOGLYME-MEA polymer compared to the SOGLYME-MEA-ST polymer showing that the higher cross-link density of the former polymer also has a significant effect on broadening the glass transition.

The $E′$ values, T_g values (tan δ maxima) as determined from DMA, flexural moduli, and flexural strengths of the SOGLYME-MEA, SOGLYME-MEA-ST (33 wt% styrene), and SOGLYME-ST (33 wt% styrene) polymers are listed in Table 7.9. As can be seen, the SOGLYME-MEA-ST polymer has the highest modulus and strength followed by the SOGLYME-MEA polymer. The T_gs of these two polymers do not show a significant difference. The SOGLYME-ST

Table 7.9 The 30°C E' values, T_g values (tan δ maxima), flexural modulus and flexural strength of the SOGLYME-MEA, SOGLYME-MEA-ST (33 wt% styrene) and SOGLYME-ST (33 wt% styrene) polymers

Resin	E'(30°C) (GPa)	T_g (°C)	Flexural modulus (GPa)	Flexural strength (MPa)
SOGLYME-MEA	0.86	131.7	0.80	26.5
SOGLYME-MEA-ST	1.15	132.6	1.04	49.0
SOGLYME-ST	0.23	65.5	0.26	1.0

Table 7.10 The cross-link densities (v) of the SOGLYME polymers

Resin	v (moles/m³)
SOGLYME-MEA	701
SOGLYME-MEA-ST	609
SOGLYME-ST	520

polymer, on the other hand, exhibits considerably lower modulus, strength, and T_g values compared to the other two polymers. The properties of the individual monomers in these polymer systems as well as the cross-link density are both detrimental to the mechanical properties. The cross-link densities as determined by using the modulus values in the rubbery region of these polymers are shown in Table 7.10.

The cross-link densities of the SOGLYME-MEA and SOGLYME-MEA-ST polymers are in the neighborhood of the SOPERMA-styrene polymer at 50 wt% styrene concentration. The mechanical properties observed for these two polymers, however, are much lower than those observed for the SOPERMA-styrene polymers. The SOPERMA-styrene polymer at 50 wt% styrene has a flexural modulus of 1.65 GPa and a flexural strength of 62 MPa. This fact clearly shows that the maleate comonomer compared to the methacrylates, and the styrene comonomer compared to the methacrylic acid, bring more rigidity and strength to the triglyceride-based polymers. The cross-link density of the SOGLYME-MEA-ST polymer is lower than that of the SOGLYME-MEA polymer, as expected due to the increase in comonomer content. The SOGLYME-MEA-ST polymer, despite its lower cross-link density, still shows superior properties compared to those of the SOGLYME-MEA polymer due to the presence of the rigid styrene molecules in the polymer matrix. The significantly lower modulus, strength, and T_g observed for the SOGLYME-ST polymer is unexpected, especially when considering the properties of the SOGLYME-MEA polymer, and can only be attributed to its lower cross-link density compared to the other polymers. It is shown that the fracture strength σ of all bio-based polymers depends on modulus and cross-link density v, as $\sigma \sim [Ev]^{1/2}$.

Castor Oil-Based Polymer Properties

The basic fatty acid constituent of castor oil is ricinoleic acid, which is a hydroxy monounsaturated fatty acid (12-hydroxy *cis* 9-octodecenoic acid) ($\sim$87%). Castor oil was thus first alcoholized with pentaerythritol, glycerol, and an aromatic diol BPA propoxylate, and the alcoholysis products were then malinated with maleic anhydride, introducing maleate functionality to both the polyol hydroxyls and fatty acid hydroxyls. The resulting resins were labeled COPERMA, COGLYMA, and COBPAPRMA, respectively. The reaction schemes and the structures of the final malinated products for the COGLYMA, COPERMA, and COBPAPRMA products were shown earlier in Figs. 7.9–7.11, respectively. Additionally, castor oil was directly malinated and castor oil maleates (COMA) were also prepared. In this section, the properties of these castor oil-based polymers are introduced and analyzed with reference to the network structure. The effect of styrene concentration on mechanical properties of the COPERMA-styrene polymers is also explored and compared to the observed trend for the SOPERMA-styrene polymers.

Effect of Styrene Concentration on the COPERMA-Styrene Polymer Properties

Figure 7.33 shows the storage modulus values, E', of the COPERMA-styrene polymers as a function of temperature at increasing styrene concentrations as determined from DMA. As can be seen, the room temperature modulus values increase significantly, going from 20 wt% styrene to 30 wt% styrene. The increase in styrene concentration has a much less pronounced effect on the modulus at higher concentrations.

The changes in flexural modulus and flexural strength of the COPERMA-styrene polymers at increasing styrene concentrations are shown in Fig. 7.34a and b, respectively. The increase in flexural modulus and strength with increasing styrene concentrations follows a trend similar to that observed for the storage modulus. A significant increase in both the modulus and strength is observed while going from 20 to 30 wt% styrene, but this increase levels off rapidly at higher concentrations. As discussed in section "Properties of Plant Oil Resins," the SOPERMA-styrene polymers showed a continuous increase in both the modulus and strength, with styrene content in similar concentrations.

To explain the difference in the effect of styrene concentration on the mechanical properties of these two polymers, it is useful to determine the cross-link densities of the COPERMA-styrene polymers. Figure 7.35 shows the cross-link densities (ν) of the COPERMA-styrene polymers at increasing styrene concentrations as determined from the rubbery modulus by DMA. The cross-link

Fig. 7.33 The change of storage modulus (E′) values with temperature for the COPERMA polymers at increasing styrene concentrations

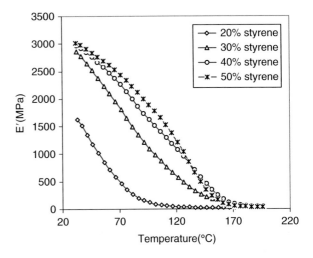

Fig. 7.34 The change in (**a**) flexural modulus and (**b**) flexural strength of the COPERMA polymers with increasing concentrations of styrene

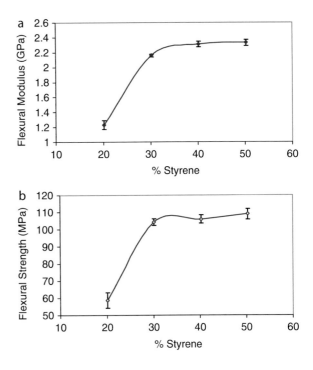

densities (v) of the SOPERMA-styrene polymers at the same styrene concentrations are also shown in the same figure, for comparison. For the COPERMA resin, which has a much higher maleate content per triglyceride than the SOPERMA resin, at 20 wt% styrene, the styrene concentration is too low to incorporate all of the maleates into polymerization, since the maleate half-esters do not homopolymerize.

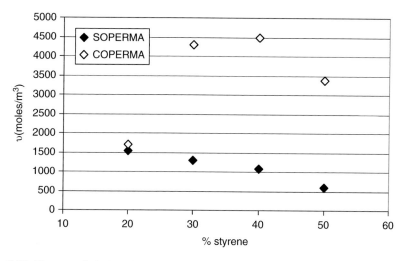

Fig. 7.35 The cross-link densities of the SOPERMA and COPERMA polymers at increasing styrene concentrations

The molar ratio of styrene double bonds to maleate double bonds for the 20 wt% styrene COPERMA resin is approximately 0.8, as determined from [1] H-NMR analysis. Thus, the 20 wt% styrene COPERMA polymer has the lowest cross-link density. At 30 and 40 wt% styrene, the molar ratio of styrene double bonds to maleate double bonds is 1.3 and 1.8, respectively, thus a significant increase in cross-link density is observed at 30 wt%, and at 40 wt% the cross-link density reaches its optimum value where all the available reactive groups of the COPERMA monomer can react with styrene. At higher concentrations, the cross-link density starts to decrease again since the added styrene increases the length of the segments between the cross-links. Thus, the modulus and strength of the COPERMA polymers at 20 wt% styrene are especially low and show a big increase at 30 and 40 wt% styrene due to the significant increase in cross-link density. After this point, the styrene content does not seem to have a significant effect on the modulus and strength. This behavior is different from that of the SOPERMA polymers, which showed a continuous increase in modulus and strength in the same styrene concentrations, despite the decrease in cross-link density. The introduction of the rigid aromatic rings of the styrene comonomer into the SOPERMA monomer with long flexible fatty acid chains results in a higher net increase in both the modulus and strength of the network than that observed for the COPERMA polymers because the fatty acids present in the COPERMA monomer are malinated and therefore incorporated into the network.

The tan δ curves for the COPERMA-styrene polymers at increasing styrene concentrations, shown in Fig. 7.36, also reflect the trend observed in cross-link density. As can be seen, the 20 wt% styrene polymer with the lowest cross-link density shows the tan δ maximum at the lowest temperature and therefore has the lowest T_g. The tan δ maximum shifts to higher temperatures with increases up to 40 wt% styrene due to the increase in cross-link density. After this point the

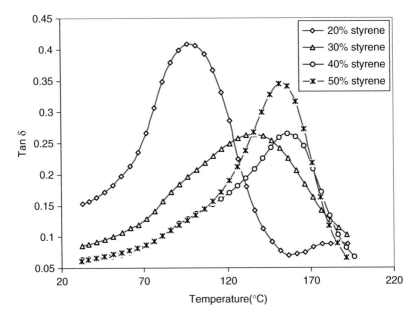

Fig. 7.36 The change of tan δ values with temperature for the COPERMA polymers at increasing styrene concentrations

increase in styrene concentration decreases the cross-link density and the T_g starts to decrease again. Thus the highest T_g is observed with 40 wt% styrene at 156°C, for the COPERMA-styrene polymers.

Comparison of COPERMA- and SOPERMA-Styrene Polymer Mechanical Properties

As can be seen in Fig. 7.35, the COPERMA-styrene polymers show significantly higher cross-link densities than those of the SOPERMA-styrene polymers, especially at 30 wt% and higher weight percentages of styrene. Table 7.11 shows the properties of the 30 wt% styrene SOPERMA and COPERMA polymers for a direct comparison. It can be seen that the modulus value nearly doubles and the flexural strength shows even a larger increase with the change from soybean oil to castor oil in the formulation. The glass transition temperature of the COPERMA polymer is about 9°C higher than that of the SOPERMA polymer. The incorporation of the fatty acid chains into the polymerization both increases the cross-link density and reduces the plasticization effect of the fatty acid chains in the COPERMA polymer, which in turn leads to a considerable increase in modulus, strength, and T_g compared to those properties of the SOPERMA polymer.

Table 7.11 The mechanical properties of COPERMA and SOPERMA polymers at 30 wt% styrene

Property	COPERMA (30% styrene)	SOPERMA (30% styrene)
Flexural strength (MPa)	101.23	43.86
Flexural modulus (GPa)	1.95	1.10
T_g (°C)	144	135
Storage modulus (GPa)	2.88	1.24

For the COPERMA resin, which has the highest maleate content among all the castor oil-based resins, the styrene concentration should be more than 30 wt% to fully incorporate all of the maleates during polymerization. This level should be lower for the other resins, which show lower maleate contents than the COPERMA resin. Thus 33 wt% styrene concentration is used for the preparation of other castor oil-based polymers, which keeps the renewable content of the resin within a reasonable range and also gives a chance to compare the properties of these materials to the commercially available unsaturated polyesters that use similar formulations.

The ratio of the flexural strength of COPERMA/SOPERMA is $\sigma_1/\sigma_2 = 101.2/43.9 = 2.1$. If the square root law for strength is applied, where

$$\frac{\sigma_1}{\sigma_2} = \left[\frac{E_1 v_1}{E_2 v_2}\right]^{1/2},$$

then, theory can be readily compared with experiment. Using $E_1 = 1.95$ GPa and $E_2 = 1.1$ GPa (Table 7.11), $v_1 = 4{,}300$ mol/m^3 and $v_2 = 1{,}300$ mol/m^3 (Fig. 7.35 at 30% styrene), then the predicted ratio for strength is $\sigma_1/\sigma_2 = 2.4$, which is in excellent accord with the experimental ratio.

Thermomechanical Properties of Castor Oil-Based Polymers

All of the malinated castor oil-styrene-based polymers exhibited broad transition profiles from the glassy state to the rubbery state, similar to the SOPERMA-styrene polymers. The broad transitions observed were attributed to the phase separation observed on the microscale and the high cross-link density exhibited by these polymers. The plasticization effect of the fatty acids should have a less pronounced effect for the castor oil-based polymers because the hydroxy fatty acids that constitute the majority of the fatty acids (87%) in castor oil were malinated and therefore connected to the network structure. Figure 7.37 shows the tan δ curves of the castor oil-based polymers as determined by DMA.

As discussed above, the damping peak position is sensitive indicator of cross-linking. As the cross-link density increases, the tan δ maximum shifts to higher temperatures, the peak broadens and a decrease in the tan δ value is observed. The cross-link densities of the malinated castor oil-based polymers (33 wt% styrene)

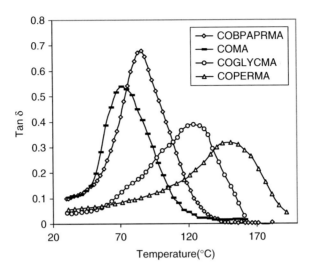

Fig. 7.37 The tan δ values of the castor oil polymers as a function of temperature

Table 7.12 Crosslink densities of castor oil polymers

Resin type	Mc(g/mole)	v(moles/m^3)
COPERMA	255	4,310
COGLYMA	441	2,494
COMA	732	1,511
COBPAPRMA	783	1,418

determined by using the value of E in the rubbery region of the polymers as determined by DMA are shown in Table 7.12. As can be seen, the COPERMA polymer has the highest cross-link density, followed by COGLYMA, COMA, and COBPAPRMA polymers. The COBPAPRMA polymer shows the lowest cross-link density, despite the higher maleate content than both COMA and COGLYMA which can be attributed to the bulkiness of the Bisphenol A propoxylate moiety. As can be seen in Fig. 7.37, the COPERMA polymer with the highest cross-link density exhibits the highest T_g and shows the broadest peak, with the lowest tan δ value. The COBPAPRMA, which has the lowest cross-link density, shows the highest tan δ values as expected; however, its tan δ max temperature is about 14º C above than that of the COMA polymer. The higher T_g observed for the COBPAPRMA may be explained by the presence of the rigid aromatic backbone of BPA propoxylate, as compared to the aliphatic fatty acid backbone of the COMA polymer. Thus, the monomer chemical structural influence dominates the cross-link density effect on T_g.

Table 7.13 The mechanical properties of castor oil polymers

Resin type	E′ (30°C) (GPa)	T_g (°C)	Flexural modulus (GPa)	Flexural strength (MPa)	Surface hardness (D)
COPERMA (1:2:10.7)	2.94	150	2.21	101.60	89.3
COGLYMA (1:2.2:9.2)	2.40	124	1.76	78.89	86.1
COBPAPRMA (1:2:6.7)	2.69	86	2.17	83.20	88.5
COMA (1:3)	1.15	72	0.78	32.83	77.1
Ortho-UP	–	135–204	3.45	80	–
Iso-UP	–	135–204	3.59	100	–

Mechanical Properties of Castor Oil-Based Polymers

The storage modulus values at 30°C and the T_g's as determined from DMA, as well as the flexural modulus, flexural strength, and the surface hardness values of the castor oil polymers are given in Table 7.13. The styrene content of each resin was 33 wt%. The mechanical property hardness is the ability of the material to resist indentation, scratching, abrasion, cutting, and penetration. This property may be important for structural materials that require a high resistance to indentation or abrasion. The hardness of a polymer reflects such other qualities as resilience, durability, uniformity, strength, and abrasion resistance. As can be seen in Table 7.13, the surface hardness of the castor oil-based polymers changes proportionately with the strength of the polymers.

The observed mechanical properties of the castor oil-based polymers can be explained both in terms of the cross-link density and the chemical structures of the polyols used. The COPERMA polymer, which has the highest cross-link density, shows superior properties to the other castor oil-based polymers. The COPERMA polymer with its T_g around 150°C and flexural modulus of 2.2 GPa and flexural strength of 105 MPa exhibits the highest T_g and strength obtained from any triglyceride-based thermoset resins. The COBPAPRMA polymer's modulus, strength and surface hardness values are higher than those of both COMA and COGLYMA polymers and approach those of COPERMA polymer, although its cross-link density is slightly lower than these two polymers. The aromatic structure of the BPA propoxylate moiety brings both rigidity and strength to the polymer network. Thus, this resin shows both high modulus and strength with a reasonable T_g, despite its lower maleate content, which is beneficial for the formulation since it decreases the nonrenewable content of the polymer. The COMA polymer shows the lowest modulus, strength, surface hardness, and T_g values due to its low cross-link density and also shows that a multifunctional unit at the center of the triglyceride monomer structure is essential for improved properties for these polymers.

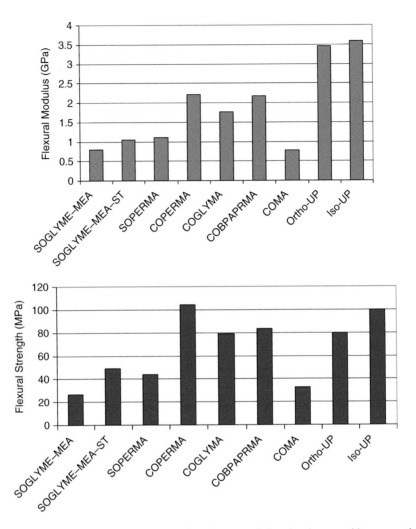

Fig. 7.38 A property comparison of soyoil and castor oil based polymers with commercial petroleum based Ortho and Iso-Unsturated Polyester resins

Figure 7.38 gives a comparison between the mechanical properties of these bio-based resins and the properties of two of the most commonly used UP resins: Orthophthalic (Ortho-UP) and Isophthalic (Iso-UP) UP resins. As can be seen, the properties of castor oil-based polymers are in a comparable range with those of the commercially successful UP resins. The properties of these bio-based resins can be significantly improved by the addition of lignin, which introduces the required aromatic groups for high stiffness and high T_g polymers and further increases their bio-based content with low-cost renewable material.

Future Directions

Triglyceride oils are an abundant natural resource that has yet to be fully exploited as a source for polymers and composites. The different chemical functionalities allow the triglyceride to be converted to several promising monomers. These are being commercialized for applications in composite resins, foams, pressure sensitive adhesives, elastomers, coatings with significantly reduced toxicity, and carbon footprint compared to their petroleum-based counterparts. When blended with comonomers, these monomers form polymers with a wide range of physical properties. They exhibit flexural strength up to 100 MPa and tensile moduli in the 1–3 GPa range, with T_g ranging from 70°C to 150°C, depending on the particular monomer and the resin composition. DMA shows that the transition from glassy to rubbery behavior is extremely broad for these polymers as a result of the triglyceride molecules acting both as cross-linkers and as plasticizers in the system. Saturated fatty acid chains are unable to attach to the polymer network, causing relaxations to occur in the network. However, this transition can be sharpened by reducing the saturation content, as demonstrated with the genetically engineered oil and pure triolein oil. The new design rules for controlling the molecular architecture based on the Twinkling Fractal Theory of the Glass Transition [67–69] will allow the facile development of these materials in future years. When combined with natural fibers (chicken feathers, hemp, flax, straw, etc.), useful high volume applications are made possible in hurricane resistant housing, positive net energy housing, earthquake resistant structures, eco-leather, and wind turbines in support of renewable energy.

This area of research sets a foundation from which completely new materials can be produced with novel properties. Work continues on optimizing the properties of these green materials and understanding the fundamental issues that affect them. Computer simulation can be used to optimize the choice of the fatty acid distribution function and determine the resulting architecture and mechanical properties for the particular chemical pathways shown in Fig. 7.2. Use of a computer significantly reduces the number of chemical trials required for a system with a large number of degrees of freedom and suggests the optimal oil most suited to a particular type of resin. In this manner, more renewable resources can be used to meet the material demands of many industries.

Bibliography

1. Wool RP, Kusefoglu S, Zhao R, Palmese G, Khot S (2000) High Modulus Polymers and Composites from Plant Oils. U.S. Patent 6,121,398
2. Wool RP (1999) Chemtech 29:41
3. Liu K (1997) Soybeans: chemistry, technology, and utilization. Chapman and Hall, New York
4. Gunstone F (1996) Fatty acid and lipid chemistry. Blackie Academic and Professional, New York
5. Cunningham A, Yapp A (1971) U.S. Patent 3,827,993

6. Bussell GW (1971) U.S. Patent 3,855,163
7. Hodakowski LE, Osborn CL, Harris EB (1975) U.S. Patent 4,119,640
8. Trecker DJ, Borden GW, Smith OW (1976) U.S. Patent 3,979,270
9. Trecker DJ, Borden GW, Smith OW (1976) U.S. Patent 3,931,075
10. Salunkhe DK, Chavan JK, Adsule RN, Kadam SS (1992) World oilseeds: chemistry, technology, and utilization. Van Nostrand Reinhold, New York
11. Force CG, Starr FS (1988) U.S. Patent 4,740,367
12. Barrett LW, Sperling LH, Murphy CJ (1993) J Am Oil Chem Soc 70:523
13. Qureshi S, Manson JA, Sperling LH, Murphy CJ (1983) In: Carraher CE, Sperling LH (eds) Polymer applications of renewable-resource materials. Plenum Press, New York
14. Devia N, Manson JA, Sperling LH, Conde A (1979) Polym Eng Sci 19:878
15. Devia N, Manson JA, Sperling LH, Conde A (1979) Polym Eng Sci 19:869
16. Devia N, Manson JA, Sperling LH, Conde A (1979) Macromolecules 12:360
17. Sperling LH, Carraher CE, Qureshi SP et al (1991) In: Gebelein CG (ed) Polymers from biotechnology. Plenum Press, New York
18. Sperling LH, Manson JA, Linne MA (1984) J Polym Mater 1:51
19. Sperling LH, Manson JA (1983) J Am Oil Chem Soc 60:1887
20. Fernandez AM, Murphy CJ, DeCosta MT et al (1983) In: Carraher CE, Sperling LH (eds) Polymer applications of renewable-resource materials. Plenum Press, New York
21. Sperling LH, Manson JA, Qureshi SA, Fernandez AM (1981) Ind Eng Chem 20:163
22. Yenwo GM, Manson JA, Pulido J et al (1977) J Appl Polym Sci 21:1531
23. Frischinger I, Dirlikov S (1991) Polymer Comm 32:536
24. Frischinger I, Dirlikov S (1994) In: Sperling LH, Kempner D, Utracki L (eds) Interpenetrating polymer networks. Advances in chemistry series 239. American Chemical Society, Washington, DC, p 517
25. Rosch J, Mulhaupt R (1993) Polymer Bull 31:679
26. Meffert A, Kluth H (1989) Denmark Patent 4,886,893
27. Rangarajan B, Havey A, Grulke EA, Culnan PD (1995) J Am Oil Chem Soc 72:1161
28. Zaher FA, El-Malla MH, El-Hefnawy MM (1989) J Am Oil Chem Soc 66:698
29. Friedman A, Polovsky SB, Pavlichko JP, Moral LS (1996) U.S. Patent 5,576,027
30. Swern D, Billen GN, Findley TW, Scanlan JT (1945) J Am Oil Chem Soc 67:786
31. Sonntag NOV (1982) J Am Oil Chem Soc 59:795
32. Solomon DH (1967) The chemistry of organic film formers. Wiley, New York
33. Can E (1999) M.S. Thesis, Bogazici University, Turkey
34. Bailey AE (1985) In: Swern D (ed) Bailey's industrial oil and fat products. Wiley, New York
35. Hellsten M, Harwigsson I, Brink C (1999) U.S. Patent 5,911,236
36. Cain FW, Kuin AJ, Cynthia PA, Quinlan PT (1995) U.S. Patent 5,912,042
37. Eckwert K, Jeromin L, Meffert A, et al (1987) U.S. Patent 4,647,678
38. Khot SN (1998) M.S. Thesis, University of Delaware
39. Wypych J (1986) Polyvinyl chloride stabilization. Elsevier, Amsterdam
40. Sears JK, Darby JR (1982) The technology of plasticizers. Wiley, New York
41. Carlson KD, Chang SP (1985) J Am Oil Chem Soc 62:931
42. Raghavachar R, Letasi RJ, Kola PV et al (1999) J Am Oil Chem Soc 76:511
43. Pashley RM, Senden TJ, Morris RA, et al (1991) U.S. Patent 5,360,880
44. Likavec WR, Bradley CR (1999) U.S. Patent 5,866,628
45. Bordon GW, Smith OW, Trecker DJ (1971) U.S. Patent 4,025,477
46. La Scala JJ, Wool RP (2002) J Am Oil Chem Soc 79:59
47. Bunker SP (2000) M.S. Thesis, University of Delaware
48. Chu TJ, Niou DYJ (1989) Chin Inst Chem Eng 20:1
49. Betts AT (1975) U.S. Patent 3,867,354
50. Mitch EL, Kaplan SL (1975) In: Proceedings of 33 rd annual SPE technical conference, Atlanta

51. Can E, Kusefoglu S, Wool RP (2001) Rigid thermosetting liquid molding resins from renewable resources: I. Copolymers of soyoil monoglycerides with maleic anhydride. J Appl Polym Sci 81:69
52. Gardner HC, Cotter RJ (1981) European Patent 20,945
53. Thomas P, Mayer J (1971) U.S. Patent 3,784,586
54. Lee SH, Park TW, Lee SO (1999) Polymer (Korea) 23:493
55. ShioneH, Yamada J (1999) Japanese Patent 11,147,222
56. Hasegawa H (1999) Japanese Patent 11,240,014
57. Johnson LK, Sade WT (1993) J Coat Tech 65:19
58. Solomons TWG (1992) Organic chemistry. Wiley, New York
59. La Scala J, Wool RP Polymer. in preparation
60. Flory PJ (1975) Principles of polymer chemistry. Cornell University, Ithaca
61. Nielsen LE, Landel RF (1991) Mechanical properties of polymers and composites. Marcel Dekker, New York
62. Wool RP, Khot SN (2000) In: Proceedings of ACUN-2, Sydney
63. Can E, Kusefoglu S, Wool RP (2001) Rigid thermosetting liquid molding resins from renewable resources: (I) copolymers of soyoil monoglycerides with maleic anhydride. J Appl Polym Sci 81:69
64. Khot SN, La Scala JJ, Can E et al (2001) J Appl Polym Sci 82:703
65. Can E, Kusefoglu S, Wool RP (2002) Rigid thermosetting liquid molding resins from renewable resources. II. Copolymers of soybean oil monoglyceride maleates with neopentyl glycol and bisphenol A maleates. J Appl Polym Sci 83(5):972–980
66. Can E (2004) PhD Thesis, University of Delaware
67. Wool RP (2008) J Polym Sci Part B: Polym Phys 46:2765
68. Wool RP, Campanella A (2009) J Polym Sci Part B: Polym Phys 47:2578
69. Stanzione JF III, Strawhecker KE, Wool RP, doi:10.1016/j.jnoncrysol.2010.06.041

Chapter 8
Organic Batteries

Hiroyuki Nishide and Kenichi Oyaizu

Glossary

C Rate	1C (or $1\ I_t$) Discharging corresponds to the constant-current discharging of rated capacity at 1 h. For example, the 1 C discharging of a lead acid battery with a rated capacity of 12 Ah corresponds to the discharging at 12 A, and 2 C corresponds to the discharging at 24 A. The C rate is also used to show the charging current.
Capacity	The quantity of electricity (Ah) that can be obtained by discharging until the cell reaches the end of life. Electricity for 1 mol of e^- = 1 Faraday (F) = 96,485 C = 26.8 Ah.
Cyclability	Capability of charging/discharging cycles without loss of capacity for secondary batteries. A commercial cell must be capable of completely discharging and then fully recharging for >300 times and not losing >20% of its capacity, which requires a robust system and reversible electrode reactions. There should be no side reactions that result in the loss of the active materials during the cycle. High cyclability secondary batteries reduce their environmental load.

This chapter was originally published as part of the Encyclopedia of Sustainability Science and? Technology edited by Robert A. Meyers. DOI:10.1007/978-1-4419-0851-3

H. Nishide (✉) • K. Oyaizu
Department of Applied Chemistry, Waseda University, Tokyo 169–8555, Japan
e-mail: nishide@waseda.jp; oyaizu@waseda.jp

P.T. Anastas and J.B. Zimmerman (eds.), *Innovations in Green Chemistry and Green Engineering*, DOI 10.1007/978-1-4614-5817-3_8,
© Springer Science+Business Media New York 2013

Electrode-active material	The material that generates electrical current by means of a chemical reaction within the battery. The *anode*- and *cathode*-active materials are used in the negative and positive electrodes associated with oxidative and reductive chemical reactions that release and gain electrons through the external circuit during discharging, respectively. Organic batteries use organic redox-active molecules, which are typically polymers, for the electrode-active materials.
Electrolyte	Salt solution (or ionic liquid) capable of electric conduction by ionic transport. The electrolyte layer in batteries serve as a simple buffer for ionic flow between the electrodes as in Li-ion and Ni-Cd cells, or it participates in the electrochemical reaction as in Pb-acid cells. For battery sustainability and safety, the low flammability and leakage are required so that (semi)solid electrolyte is often employed such as those with poly(ethylene glycol).
Electromotive force (emf)	A difference in potential that tends to give rise to electric current. For a cell with a configuration of

$$\text{Anode}|\text{Ox}_2, \text{Red}_2|\text{Electrolyte}|\text{Ox}_1, \text{Red}_1|\text{Cathode},$$

the overall reaction is $n_2\text{Ox}_1 + n_1\text{Red}_2 \ \square \ n_2\text{Red}_1 + n_1\text{Ox}_2$ and the emf is given by

$$E = E_1 - E_2 = \Delta E^\circ - RT/(n_1 n_2 F)$$
$$\ln\{(a_{\text{Red}_1}{}^{n_2} a_{\text{Ox}_2}{}^{n_1})/(a_{\text{Ox}_1}{}^{n_2} a_{\text{Red}_2}{}^{n_1})\}.$$

The cell voltage is $V = E - IR$ where R is the resistance of the cell. The power is P (W) = current I (A) $\times$ cell voltage V (V). Thus, large emf is advantageous to produce large power, although special attention has to be paid to the safety of the battery.

Faraday efficiency (Coulomb efficiency)	Efficiency of active mass = (actual capacity)/(theoretical capacity).
Rechargeable (secondary) battery	An electrochemical cell for the generation of electrical energy in which the cell, after being discharged, is restored to its original charged condition reversibly by an electric current flowing in the direction opposite to the flow of current when the cell was discharged. The reusability lowers the price and the environmental load per supplied electrical power.

Self-discharge	The loss of performance when a battery is not in use, which is related to the loss of energy accompanied by the energy storage with the battery. $Li - MnO_2$ primary cells will deliver 90% of their energy even after 8 years on the shelf; that is, their self-discharge is low. Rechargeable batteries generally have more rapid loss of capacity on storage. The rechargeable $Ni - MH$ cell, for instance, will lose up to 30% of its capacity in a month. In organic batteries, immobilization of the organic electrode-active materials is crucial to prevent or suppress the self-discharge. Cross-linking is often applied to the polymeric materials to avoid dissolution into electrolyte solutions.
Separator	A substance, such as microporous polyethylene, that separates the anode and the cathode. An organic shutdown separator, which is molecularly designed to shut down the battery on over-temperature or overcharging and discharging, is desired for the safety concerns, especially in the case of Li-ion batteries.
Specific capacity (in mAh/g or Ah/kg)	1,000/(electrochemical equivalent $\times$ 3,600), where electrochemical equivalent is the mass of element or molecule (in g) transported by 1 C of electricity. Examples: Li (6.941 g/mol), $Li \rightarrow Li^+ + e^-$, electrochemical equivalent = 6.941/96,485 = 7.19×10^{-5} g,

specific capacity = $1,000/(7.19 \times 10^{-5} \times 3,600)$ = 3,863 mAh/g. PTMA (240.3 g/mol unit), electrochemical equivalent = 2.49×10^{-3} g, specific capacity = 112 mAh/g.

Sustainable battery	Energy devices designed to store and supply electric power, characterized by clean and reduced CO_2 emission, effective utilization of alternative fuels, low environmental load, and safety, for a sustainable society. Organic batteries are inherently sustainable, owing to the environmentally benign fabrication and disposing processes, low toxicity and safety, based on the use of organic, resource-unlimited, and heavy metal-free materials.

Definition of the Subject

Batteries are composed of cathode- and anode-active materials, a separator to separate the two electrode-active materials, an electrolyte layer, and current collectors. The cathode- and anode-active materials of conventional primary and secondary batteries are usually metal oxides, such as manganese, silver, lead, nickel, and vanadium oxides, and metals, such as zinc, lead, cadmium, and lithium. All the electrodes of conventional batteries are composed of metals and metal oxides (except oxygen and carbon that are used for the air battery cathode and the lithium battery anode, respectively), many of which come from limited resources.

Recent development of organic batteries aiming at increasing the energy density has made them promising candidates of energy-storage devices to complement the conventional inorganic-based batteries. Definition of organic battery in this entry is based on the use of organic materials designed for the electrode-active materials. (Note that those using organic materials for electrolyte layers have often been called likewise.) Organic materials for the electrode-active materials are π-conjugated polymers using their doping/de-doping behaviors, and redox polymers, either conjugated or nonconjugated, involving polyaniline-based disulfides and other organic redox sites such as tetracyanoquinodimethane. The most typical and recently developed organic battery includes a radical battery, using organic radical polymers which are a class of aliphatic or nonconjugated redox polymers with organic robust radical pendant groups as the redox site. The advantage of organic battery in terms of sustainability also comes from the excellent recyclability and safety as the inherent virtue of the organic materials.

Introduction

Development of sustainable processes for energy storage and supply is one of the most important worldwide concerns today. Primary batteries, such as alkaline manganese and silver oxide batteries, produce electric current by a one-way chemical reaction and are not rechargeable and hence useless for reversible electricity storage. Portable electronic equipments, electric vehicles, and robots require rechargeable secondary batteries. Li-ion, lead acid, and nickel-metal hydride batteries are generally used at the present to power them. Solar cells and wind-power generators expect a parallel use of rechargeable batteries for leveling and preserving their generated electricity. Ubiquitous electronic devices such as integrated circuit smart cards and active radio-frequency identification tags need rechargeable batteries that are bendable or flexible and environmentally benign for durability in daily use. Designing of soft portable electronic equipments, such as roll-up displays and wearable devices, also require the development of flexible batteries. It is essential to find new, low-cost, and environmentally benign electroactive materials based on less-limited resource for electric energy storage and supply [1]. Reversible storage materials of electric energy or charge that are currently under use in electrodes of rechargeable batteries are entirely inorganic materials, such as Li-ion-containing cobalt oxide, lead acid, and nickel-metal hydride.

Among the secondary batteries, Li-ion batteries are becoming the most popular [2, 3]. Almost eight billion batteries which are worth US$9.5 billion were produced worldwide in 2009 [4]. Typical Li-ion battery is configurated with the cathode composed of Li-ion-containing metal oxides, such as lithium cobalt oxide, and the anode of graphite carbon [5]. During the charging process, Li ions are eliminated from the lattice of the metal oxide cathode and intercalated into the carbon anode. During the discharging process, the cobalt oxide is regenerated by the elimination of the Li ions. Li is one of the smallest elements to yield large specific capacity ($Li_6C \rightarrow 6Li^+ + 6e^- + C$: 372 mAh/g) and the very large electromotive force generated by the Li ion ($0.5C_6Li + Li_{0.5}CoO_2 \square 3 C + LiCoO_2$: 3.6 V), which result in a significantly high energy density and the highest given voltage of the battery. Another feature of the Li-ion battery is the rocking-chair-type intercalation reactions of the Li ion between the cathode- and the anode-active materials [3, 5]. The intercalation of Li ions into the electrodes suppresses metallic Li (dendrite) formation by separating and immobilizing each Li ion in the lattice and the layered structure. However, the intercalating diffusion of Li ions accompanied by transformation of the lattice and the layer structure during the electrode reactions results in slow kinetics and heat generation in the charging and discharging processes, often producing overheating and occasional ignition. Indeed, battery makers often issued recalls of their million-scale Li-ion batteries for mobile phones and laptop computers, due to overheating accidents.

Tedious waste processing of used batteries is difficult and crucial issue of the metal- and metal oxide-based conventional batteries. Used batteries are, in these days of environmental concern, being collected and recycled. Among the recycling of used batteries, the Li-ion batteries are highly valued for their excellent collection yield (e.g., 60% in Japan [6, 7]) from personal computers and mobile phones. The typical recycle processing of used Li-ion batteries includes several steps of calcination, sieving, dissolving in acid, and solvent extraction, and is reasonably working at least in Japan [6, 7]. However, the collection yields of used Li-ion batteries are somewhat decreasing because the batteries are being equipped in many varieties of small devices. Additionally, other used batteries are often found in landfill disposal or simply stand at the level at 60 kt/year even in Japan. The conventional metal-based batteries involve several inherent unsolved issues from the standpoint of sustainable technology.

Organic functional polymers have been developed as alternatives of inorganic functional materials because of their lightweight, flexibility, thin film–forming ability, processability, metal-free and benign environmental aspects, and fewer limitations of organic resources. A new trend in materials research is emerging recently, focusing on the fabrication of organic electrodes or electrodes composed of electrode-active organic compounds, characterized by the redox (oxidative/

reductive or electron-releasing/gaining reaction) activity and/or electric conduction of organic molecules that typically consist of redox polymers and/or conducting polymers. The exploration of properties of such organic materials leads to new technological perspectives of various totally organic electrochemical devices, not only batteries and hybrid capacitors but sensors and displays as well. The organic batteries and devices are free from using (heavy) metals and limited resources, and are potentially involved in the carbon cycle. In this entry, the concept and outline of organic electrodes for use in rechargeable batteries is reviewed, emphasizing the material science of the organic polymers for charge transport and storage for the development of sustainable science and engineering.

Conjugated Polymers and Redox Polymers for Electrode-Active Materials

Research on organic electrode-active materials was prompted by the discovery of the electric conductivity of doped polyacetylene, which led to explore organic batteries by employing the redox capability of polyacetylene based on the reversible p-type and n-type doping/de-doping processes for the cathode and the anode, respectively, as the principal electrode reactions [8]. Electrically conducting polymers with π-conjugated skeleton structures, such as polyaniline, polypyrrole, polythiophene, and their related derivatives (Scheme 8.1a), have been similarly examined as electrode-active materials, based on their reversible electrochemical doping behaviors [9]. The polymers are conveniently placed on the surface of a current collector by a solution-based wet process such as spin-coating and electropolymerization methods. The limitations of the π-conjugated polymers as the electrode-active materials are based on their insufficient doping levels, the resulting low redox capacities, and fluctuation of the cell voltage through the doping/de-doping process. Chemical instability of the doped states of the π-conjugated polymers is the fatal flaw to practical application because it frequently leads to the self-discharge and degradation of the rechargeable properties of the resulting batteries.

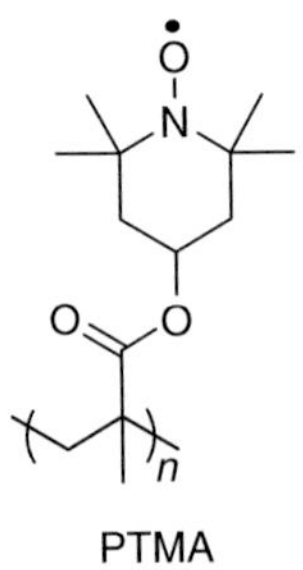

Scheme 8.1 Chemical structures of the (**a**) π-conjugated polymers and (**b**) the dithiol/disulfide-based electrode-active materials

PTMA

A pair of p- and n-type redox-active organic π-conjugated compounds, such as indigo derivatives bearing two sulfonic acid groups, was incorporated as dopants into polypyrrole [10]. The doped polypyrrole held the redox-active dopants through electrostatic interaction and worked as the electroconductive cathode and anode for a rechargeable battery. However, amount of the loadable redox-active dopants was limited by the doping degree and the battery lacked in stability to elute the dopant through the discharging process.

Organodisulfide compounds such as those derived from [1,3,4]thiadiazole-2,5-dithiol (362 mAh/g) (Scheme 8.1b) act as a cathode-active material with a high charge capacity through their two-electron redox reaction according to

$$n \; {}^{\ominus}\text{S-R-S}^{\ominus} \; \underset{+2ne^-}{\overset{-2ne^-}{\rightleftharpoons}} \; \{\text{S-R-S}\}_n$$

Emf of the redox reaction versus Li is 2.7–3.0 V to give significantly high energy density [11]. However, the redox reaction involves the formation of the disulfide polymers (Scheme 8.1) with the formation and cleavage of the chemical bond or the formation and the scission of the polymer chain, which renders the rate performance of the cathode low. Blending of the organothiolate with polyaniline as a conducting polymer has been examined to improve the rate performance, where the polyaniline served as a conducting pathway for the redox reaction of the disulfide. Nasty odor of the sulfur compounds was still a problematic issue. The test battery in conjunction with the lithium anode performed the output voltage of ca 3.5 V with discharging capacity of ca. 150 Ah/kg.

Organodisulfide residue has been intramolecularly combined in the repeating unit of polyaniline [12], as represented in Scheme 8.1. Cyclability in the redox reactivity and the smelly problem were much improved, while there still remained limitation of slow kinetics in the charging/discharging processes.

Organic Radical Polymers as Electrode-Active Materials

Another avenue toward the organic electrode-active materials is based on the sufficiently large redox capacity of aliphatic radical polymers, that is, organic polymers densely populated with pendant redox-isolated sites. The principal finding to permit the use of the nonconjugated radical polymers is the capability of organic robust radicals, or open-shell molecules, to allow fully reversible one-electron redox reactions featuring fast electrode kinetics and reactant cyclability [13]. Organic radicals are stabilized thermodynamically by effective delocalization of the unpaired electron and/or kinetically by bulky substituents to employ as the redox sites. The nitroxide radicals such as 2,2,6,6-tetramethylpiperidine-1-oxyl (TEMPO) derivatives and nitronyl nitroxides, the nitrogen radicals such as triarylaminium cation radicals, diphenylpicrylhydrazyl derivatives and verdazyls, and the oxygen radicals such as phenoxyls and galvinoxyls are typically examined

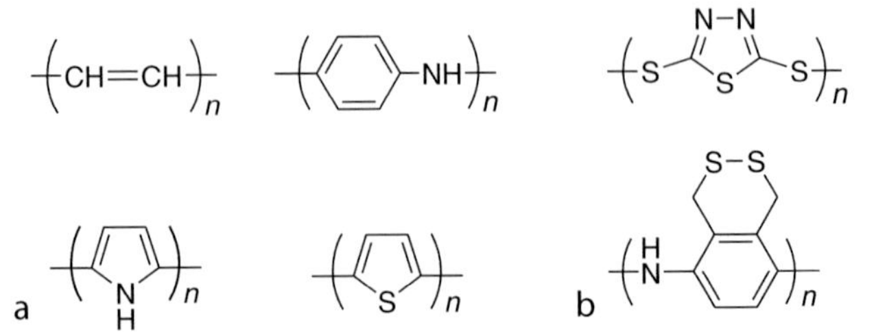

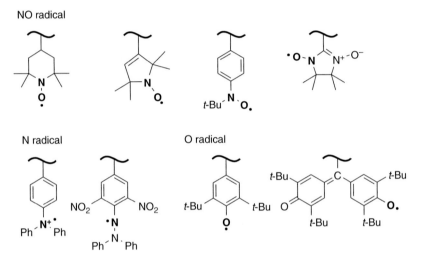

Scheme 8.2 Chemical structures of organic robust radicals in the radical polymers as the electrode-active materials in organic batteries

NO radical

N radical   **O radical**

Scheme 8.3 Redox couples of TEMPO and phenoxyl

as the pendant group of the radical polymers (Scheme 8.2). These radicals are persistent at ambient temperatures under air and characterized by a rapid and reversible $1e^-$ electrode reactions, typically with heterogeneous electron-transfer rate constants in the order of $k_0 = 10^{-1}$ cm/s [14, 15].

Nitroxide radicals are oxidized to oxoammonium cations near 1 V and reduced to aminoxy anions near -0.5 V by $1e^-$ transfer processes. The negative charging of the neutral radicals usually takes place at more negative potentials than those for the positive charging. Attempts have been made to use these reactions for the electrode-active materials. The most typically used radicals for the cathode and the anode are TEMPO and phenoxyl (or galvinozyl), respectively, which give the corresponding oxoammonium cation and the phenolate (or galvinolate) anion upon the electro-chemically reversible $1e^-$ reactions (Scheme 8.3).

A variety of polymer backbones have been employed to bind the radicals, such as poly(meth)acrylates, polystyrene derivatives, poly(vinyl ether)s, polyethers, and

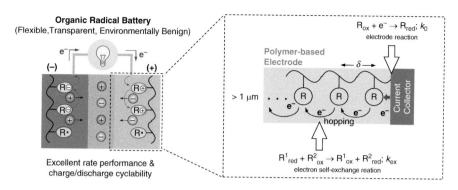

Fig. 8.1 Discharging processes of the totally organic radical battery, showing the redox gradient-driven charge transport in the radical polymer layer at the current collector surface. Charge propagation is accomplished within redox polymer layers, swollen in electrolyte solutions, by an electron self-exchange (hopping) mechanism. R represents the redox site in the polymer

polynorbornenes. The radical sites must be bound to polymer chains because of the need to immobilize them at current collectors for exclusion of elution into electrolyte solutions that lead to discharging, to accomplish appropriate mobility of the counterions, to make the electrode-active materual amorphous and plasticized to accommodate deformation without heat generation, and to allow wet fabrication processes. Unpaired electron density of the polymers, determined by SQUID measurements, has revealed the presence of the radicals substantially per repeating unit. Specific capacities based on the formula weight of the repeating unit are in the range of 50–150 mAh/g.

The unpaired electron density of the radical polymers amount to several molars in the bulk of the swollen polymer equilibrated in electrolyte solutions. Under these conditions, charge propagation within the polymer layer is sufficiently fast leading to high-density charge storage, because the redox sites are so populated that electron self-exchange reaction goes to completion within a finite distance of the polymer layer. The concentration gradient-driven charge propagation is accomplished by the electron self-exchange reaction. The exchange reaction is sufficiently fast for the diffusion-limited outer-sphere redox reactions, resulting in the efficient transfer of charge produced at the surface of the current collector (Fig. 8.1). Combination of the fast electrode process ($k_0 \sim 10^{-1}$ cm/s) and self-exchange reaction ($k_{ex} \sim 10^{6-8}$ $M^{-1}s^{-1}$) lead to the high rate performance of the radical-based organic battery.

The radical polymers act as both cathode- and anode-active materials, depending on the redox potential. A couple of polymers, different in redox potentials, are used as the electroactive materials in the organic radical battery (Fig. 8.1). The charging process corresponds to the oxidation of the radical to the cation at the cathode, and the reduction of the radical to the anion at the anode. The electromotive force is close to the potential gap between the two redox couples, which typically amounts to 0.5–1.5 V. The radical polymers are reversibly converted to the corresponding polyelectrolytes during the charging process. The rapid electrode reaction of the radicals and the efficient charge propagation within the polymer layer lead to a high rate performance, allowing rapid charging and large discharge currents without substantial loss of output voltages.

Performance of Organic Radical Batteries

The amorphous radical polymers allow fabrication of flexible thin-film devices (Fig. 8.2) [16]. A curious feature of the radical battery is the excellent rate performance to produce burst power (50 mA/cm^2 and >95% capacity at 60 C discharging), which also allow instant full charging (i.e., in a few seconds). Optimization of the battery has allowed retention of >90% capacity even after 1,000 charging/discharging cycles. The maximum energy and power densities are 120 Wh/kg and 20 kW/kg, respectively. The resulting radical battery has allowed the rated performance of the power density of 5 kW/L for more than 10^4 times pulse charging/discharging at 100 mA, and 1 A discharging and 2 W power from only <1 mm-thick flexible battery, that is suitable for IC cards, electronic papers, and active Radio Frequency Identification (RFID) [17, 18].

Radical batteries have proved to be a new class of organic rechargeable devices, which are characterized by the excellent rate performance and the moderate energy density as illustrated by the Ragone plots (Fig. 8.3). The mass-specific energy density is placed between those of polyacetylene- and disulfide-based organic batteries, but the power density is much larger and comparable to those of supercapacitors. Efforts have been directed toward further increasing the theoretical energy density of the radical batteries. For this purpose, radical polymers with even smaller formula weights per repeating unit are designed.

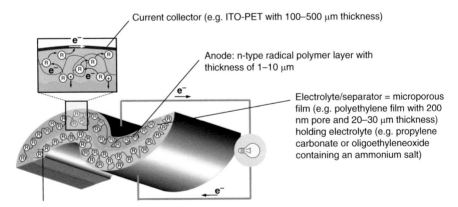

Current collector (e.g. ITO-PET with 100–500 μm thickness)

Anode: n-type radical polymer layer with thickness of 1–10 μm

Electrolyte/separator = microporous film (e.g. polyethylene film with 200 nm pore and 20–30 μm thickness) holding electrolyte (e.g. propylene carbonate or oligoethyleneoxide containing an ammonium salt)

Cathode: p-type radical polymer layer with thickness of 1–10 μm

Fig. 8.2 Charge-storage configuration of a flexible organic radical battery

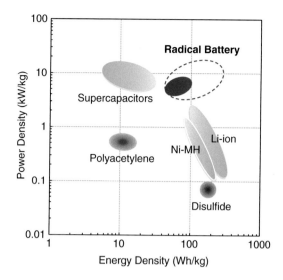

Fig. 8.3 Energy versus power density plots for batteries and supercapacitors

Future Directions

The organic batteries have several advantages over the Li-ion batteries, such as higher safety, adoptability to wet fabrication processes, easy disposability, and capability of fabrication from less-limited resources. While organic batteries have intrinsically lower volumetric energy density, this limitation can be overcome in the near future so that they can be designed to be compatible with and installable in various electronic equipments.

Bibliography

1. Service RF (2006) New 'Supercapacitor' promises to pack more electrical punch. Science 313:902
2. Pistoia G (2005) Batteries for portable devices. Elsevier, Amsterdam
3. Scrosati B, Schalkwijk WAV (2002) Advances in lithium-ion batteries. Plenum, New York
4. Report of Battery Association of Japan. http://www.baj.or.jp/statistics/01.html. Accessed 19 March 2011
5. Balakrishnan PG, Ramesh R, Kumar TP (2006) Safety mechanisms in lithium-ion batteries. J Power Sources 155:401–414
6. Report from Ministry of Economy, Trade and Industry, Japan. Year-book of Machinery Statistics (2008)
7. Report from Japan Portable Rechargeable Battery Recycling Centor. http://www.jbrc.net/hp/contents/recycle/index.html. Accessed 19 March 2011

8. Nigrey PJ, MacDiarmid AG, Heeger AJ (1979) Electrochemistry of polyacetylene, (CH)$_x$: electrochemical doping of (CH)$_x$ films to the metallic state. J Chem Soc Chem Commun (14):594–595
9. Roncali J (1997) Synthetic principles for bandgap control in linear π-conjugated systems. Chem Rev 97:173–205
10. Song HK, Palmore GTR (2006) Redox-active polypyrrole: toward polymer-based batteries. Adv Mater 18:1764–1768
11. Oyama N, Tatsuma T, Sato T, Sotomura T (1995) Dimercaptan–polyaniline composite electrodes for lithium batteries with high energy density. Nature 373:598–600
12. Takemura I, Iwasaki T, Takeoka S, Nishide H (2004) 3-Dithiolthione-substituted polythiophene and its redox activities. Chem Lett 33:1482–1483
13. Oyaizu K, Nishide H (2009) Radical polymers for organic electronic devices: a radical departure from conjugated polymers? Adv Mater 21:2339–2344
14. Yonekuta Y, Oyaizu K, Nishide H (2007) Structural implication of oxoammonium cations for reversible organic one-electron redox reaction to nitroxide radicals. Chem Lett 36:866–867
15. Suga Y, Pu YJ, Oyaizu K, Nishide H (2004) Electron-transfer kinetics of nitroxide radicals as an electrode-active material. Bull Chem Soc Jpn 77:2203–2204
16. Nishide H, Oyaizu K (2008) Toward flexible batteries. Science 319:737–738
17. http://www.nec.co.jp/press/ja/0512/0702.html. Accessed 19 March 2011
18. http://www.nec.co.jp/press/ja/0902/1302.html. Accessed 19 March 2011

Chapter 9
Oxidation Catalysts for Green Chemistry

Colin P. Horwitz

Glossary

Catalyst	A substance that initiates or increases the rate of a chemical reaction by reducing the activation energy for the reaction without itself undergoing any permanent chemical change.
Commodity chemicals	Chemicals that are produced on the millions of tons per year scale that are used as building blocks for more sophisticated chemicals.
Green chemistry	The design of chemical products and processes that reduce or eliminate the use and generation of hazardous substances.
Oxidation	The loss of electrons or the increase in oxidation state of a molecule, atom, or ion.
Self-oxidation	A situation where the active oxidation catalyst oxidizes itself, usually resulting in degradation and loss of activity.
Speciality chemicals	Chemicals that are produced on a relatively small scale often for a particular application.

This chapter was originally published as part of the Encyclopedia of Sustainability Science and Technology edited by Robert A. Meyers. DOI:10.1007/978-1-4419-0851-3

C.P. Horwitz (✉)
GreenOx Catalysts, Inc, 4400 Fifth Avenue, Pittsburgh, PA 15213, USA
e-mail: chorwitz@greenoxcatalysts.com

P.T. Anastas and J.B. Zimmerman (eds.), *Innovations in Green Chemistry and Green Engineering*, DOI 10.1007/978-1-4614-5817-3_9,
© Springer Science+Business Media New York 2013

Definition of the Subject and Its Importance

Green Oxidation Catalysts

The term "green catalyst" has no single definition. Currently, it is most commonly associated with catalysts that are recoverable or prepared from readily available starting materials. A truer definition, although circular, is that a green oxidation catalyst, or any catalyst for that matter, is one that conforms to green chemistry and green engineering principles. Creating a green oxidation catalyst *a priori* is a complex task because every aspect of the catalyst needs examination and to be of practical value it must provide a cost benefit to the end-user. Here, some general guidelines for what a green oxidation catalyst might be are presented. One of the foremost factors is the catalyst's elemental composition. The chart in Fig. 9.1 shows the relative abundance of chemical elements in the earth's crust. The vast majority of man-made and nature's catalysts contain metals from the middle of the periodic table. While not a hard and fast rule, organisms have evolved to effectively utilize metals at high-ambient concentrations, while those at lower concentrations tend to be toxic; exposure to iron (Fe) is likely to be much less of a concern than ruthenium (Ru) or osmium (Os), its family members in Group 8 of the periodic table [1, 2]. Catalysts containing metals with known toxicity issues like mercury or lead cannot be considered as green because the potential hazards associated with the elements unintentionally getting into the environment outweigh the benefits of the reaction. Currently, the USEPA is developing a *Framework for Metals Toxicity* that is

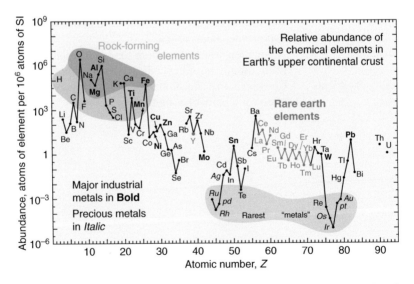

Fig. 9.1 Abundance of the chemical elements in Earth's upper continental crust as a function of atomic number (Figure courtesy of the U.S. Geological Survey)

intended to provide guidance regarding various properties of metals, such as environmental chemistry, bioavailability, and bioaccumulation [3]. Next, it has to be ascertained where and how the metal was sourced. For example, new and exciting developments are occurring using gold catalysts [4], but this needs to be tempered with the cost to humans and the environment that has long been associated with mining gold. Mining is always a hazard minimization operation. If organic molecules are used as the catalyst or as an integral part of the catalyst, it is needed to at least ask how were they synthesized, where were the raw materials sourced, and what is being done with the waste generated during their production. Others factors to include in assessing a green oxidation catalyst are determining whether the catalyst is a general or specific one (can one catalyst handle multiple steps, or do intermediates need to be isolated), the impact of the choice of oxidizing agent (e.g., potential chlorinated by-products if hypochlorite is used), the energy metrics of the process, and the potential for nonproductive use of the oxidizing agent. It is essential that these questions and others become a part of the culture of catalyst design. This is basically a restatement of a life cycle assessment (LCA) approach but focused on the catalyst itself. For now, less than ideal catalysts must be used to achieve desired goals as the means to make and implement these designs and then test them are still being developed. The challenge will be to find ways to rapidly adapt and adopt new technologies that bring us closer to the ideals.

Introduction

Catalysis -> Green Chemistry/Engineering -> Sustainable development These three concepts and practices are inexorably connected. *Catalysis* is the process whereby materials and energy are saved in a chemical transformation. *Green Chemistry and Green Engineering* are the pursuit of products and processes that reduce or eliminate hazard (physical, toxicological, ecotoxicological, and others) [5, 6]. *Sustainable development* is development that "meets the needs of the present without compromising the ability of future generations to meet their own needs" [7].

The imperative of sustainable development is leading to a wholesale rethinking of the way that humankind needs to approach its continued existence on earth. The green chemistry movement is bringing guidance as to how chemistry should be practiced now and in the future [8]. Over the last nearly two decades, there has been the gradual development and increasing acceptance of green chemistry both in academia and industry. In 2005, the Nobel Prize was awarded to Dr. Yves Chauvin, Professor Richard Schrock, and Professor Robert Grubbs for their work on the olefin metathesis reaction with the Nobel committee clearly stating that their work was a *"great step forward for "green chemistry," reducing potentially hazardous waste through smarter production."* The general concepts that define green chemistry are that the entire life-cycle of a chemical should do no harm and that all aspects of the chemistry should be sustainable. These concepts are embodied in the 12 principles of green chemistry that were first put forward by Professor Paul Anastas and Dr. John Warner

Fig. 9.2 Reaction coordinate diagram showing the different activation energies for catalyzed and uncatalyzed reactions. The catalyst opens a lower-energy pathway to the desired products

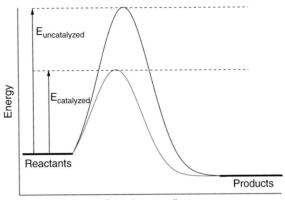

in their landmark 1998 book "Green Chemistry: Theory and Practice" [9]. These principles are augmented by the green engineering principles espoused in 2003 by Professor Anastas and Professor Julie Zimmerman [6]. When current and future chemical products and processes incorporate these principles, then chemistry will flourish for the foreseeable future.

Catalysis is at the heart of Green Chemistry as it is the means to increase efficiency and efficacy of chemical and energy resources while promoting environmental friendliness and intensifying time and cost savings in chemical synthesis [10]. A catalyst's function simply is to provide a pathway for chemicals (reactants) to combine in a more effective manner than in its absence. This is generally depicted as a reaction coordinate diagram like the one shown in Fig. 9.2. In the absence of a catalyst, heat is usually the way to overcome the energy barrier, but this increases energy consumption and often results in unwanted side reactions. A catalyst cannot make an energetically unfavorable reaction occur or change the chemical equilibrium of a reaction because the catalyzed rate of both the forward and the reverse reactions are equally affected.

The concept of catalysis traces its birth to the 1830s when Jöns Jacob Berzelius [11] rationalized the results of other investigators into the concept of "catalytic power" and coined the word catalysis [12]. Interestingly, it was Sir Humphry Davy in 1817 who first noted the catalytic properties of a metal when he described the reaction of coal gas and oxygen over platinum and palladium wires. In that report, he noted that the reaction did not proceed in the presence of copper, silver, gold, and iron. Roberts proposes in his brief history of catalysis [13] that this may have been the earliest recorded pattern of selectivity in catalysis.

Oxidation Chemistry

Oxidation chemistry – what is it and what is it good for? Oxidation is defined as the loss of electrons or the increase in oxidation state of a molecule, atom, or ion. Combustion of organic matter like coal, natural gas, or gasoline is a familiar oxidation reaction. This is a destructive or nonselective form of oxidation chemistry. Here, a "reduced" form of carbon is being taken and combined with oxygen and then oxidized to CO_2 with water as the other product. In the constructive or selective regime of oxidation chemistry, the one practiced by most chemists, the organic matter is transformed either using oxygen or some other electron acceptor into a new chemical that is of higher value. Oxidation chemistry is probably the largest branch of chemistry pursued by industry, human activity and diverse natural systems.

- Oxidation chemistry is being used when doing laundry, where bleach (sodium hypochlorite) or hydrogen peroxide is the oxidizing agent that destroys or degrades the molecules responsible for the stains on clothes.
- The society benefits from the application of oxidation chemistry at the wastewater treatment plant that treats sewage and the drinking water treatment plant that provides clean drinking water. When the chlorination of drinking water was developed, many of the water-borne pathogens that caused so much suffering and misery were removed and human health took a great leap forward. However, it has been known for many years that some of the by-products from water chlorination, which are referred to as disinfection by-products (DBPs), are human health hazards. In order to avoid DBPs, alternative oxidants such as chloramine and ozone are being implemented, but these too have some negative consequences associated with their application.
- The catalytic converter on automobiles might be considered one of the most successful and detrimental oxidation catalysts from the perspective of human health and the environment. The modern 3-way catalytic converter oxidizes carbon monoxide and unburnt hydrocarbons to carbon dioxide and water while reducing nitrogen oxides to nitrogen gas and water. This catalyst system has made the environment vastly better by reducing smog causing chemicals and eliminating the use of leaded gasoline. Among its negatives are the production of carbon dioxide that contributes to global climate change, expansion of the automobile industry because smog producing chemicals are less prevalent in automobile exhaust, loss of land as a result of road building, and the catalyst materials are precious metals (platinum, palladium, and rhodium) that are obtained by mining.
- Oxidation reactions are constantly occurring in biological systems. The cytochrome P450 enzyme in the liver detoxifies chemicals that are ingested by oxidizing them to less toxic substances. Cytochrome P450 is an exquisitely designed oxidation catalyst system that uses oxygen in a precise and controlled manner to destroy the toxin. Chemists and biologists have been fascinated by the cytochrome P450s, resulting in hundreds if not thousands of research publications describing their chemical, physical, and biological properties.
- In the environment, many different kinds of organisms are essential for the degradation of natural and unnatural substances that constantly enter soils and waterways. For example, a decaying tree in the forest might be under attack from the enzymes lignin peroxidase and manganese peroxidase in white rot fungi that

utilize hydrogen peroxide to degrade the lignin. In this case, the degraded material ends up as new fertilizer for the forest floor. Interestingly, lignin peroxidase also has been found to degrade some recalcitrant, man-made pollutants.

These examples of oxidation chemistry only scratch the surface of the field. Oxidation processes are used for odor control, bleaching of pulp for paper production, wastewater treatment, disinfection, bulk and specialty chemical production, aquatics and pools, food and beverage processing, cooling towers, agriculture/ farming, and many others. It is often noted in reviews about oxidation chemistry that catalysts would make the field substantially more efficient. In fact, there are a great many oxidation reactions that are catalytically driven. A search of the Chemical Abstracts Services database using *chemical oxidation catalyst* as the search criteria and limiting results to patents and journal article returned more than 500 entries for the year 2009, likewise the Royal Society of Chemistry's RSC Catalysis and Catalysed Reactions database resulted in more than 100 entries using *Reaction type – oxidation*, and a search of patent application titles at the US Patent and Trademark Office with *oxidation* and *catalyst* in the title yielded 230 entries since 2001 – undoubtedly, there were many more.

Oxidation reactions whether they be catalytic or not remain an enormous undertaking from the laboratory scale to industrial production. An oxidation reaction might be contemplated on a molecule that already has one or more functional groups that might be affected by the oxidation process. Thus, a successful oxidation process involves a deep understanding of inherent chemical and physical properties of the target molecule and how these might be impacted by the choice of oxidation catalyst, the oxidant, and the reaction medium [14].

The definition of oxidation gives wide latitude to what is desired from an oxidation catalyst. Indeed, one of the great challenges in oxidation chemistry is the breadth of the field. The most recent edition of *March's Advanced Organic Chemistry: Reactions, Mechanisms, and Structure, 6th Edition*, a standard reference source for the organic chemist lists six groups of oxidations that comprise 32 different types of oxidation reactions [15]. Entire libraries of catalysts have been developed to achieve a single type of oxidation process. For example, 2001 Nobel Prize winner Professor K. Barry Sharpless developed a catalytic system for the enantioselective epoxidation of allylic alcohols, Fig. 9.3 [16]. However versatile these catalysts might be within their class of oxidations, they are not generally applicable to the epoxidation of other olefin containing compounds.

Catalytic oxidation processes are used in the production of some extremely large volume industrial chemicals. For example, a cobalt/manganese catalyst with a bromine promoter is used to oxidize *p*-xylene to terephthalic acid, Fig. 9.4 [17], a roughly 44 million ton/year chemical. The terephthalic acid ultimately is used for polyethylene terephthalate, the polymer used in many drinking bottles. Large-scale chemical manufacturers know that they have to adapt their processes as feedstock resources become more limited and that they will have to comply with ever more stringent environmental guidelines [18].

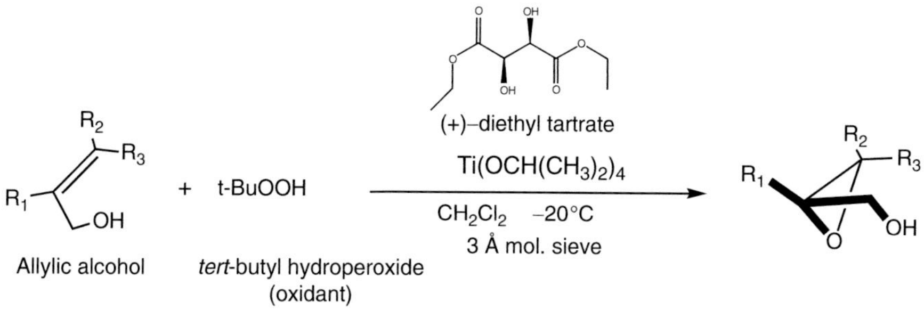

Fig. 9.3 Typical Sharpless epoxidation conditions. Substitution of the nonnatural (-)-diethyl tartrate results in oxygen addition on the upper face of the allylic alcohol

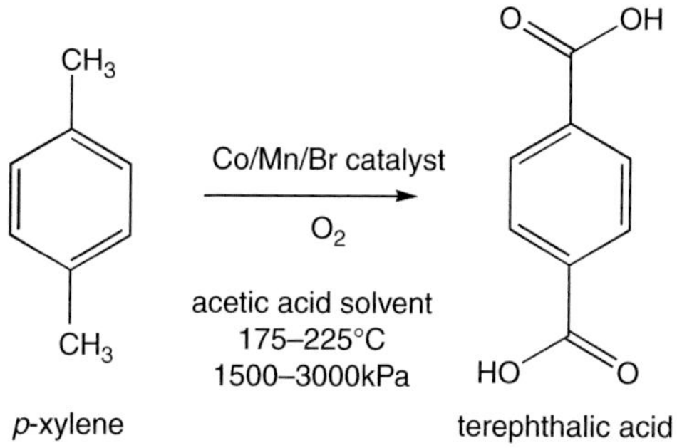

Fig. 9.4 The Amoco industrial synthesis of terephthalic acid

Finally, oxidation catalysts can play a crucial function cleaning organic pollutants from the environment. There are roughly 30,000 chemicals used on a regular basis throughout commerce, approximately 84,000 are registered with the USEPA under the Toxic Substances Control Act. Some of these ultimately find their way into the water that sustains life. The effects of many of these chemicals on human health are not well known although the President's Cancer Panel recently suggested that the role of environmental toxins (pollutants) as cancer causing agents is substantially underestimated. Fortunately, there are many technologies to remove toxic compounds from water. Oxidation catalysts can play a pivotal role in breaking down the recalcitrant compounds that are not degradable by biological means. A number of different oxidation technologies are being investigated, but only a few employ catalysts [19]. Fenton's reagent, an iron-based catalytic process that utilizes H_2O_2, is a popular option; photocatalytic processes using titanium dioxide is another approach; catalytic wet-air oxidations for waste streams containing high concentrations of organic materials is a powerful methodology, and an emerging technology is based on using oxidatively robust small molecule catalysts. The real test is determining which catalysts are green and getting them adopted as widely as possible by moving them from the chemical literature to commercial ventures.

Industrial Catalysts

Industrial, bulk, or commodity chemicals are those chemicals that are produced on a vast scale (millions of tons per year, Mt a^{-1}), often as building blocks for more sophisticated chemicals. As such, they generally cost less than \$0.50/lb. In the industrial chemical sector, there is tremendous pressure to make chemicals as inexpensively as possible – economics drives production of large-volume chemicals and the application of green chemistry principles enhances the bottom line. Catalysis plays a vital role in providing pathways to these chemicals. Table 9.1 provides examples of some bulk chemicals produced by catalytic oxidation processes on the Mt a^{-1} scale.

Many, if not most, of the organic bulk chemicals in Table 9.1 are produced from nonrenewable resources such as petroleum or natural gas so their use and preparation violate Green Chemistry principles. It is often proposed that careful stewardship of the nonrenewable resources is what is needed to sustain the chemical industry. However, if the chemical industry truly needs to continue supplying these chemicals, renewable resources will have to be used and transformations of these natural materials based on Green Chemistry principles developed [20]. This is clearly as exciting a goal for current and future chemists and biologists as the one that led prior generations to the current portfolio of chemicals [21]. It is heartening to know that substantial advances are taking place to include renewable resources among these feedstocks [22, 23]. Still, the vast scale of production of "bulk" chemicals means that it will be decades before renewable resource-derived products and the associated processes can replace current practices on the industrial scale. Until such time, catalytic oxidation of petroleum and natural gas-derived feedstocks will prevail.

The vast majority of catalysts for bulk chemicals is heterogeneous and are either of the supported or bulk type. Ease of separating the catalyst from reactants and products, thermal stability, and recyclability of the catalyst are among the most important

Table 9.1 Examples of bulk chemicals produced by catalytic oxidation processes

Chemical	Chemical precursor	Production (Mt a^{-1})	Example end uses
Terephthalic acid	p-Xylene	44	Polyethylene terephthalate (PET) drink bottles
Formaldehyde	Methanol	19	Resin component
Ethylene oxide	Ethylene	18	Polyester fiber, automotive engine antifreeze, production of polyethylene terephthalate
Propylene oxide	Propylene	8	Polyurethane systems
Cyclohexanone	Cyclohexane	6	Nylon 6,6
Phthalic anhydride	o-Xylene	5	Plasticizers
Maleic anhydride	Butane	2	Polyester and alkyd resins, lacquers, plasticizers, copolymers, and lubricants

reasons for using heterogeneous systems. Virtually, all of the heterogeneous catalysts contain transition metals from the middle of the periodic chart as the catalyst or as additives that promote the catalytic reaction. Chemists have done an amazing job over the years in understanding how these metals function and how to manipulate their activity and reactivity. However, some are also well known for their toxicity [24] and for the environmental [25] and societal damage that extracting them from the earth causes. Nevertheless, for the foreseeable future, metal-containing catalysts will remain the first choice for the chemical industry. Fortunately, there are many catalyst recycling businesses, most of which claim to recover the metals in an environmentally friendly manner.

Challenges in Industrial Oxidation Catalysis

There have been many recent advances in catalysts and catalytic reactions for some of the chemicals listed in Table 1. Excellent recent reviews cover selected "dream" reactions for oxidative catalysis that are currently under investigation or have become commercial processes [22, 26]. The incredible volume of activity in the industrial oxidation area makes it impossible to assess the greenness of each catalyst here. To provide illustrative examples, two of these processes, oxidation of propene to propylene oxide and adipic acid synthesis, will be examined in some detail with respect to how the catalysts being evaluated conform to Green Chemistry/Engineering principles.

As already mentioned, the chemicals in Table 1 must be produced with an eye to economics. This leads to the use of oxygen (or air) as the primary oxidant. Oxygen is certainly a green reagent; however, there is risk of explosion and fire, so risk management in terms of oxygen concentration and equipment construction always must be considered. Hydrogen peroxide is the next best oxidant as its by-product is water. However, its cost can be prohibitive especially when the reaction consumes multiple equivalents of hydrogen peroxide. Hydrogen peroxide is also inherently a water-based process, so the chemistry must be compatible with this type of environment.

Propene to Propylene Oxide

The first reaction described is conversion of propene to propylene oxide, which is becoming recognized as a green process. Propylene oxide is 6 Mt a^{-1} bulk chemical used primarily for the production of polyether polyols, polyurethane precursors, as well as for propylene glycol and glycol ethers. The ideal reaction is direct epoxidation of propene with oxygen, but when allylic hydrogen atoms are present, as they are in propene, selectivity becomes a factor as over-oxidation occurs. The chlorohydrin process for propylene oxide production uses chlorine gas in the reaction, Fig. 9.5, and it is this process that is targeted for replacement. In addition to using

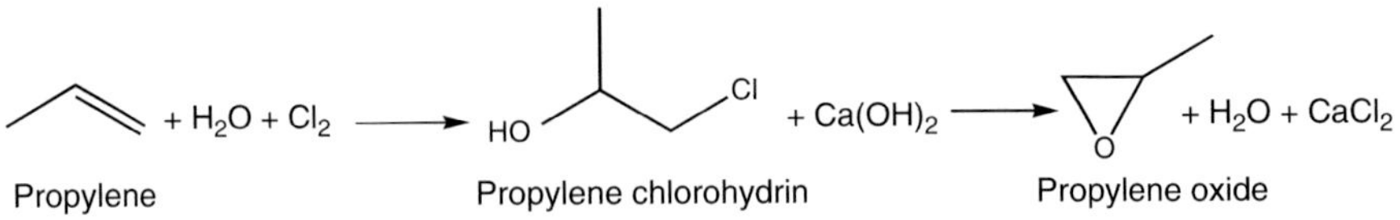

Fig. 9.5 Chlorohydrin process for producing propylene oxide

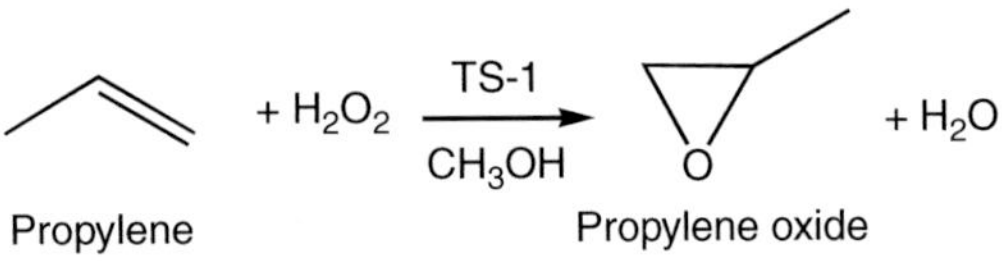

Fig. 9.6 Direct oxidation of propylene to propylene oxide using a catalyzed HPPO process

the highly toxic and dangerous gas chlorine, the chlorohydrin process requires for each ton of propylene oxide, 1.4 t of chlorine, and 1.0 t of calcium hydroxide. In terms of waste, approximately 2.0 t of $CaCl_2$ are obtained and a large volume of salt-containing wastewater is generated [18].

Catalytic oxidations of propylene to propylene oxide using H_2O_2 or cumyl hydroperoxide are recent commercial successes. Work began in the early 1990s on catalytic oxidation of propylene using H_2O_2 and a titanium silicalite (TS-1) as the catalyst [27–29]. The generalized reaction is shown in Fig. 9.6. The practical potential for this chemistry is enormous since it is direct conversion to product, only one equivalent of H_2O_2 is needed, and the waste product is water. This chemistry has evolved into the so-called HPPO process (Hydrogen Peroxide Propylene Oxide), where propylene is oxidized to propylene oxide under mild reaction conditions. It is run at $<60°C$ in aqueous methanol as solvent and a propylene pressure of ca. 4 bar (3.95 atm or 58 psi). The methanol is recycled. Selectivity to product is $>90\%$ and can be as high as 97% based on hydrogen peroxide [30].

An Evonik–Uhde partnership has licensed the TS-1/H_2O_2 technology to the Korean firm SKC of Seoul that commissioned a 100,000 t/year propylene to propylene oxide plant in 2008. This plant also uses a platinum/palladium-based catalyst technology developed by Headwaters Technology Innovation that makes H_2O_2 directly from H_2 and O_2. The catalyst, NxCat™ lowers energy use, uses renewable resources and generates no toxic waste during the H_2O_2 production. This technology was the recipient of a 2007 USEPA Presidential Green Chemistry Challenge Award. A BASF – Dow Chemical partnership recently commissioned a 300,000 Mt a^{-1} propylene to propylene oxide plant using a titanium silicate oxidation catalyst for H_2O_2 activation. According to BASF/Dow, wastewater is reduced by 70–80%, and energy usage by 35%, compared with existing PO technologies. The BASF-Dow Chemical process received a 2010 USEPA Presidential Green Chemistry Challenge Award.

Sumitomo has commercialized a process that also utilizes a proprietary titanium-silicate catalyst. The Sumitomo method is interesting as it couples the non-catalytic

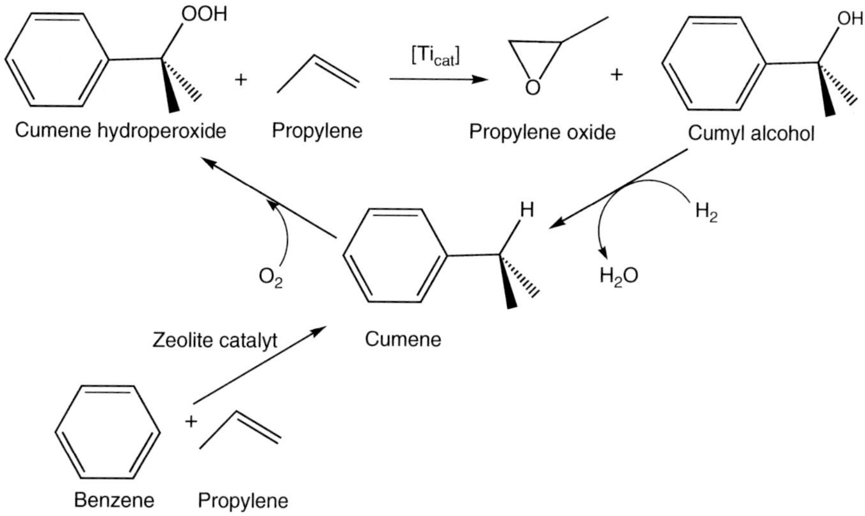

Fig. 9.7 Sumitomo catalytic oxidation of propylene to propylene oxide using a proprietary titanium-based catalyst Ti_{cat}

oxidation of cumene to cumene hydroperoxide with the catalytic oxidation of propylene to propylene oxide, Fig. 9.7. This oxidation route is viable in part because of the high stability of cumyl hydroperoxide that results in high recyclability of cumene.

TS-1 is a crystalline zeolite type material containing tetrahedral TiO_4 and SiO_4 units with a three-dimensional system of channels. It is particularly good at activating H_2O_2, and it has been a significant addition to the industrial catalysis field where many reactions now use it alone or in modified form [31]. TS-1 contains less than 2% titanium, and these appear to be the catalytic sites as the Ti-free material does not show a comparable activity.

Current technologies to prepare TS-1 and many other mesoporous materials can be categorized as being modestly green. TS-1 is synthesized by controlled hydrolysis of a synthesis gel comprised of a tetraalkyltitanate, tetraalkylsilicate, and a tetrapropylammonium hydroxide [32]. An excellent recent review that tackles many of the green issues associated with sol-gel methods for preparing silicates is available [33]. For TS-1, the proportions of the different reagents are important for controlling catalyst activity. From a green catalyst perspective, the mass yield of catalyst is relatively low. In 1992, Professor Roger Sheldon [34–37] codified the E-factor ($E =$ Environmental) which takes into account the actual amount of waste produced in a given process and includes reagents, solvent losses, process aids and fuel. In simplest terms, it is the ratio of kilograms of by-product/kilograms product [38]. The TS-1 synthesis from a 2005 Degussa patent US 6,841,144 states:

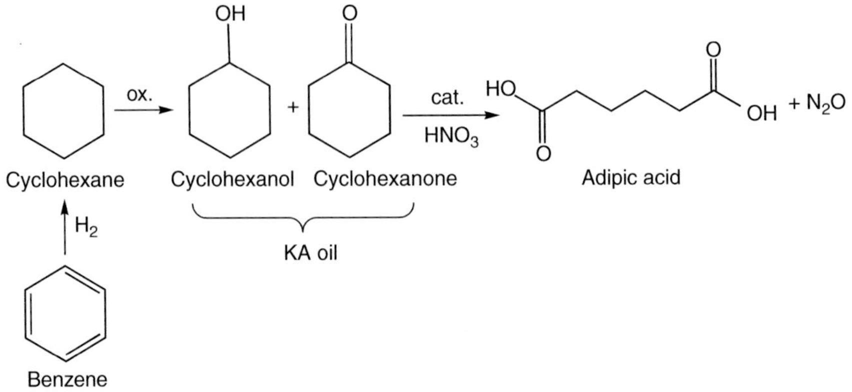

Fig. 9.8 Adipic acid synthesis from benzene. Nitric acid (HNO$_3$) is the oxidant and generates N$_2$O a greenhouse gas. The catalyst (cat.) is homogeneous, containing Cu^{II} salts and an ammonium metavanadate

"This synthesis method resulted in a consumption of *tetra-n*-propyl ammonium hydroxide (40% solution) of 1.6 kg per kg TS-1, a yield of 110 g TS-1 per kg synthesis gel …."

This results in an *E*-factor for TS-1 of at least 8 (*E*-factor = (1.0 kg synthesis gel − 0.110 kg TS-1)/(0.110 kg TS-1), and it could be higher if purification methods are needed. During the hydrolysis process, the alcohol formed must be distilled and then either recycled or removed as waste. Finally, tetraalkyltitanates and tetraalkylsilicates are flammable and thus require careful handling. Attention is being given in many laboratories that prepare these materials to find alternative, efficient synthesis methods [39].

Adipic Acid Synthesis

There are significant efforts underway to change adipic acid production to a catalytic process using what appear to be mostly green catalysts. Adipic acid is used in the synthesis of nylon 6-6. An overview of the current practice for adipic acid synthesis is given in Fig. 9.8. It is essentially a two-step process, where cyclohexane is first oxidized to a mixture of cyclohexanol and cyclohexanol (KA oil) [40], and then this mixture is oxidized with nitric acid, HNO$_3$, to adipic acid. The oxidation is catalyzed with a homogeneous catalyst containing Cu^{II} nitrate and ammonium metavanadate [18]. The objective is to remove HNO$_3$ as the oxidant because it is a serious hazard from a chemical safety point of view, and its reduction product N$_2$O is a powerful greenhouse gas (300 kg of N$_2$O are produced per ton of adipic acid). Most adipic acid producers have means to decompose, trap, reuse, or recycle the N$_2$O.

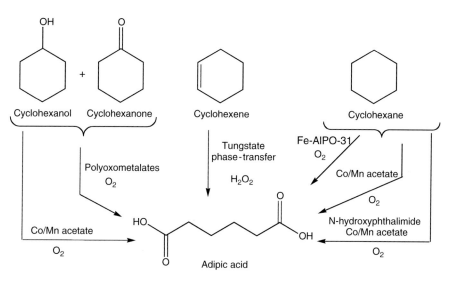

Fig. 9.9 Potential new catalytic routes to adipic acid. Information for figure adapted from Cavani [18]

Depicted in Fig. 9.9 are some of the catalytic routes and catalysts being contemplated. An excellent recent review detailing efforts ongoing in this area is available [19]. Among the limitations of the catalysts are poor conversion levels, moderate to low selectivity, challenging processability, and some demand a specific oxidizing agent. From a pure materials perspective, the $Fe^{III}AlPO$-31 catalyst is the most green. It is based on aluminophosphate molecular sieve, where Fe^{III} substitutes some of the Al^{III} in the framework. Interestingly, other alumino-phosphate structures are not as selective as AlPO-31. It is suggested that AlPO-31 has narrow pores which easily absorb cyclohexane, but intermediate oxidation products are less easily desorbed than adipic acid [41]. Benzene, a petroleum product, is the precursor to the cyclohexane derivatives used in each synthesis. Ultimately, it will be necessary to replace or sustainably produce the benzene or preferably the KA oil. There are efforts particularly in the biocatalysis field to produce adidpic acid free of a benzene source [42].

Economic viability will be the ultimate driving force for adopting new adipic acid technologies. For example, cyclohexene oxidation with the tungstate catalyst Na_2WO_4 in Fig. 9.8 is elegant chemistry that currently is not cost competitive with other technologies [43]. Na_2WO_4 might be considered a green catalyst as it has low toxicity toward fish [44], but speciation toward the metatungstate $(3Na_2WO_4 \cdot 9WO_3)$ increases toxicity [45]. The cyclohexene is catalytically oxidized with aqueous 30% H_2O_2 to adipic acid in a one-step process in the presence of the catalyst and the phase-transfer agent $CH_3(n\text{-}C_{18}H_{17}N)_3HSO_4$, there are no added solvents. Na_2WO_4 has good water solubility, but cyclohexene does not, so the phase-transfer agent is used to introduce the catalyst and cyclohexene into the same phase. Reaction temperatures are mild 75–90°C, the cyclohexene is converted directly to analytically pure, crystalline adipic acid in almost quantitative yield, and H_2O_2 consumption is nearly ideal (theoretical consumption is 4.0 mol of H_2O_2 per mol of cyclohexene and actual is 4.4 mol H_2O_2). Nevertheless, at current H_2O_2 prices, the process is not economically viable because of the relatively large amount of H_2O_2 needed [46, 47]. This example raises the point that a catalyst cannot change the stoichiometry of a reaction, so the fundamental chemical economics of a process can be determined without consideration of the catalyst. An integrated process as in polypropylene oxidation, where H_2O_2 cost is controlled, or one in which the cyclohexene cost is much less than the other starting materials might overcome economic issues, and then the issue will be changing from an entrenched multi-Mt a^{-1} process to a new one.

Long-term Industrial Challenges

Synthesis gas, which is carbon monoxide and hydrogen gas, is the current economically viable route for converting methane into commercially useful fuels, gasoline and diesel, and chemicals, principally methanol. The methanol is subsequently converted to olefins as building blocks for other chemicals [48]. The methane can come from natural gas deposits but, from a sustainability perspective, the methane produced from the decay or fermentation of organic material, biogas is the more desirable source. Steam reforming of methane, Eq. 9.1, currently is the large-scale process for natural-gas conversion to synthesis gas. This is an energy intensive process because the reaction is energetically uphill by a substantial amount ($\Delta H_{298\ K}$ = +206 kJ/mol). Temperatures in the range 700–1,000°C and steam under 3–25 bar pressure are used. The problem could be solved by oxidation, Eq. 9.2, as this is an energetically favorable process ($H_{298\ K}$ = −36 kJ/mol). Oxidation catalysts for synthesis gas production are receiving substantial research efforts in industry and academia. The main stumbling block at this time is further oxidation (over-oxidation) of the CO to CO_2 [49].

$$\text{Steam reforming } CH_4 + H_2O(+\text{heat}) \rightarrow CO + 3H_2 \tag{9.1}$$

$$\text{Partial Oxidation } CH_4 + \frac{1}{2}O_2 \rightarrow CO + 2H_2 \ (+\text{heat}) \tag{9.2}$$

Selective catalytic oxidation of hydrocarbons to oxygen-containing chemicals like alcohols, aldehydes, and carboxylic acids are also targets [50]. This would allow the direct conversion of light hydrocarbons from refinery operations to useful chemicals or precursors to more complex products. Hydrocarbons generally are unreactive, except in the presence of metal catalysts at high temperatures and pressures or in the presence of specifically designed reaction media [51]. With clear economic and energy metrics for direct conversion to oxygenates, intense research efforts are underway to make better homogeneous and heterogeneous catalysts. The use of fossil fuel-based products is not a sustainable approach to chemical manufacturing, but the energy and resource savings make this direct conversion an acceptable interim measure until renewable resources take their place.

The oxidative dehydrogenation (ODH) of alkanes to alkenes is another oxidation reaction that continues to receive considerable attention, Eq. 9.3 [52]. Ethylene is the largest production chemical at about 95 million tons/year. It is used directly in polymers like polyethylene, chlorinated for polyvinyl-chloride production, combined with benzene to produce ethylbenzene, the precursor of polystyrene, and oxidized to other smaller volume chemicals [53]. Currently, ethylene is manufactured by steam cracking of napthas from petroleum feedstocks. Like the synthesis gas reaction, this is energy intensive because the thermodynamics of the process are unfavorable, and so high temperatures, 800–850°C, are used to overcome the barrier. The high temperatures can result in cracking of the hydrocarbons to smaller fragments and reduction to coke [54].

$$2CH_4 + O_2 \rightarrow C_2H_4 + 2H_2O \tag{9.3}$$

In an ODH process, reaction temperatures can be substantially lowered because the added O_2 reacts with the hydrogen from the alkane producing water that provides the thermodynamic driving force for the reaction [53, 55]. If the technology could be developed for production of butadiene from butane, for example, this would have high commercial value since butadiene is used in the manufacture of automobile tires, synthetic rubbers, specialty polymers, and chemicals [54]. Despite the extensive literature describing the oxidative dehydrogenation of alkanes, information about the individual reaction steps comprising alkene formation is somewhat limited. Catalyst composition focuses around a few transition metals, nickel, vanadium, and molybdenum, with different additives or promoters such as potassium and calcium, and a host of different support materials each of which can influence the chemistry. Cerium oxide-based materials are showing significant promise in ODH chemistry, and these can be considered as green catalysts as long as no inherently toxic metals are used [52]. However, like many other catalyst systems and catalyst reactions, there is still much work to be done to understand how catalyst composition and structure relate to activity and selectivity [50].

Specialty Chemical Catalysts

The birth of the synthetic chemical industry can be traced to 1856 when William Henry Perkin oxidized aniline to mauveine. Under the direction of August Wilhelm Hofmann at the Royal College of Chemistry in Oxford Street, London, Perkin had been studying the reactions of various constituents in coal tars. Hofmann's research focused on manipulating the organic chemicals present in coal tars, a by-product of the gas-manufacturing plants that were lighting the streets of London. Perkin was directed to synthesize the antimalarial drug quinine from coal-tar components.

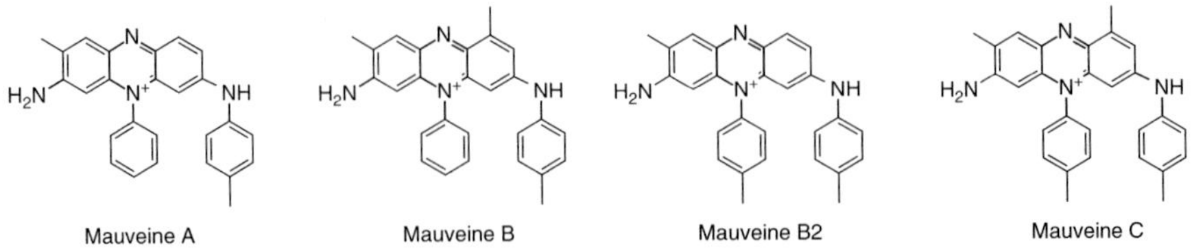

Mauveine A   Mauveine B   Mauveine B2   Mauveine C

Fig. 9.10 Major components of the dye produced by Perkin upon oxidation of aniline, *o*-toluidine and *p*-toluidine

In Perkin's home laboratory, he attempted a synthesis of quinine by oxidizing allyltoluidine with potassium dichromate. He expected a colorless product but rather obtained a dark reddish-brown solid. A repeat of this experiment using aniline rather than allyltoluidine resulted again in a dark black solid. Perkin, as any well-trained synthetic chemist would do, dug into this complex mixture and isolated a purple component, which he quickly found made an excellent fabric dye. After testing the dye in various dyehouses and achieving outstanding results, Perkin applied for and ultimately received a patent in for the synthesis of dyes. His procedure for preparing the dye powder is described in the patent as follows:

> I take a cold solution of sulphate of aniline, or a cold solution of sulphate of toluidine or a cold solution of xylidine, or a mixture of any one of such solutions with any others of other of them, and as much of a cold solution of a soluble bichromate as contains base enough to convert the sulphuric acid in any of the above-mentioned solutions into a neutral sulphate. I then mix the solutions and allow them to stand for ten or twelve hours, when the mixture will consist of a black powder and a solution of a neutral sulphate. I then throw this mixture upon a fine filter and wash it with water till free from the neutral sulphate. I then dry the substance thus obtained at a temperature of 100 degrees centigrade, or 212 degrees Fahrenheit and digest it repeatedly with coal-tar naphtha until it is free from a brown substance which is extracted by the naphtha. I then free the residue from the naphtha by evaporation and digest it with methylated spirit or any other liquid in which the colouring matter is soluble, which dissolves out the new colouring matter. I then separate the methylated spirit from the colouring matter by distillation, at a temperature of 100 or 212°F.

Buoyed by successful trials at various dyehouses, in 1857 Perkin, his father and brother established the company Perkin and Sons at Greenford Green, near Harrow, to manufacture and supply the dye. The success of the scale-up and subsequent commercialization process for Perkin and Sons is well described in many accounts of Perkin's life and career as a chemist.

Interestingly, it was not until 1994 that the structural components that comprised mauveine were conclusively identified [56, 57]. These are shown in Fig. 9.10. The components are a combination of aniline and o-toluidine and p-toluidine. Thus, the aniline that was supplied to Perkin contained toluidine impurities. Performing reproducible synthetic reactions on aniline that had varying amounts of impurities must have been challenging, especially with the crude apparatus available at the time. This makes Perkin's achievements even more remarkable.

With no disrespect to Perkin, a review of the aniline oxidation reaction in terms of green chemistry principles, reveals that very little of it would be described as "green." Aniline itself is still produced by "nongreen" methods and from nonrenewable resources, although there are ongoing efforts to green its preparation [58, 59]. Furthermore, mauveine is synthesized by the stoichiometric addition of the oxidant potassium dichromate to an impure solution of aniline in acidic water. Here, the potassium dichromate oxidizes the aniline, and its reduced form becomes a by-product of the reaction that must be separated and disposed of safely (at least in modern times). Today, there are green catalytic oxidation alternatives to stoichiometric reagents like dichromate, so it is possible that mauveine can be prepared catalytically. In fact, there is a suggestion of such possibility in a 1935 publication when a peroxidase enzyme is used with H_2O_2 to oxidize aniline, and it is proposed that mauveine is one of the products [60].

Specialty Chemicals

The development of synthetic organic chemistry began well before Perkin's time and has resulted in an enormous body of work. Today, there are numerous compendia that the chemist can turn to for insights into how to perform their particular reaction. Chemists are very skilled at devising new target molecules on paper and then mounting what frequently becomes a complex synthetic campaign covering many reaction steps and several years to make them. In fact, it can be stated with reasonable certainty that if a molecule can be envisioned, then there is a synthetic pathway that will successfully lead to its preparation. The reaction sequence often has steps with high E-factors and others having low ones. During the preparation, the chemist could be considering aspects of chemoselectivity (conversion of one functional group in the presence of another), regioselectivity (preference of reaction at one of two or more possible reaction sites), or stereoselectivity (the production of one stereoisomer of the product in preference to the other). This synthetic complexity accounts for the vast array of catalysts that have been reported and the libraries of catalysts developed to achieve a subset of reactions within a larger class. Catalysts used for bulk chemical production often employ more forcing conditions with respect to temperature and pressure than are available in many synthetic laboratories, and they tend not to have the kinds of selectivities desired by the specialty chemical producer.

In the specialty chemical area, many oxidation catalysts are based on transition metals surrounded by more or less complicated organic structures which are referred to as ligands [61]. The design of the ligands is a difficult endeavor because once reaction is initiated, then all species in solution, including the catalyst, are targets for oxidation. In an ideal scenario, the oxidation process is selective toward the target of interest. However, it is more often the case that the ligand is oxidized at nearly the same rate as the target molecule. This has lead to the development of complex protection schemes for ligands (and associated reaction pathways to make them) or alternatively the use of low catalyst to substrate ratios (higher than desirable quantities of catalyst). While both means ultimately lead to the desired oxidation product, neither exemplifies Green Chemistry principles. Nevertheless, a large amount of valuable information and useful oxidation reactions have been achieved with these systems. The most ideal approach then is to design ligands that are inherently stable under oxidizing conditions. This should result in the use of less catalyst and bring the overall process closer to Green Chemistry principles [62].

Oxidation Catalysts Inspired by Biology

One of the most logical starting points to begin the design of green oxidation catalysts is to look at how nature handles the oxidation problem. Biology has had billions of years of practice perfecting its repetitive oxidation reactions, but chemists began systematically exploring the diverse field of oxidation chemistry only little more than two centuries ago. Many biological oxidation processes are catalytic and are arranged such that a single catalyst reaction center, or a combination of reaction centers, reacts with many, many target molecules before returning to their resting state until called upon to do the reaction again. Repair and replacement mechanisms for these catalysts are an important part of the biological process, so the catalytic reaction center does not need to be robust. The chemist generally does not have the luxury of mid-process catalyst repair, and replacement means adding more of the often precious catalyst to keep the reaction proceeding.

Biocatalysis

Although enzyme systems have been intensively studied for many years, over the past decade chemists, biologists, and chemical and pharmaceutical companies have begun to aggressively embrace enzymes as new engines, *biocatalysts*, for making some of the products that are used every day [63]. The beauty of enzymes is that they are generally nontoxic, they can be grown rather than synthesized in the laboratory, they have high selectivity, their turnover frequencies are often high $(10–10,000 \text{ s}^{-1})$, and their waste products should not be an environmental burden. These attributes clearly place biocatalysts at the top level in terms of green oxidation catalysts. The discoveries and applications of biocatalysis are still in their relative infancy, but bulk chemicals, pharmaceutical and agrochemical intermediates, active pharmaceuticals, and food ingredients are being made using biocatalysts [64]. As directed enzyme evolution, genomics, and the exploration of new biological sources evolve, biocatalysis is likely to play an increasing role in the chemical enterprise [65]. Still, it will be many years before biocatalysts replace some of the traditional chemical processes because they must provide a clear cost benefit over the current process, and it is likely that there will remain some or even many chemicals that cannot be made by biocatalytic means.

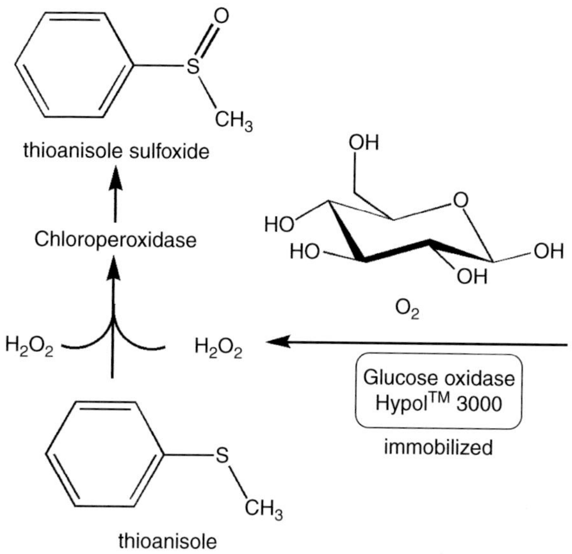

Fig. 9.11 Oxidation of thioanisole using chloroperoxidase and glucose oxidase co-immobilized in the urethane polymer derived from Hypol™ 3000

There are a plethora of interesting biocatalytic oxidations involving the oxidation of racemic alcohols, C–H bonds, Baeyer–Villager oxidations, and many others, and the details of these studies are best left to the reader to explore [66]. One clever example of a biocatalytic sulfur oxidation is provided by Professor Roger Sheldon, a leader and pioneer in the green chemistry community and movement. His group combined a heme-dependent chloroperoxidase enzyme with in situ H_2O_2 formation produced *via* the glucose/glucose oxidase reaction to enantioselectively oxidize thioanisole to the corresponding sulfoxide, Fig. 9.11. The key element in Sheldon's study was to co-immobilize the chloroperoxidase enzyme and glucose oxidase in a polyurethane foam [67]. Immobilization of the chloroperoxidase enzyme vastly improved its stability toward oxidative degradation, and in situ generation of H_2O_2 meant that supplying an external source of the oxidant (a hazard) was avoided. In a broader sense, this work brings together 11 of the 12 green chemistry principles – the only one not embodied is real-time analysis for pollution prevention.

Biocatalysis has had to overcome many obstacles during its relatively brief development time. Among the concerns voiced regarding its viability are: (a) limited substrate specificity, (b) availability of enzymes, (c) enzyme stability, and (d) the need for cofactors to regenerate the active enzyme. All of these have been addressed successfully by creative and insightful scientific approaches. One of the remaining obstacles in biocatalysis remains water demand for production of the enzyme and its use. In one recent example, an analysis of the oxidation of styrene to enantiopure (S)-styrene oxide styrene using three chemical and one biocatalytic approach revealed that the biocatalytic route required from twofold to nearly 40-fold more demand in water [68].

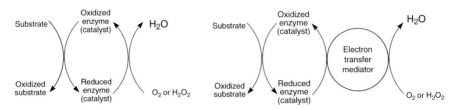

Fig. 9.12 Biomimetic catalysis based on (*left*) direct reoxidation of the enzyme (catalyst) or (*right*) electron-transfer mediator reoxidation of the enzyme (catalyst) by O_2 or H_2O_2 [71]

Biomimicry

The area of biomimicry is the examination of all aspects of nature, including its systems, processes, and elements, in order to solve human problems sustainably. A bit more narrow focus for the discussion here is biomimetics, which uses nature to inspire the creation of materials and products by reverse engineering. A biomimetic catalyst is a catalyst that carries out a reaction in a mode of action that resembles a natural enzyme. A generalized scheme for biomimetic reactions is shown in Fig. 9.12. Biomimetic oxidation reactions, especially those catalyzed by transition metal complexes, have provided a vast amount of scientific information with regard to structure, function, thermodynamics, kinetics, mechanism, synthesis, and enhanced or new analytical and spectroscopic methods [69, 70].

The biomimetic chemist has the advantage in the design of a green catalyst as nature mostly limits its metals choices to those that are readily available in the geosphere and also those that are nontoxic. Manganese, iron, and copper probably are the dominant metals involved in biological oxidation processes, although vanadium and others play prominent roles as well. As can be seen in Fig. 9.1, these are generally abundant metals. In addition, they are able to readily cycle through the electron-transfer steps that are needed for the oxidation process. Excellent reviews, books, and multivolume sets exist around the topic of biomimetic chemistry, so this subject will not be discussed in detail.

Nature's oxidations generally employ transition metals held in place by organic ligands. As mentioned earlier, one of the main problems with the design of oxidation catalysts is to avoid oxidative degradation of the ligand structure. Nature avoids the ligand degradation by elaborate design around the catalytic center to control access to it for only those chemicals that need to be oxidized, careful timing of the process, turning on and off the oxidation as needed, and living systems can rapidly repair and replace damaged catalytic centers. With few exceptions, the issue of ligand oxidation has hampered the development of highly successful commercial biomimetic oxidation catalysts. Nevertheless, even these less-than-ideal catalysts can be considered as green catalysts because they use nature's elements, and they can be useful under certain circumstances.

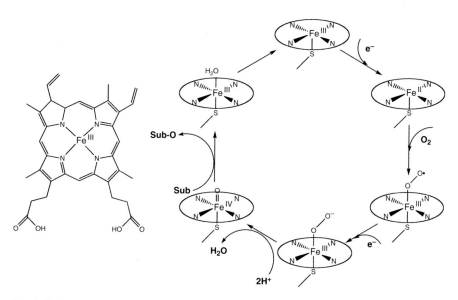

Fig. 9.13 Prosthetic heme-b group of the cytochrome P450 enzyme family and the catalytic cycle of the enzyme (Sub is the Substrate)

Cytochrome P450

The modeling of the enzyme system known as cytochrome P450 is vast both in terms of structural models and kinetic and mechanistic studies. Cytochrome P450 is a family of heme–iron enzymes, more than 2,000 are known, that activate molecular oxygen to catalyze the oxidation of organic substances incorporating one of the oxygen atoms in the substrate and the other oxygen is reduced to water, Eq. 9.4. Figure 13 shows the prosthetic group (prosthetic group is the tightly bound, nonprotein portion of an enzyme) that is conserved between the cytochrome P450's and the chemistry associated with its function. The heme-iron makes an ideal green oxidation catalyst as iron is the central metal and the heme ligand is relatively simple. In their natural environment, the P450 enzymes oxidize a large array of endogenous and exogenous compounds and most can oxidize multiple substrates. They are known to catalyze hydroxylations, epoxidations, various dealkylations, nitrogen oxidations, sulfur oxidations, and dehalogenations. Cytochrome P450 is the line of defense in the liver that metabolizes drugs, toxic compounds, and other metabolic products.

$$\text{Substrate} + O_2 + 2e^- + 2H^+$$
$$\xrightarrow{\text{cytochrome P450}} \text{Substrate} - O + H_2O \tag{9.4}$$

Even though they are exquisite in their content and fascinating in their scientific design, the kinetic and mechanistic details regarding the modeling of cytochrome

P450 are beyond the scope of this contribution [72, 73]. The key elements of the mechanism are the binding of O_2 to the iron(II) center of the prosthetic group, which is then followed by breaking of the O-O bond and formation of the active oxidant an iron-oxo species. It is this iron-oxo species that has particularly fascinated chemists and biologists for years and many attempts have been made to characterize it. Some of the issues that make its characterization difficult are its rapid reactivity and ligand decay by oxidation.

The first reports related toward mimicking P450-type catalysis were done with 5,10,15,20-tetraphenylporphyrin [(TPP)FeCl]. This catalyst suffered from rapid self-oxidation and, as a result, low efficiency [74]. Since then, considerable efforts have been invested in preparing synthetic porphyrin ligands for use as oxidation catalysts. Bulky substituents and electron withdrawing groups have been appended onto the porphyrin ligand to inhibit formation of unreactive oxo-bridged dimers and decrease the rate of self-oxidation. These approaches are somewhat successful but often rely on using functional groups like chlorine or fluorine, which severely lower the relative greenness of the catalysts as [75] many chlorinated compounds are toxic and fluorinated ones are environmentally persistent. The approach of supporting the catalysts on solids, resins, and polymers provides the catalyst more resistance toward intermolecular oxidations, but they cannot lessen the rate of intramolecular oxidation. Furthermore, many of the reactions are too slow to be of practical value [76]. Overall, synthetic porphyrins cannot be considered as ideal green oxidation catalysts because their synthesis is relatively complex, yields tend to be low, and despite the many attempts to increase their functioning lifetimes, they still remain low. Nevertheless, they may have a role to play if they are the best or only current means for oxidizing a substrate of particular value or importance [77–79].

Other Biomimetic Catalysts

An entire branch of biomimetic oxidations has developed using transition metals such as palladium, ruthenium, osmium, and others as the catalyst [71, 80]. In many cases, the catalysts are relatively simple salts of the transition metal because the problem of ligand oxidation has not been solved. Acetoxylation of cycloalkenes, oxidative carbocyclization of allene-substituted olefins, aerobic oxidation of primary and secondary alcohols, dihydroxylation of alkenes, and others are among the wide array of reactions that have been achieved using these precious metal catalysts.

A classic example using nonbiological metals is the Wacker process, Fig. 9.14, which was reported in the 1950s. This is a very well studied reaction that uses palladium to oxidize the organic substrate and copper to reoxidize the palladium. Since palladium metal is oxidized with great difficulty by O_2, this gives rise to the need for the copper portion of the catalytic cycle. The Wacker oxidation is an important industrial process and the details of its mechanism still fascinate chemists.

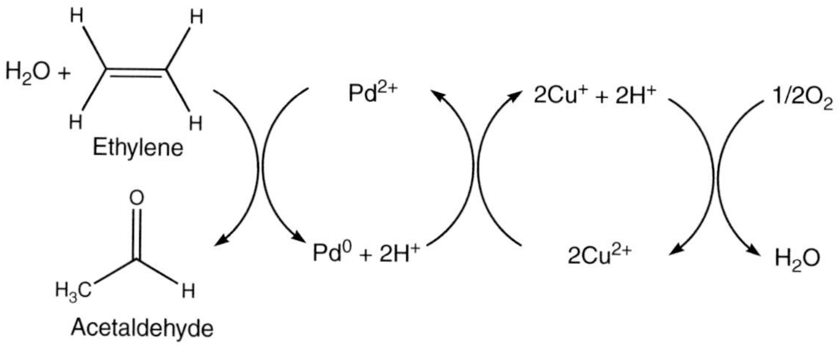

Fig. 9.14 The Wacker process for olefin oxidation using palladium as the oxidation catalyst for the olefin and copper as the co-catalyst to reoxidize the palladium

Oxidations based on these precious metals are only biomimetic in terms of mechanism as nature does not choose these metals in its chemistry. In the strictest sense then, they cannot be considered as green catalysts because they use metals that are rare, osmium is less than 0.05 ppb in the continental crust, they are toxic [81], and they are derived from mining operations that severely impact the environment. The "problem" is that these are important reactions that are an integral portion of the synthetic chemists' arsenal for the preparation of often valuable molecules. Replacement of some of these catalysts by green alternatives provides a great and exciting objective and the potential for significant economic benefits.

Polyoxometalate (POM) Catalysts

Heteropolyacids (HPAs) and the closely related polyoxometalate (POM) compounds have become valuable tools in the collection of the synthetic chemist for catalytic oxidation processes [82]. These compounds are derived from virtually any transition metal ion, oxygen, and main group elements like aluminum, phosphorous, and silicon. They are prime candidates for oxidation catalysts as they are stable toward self-oxidation because they contain no organic ligands and the transition metal and main-group elements are usually in their highest oxidation state. POMs generally have dozens of atoms arranged in well-defined structures. Examples of some classes of POMs are given in Fig. 9.15; their nomenclature is well developed [83].

A significant advantage of the POMs over many other potential oxidation catalysts is that they can self-assemble from simple, individual components by pH control. Complex molecules are thus prepared in only one synthetic step, which from a green chemistry perspective is ideal. However, in order to maintain the discrete structure during the oxidation reaction, careful pH control throughout the application might be necessary [84]. Professor Craig L. Hill has pioneered the use of POMs as oxidation catalysts. POMs have versatility well beyond oxidation chemistry and good reviews can be found on their synthesis, properties, and applications [85, 86].

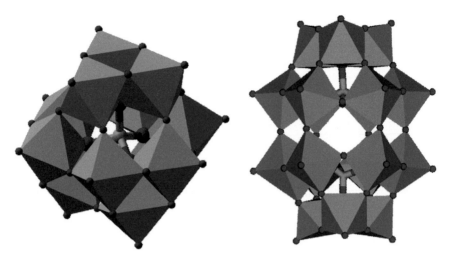

Fig. 9.15 Two polyoxometalate structures. The left is the Keggin type and the right is the Dawson. The general formula for the Keggin type is $XM_{12}O_{40}n^-$ and the Dawson type is $X_2M_{18}O_{62}n^-$, where X is typically S or P and M is Mo or W

POMs can oxidize target molecules either as electron-transfer oxidants or direct oxygen atom transfer oxidants [87]. The electron-transfer pathway depicted in Scheme 1 using the compound $H_5PV_2Mo_{10}O_{40}$ as an example is the more common [88]. In the first step, the substrate, $SubH_2$, transfers two electrons to the polyoxometalate to yield an oxidized product, Sub_{ox}, a reduced polyoxometalate ($[PV_2Mo_{10}O_{40}]^{-7}$), and two protons. In the second step, $[PV_2Mo_{10}O_{40}]^{-7}$ is oxidized by O_2. The reduced O_2 product reacts with the protons generated in the first step producing water. Here, the molecular oxygen functions as a two-electron/two-proton acceptor and the chemistry type is referred to as oxidase pathway. This mode of oxidation means that the POM must have sufficient oxidizing power such that it can remove one or two electrons from the substrate. It is often the case that organic molecules are hard to oxidize by electron-transfer pathways, so POMs must have high oxidation potentials.

Catalysis with POMs can be considered as green in the sense that they can use O_2 (or H_2O_2) for the oxidation of alcohols, amines, and phenols, and sulfur-containing compounds. In addition, their solubility can be manipulated so that they can be soluble in a variety of "green" solvents [88]. Examples are available for catalysis in aqueous biphasic reaction media, the nontoxic- and nonvolatile-solvent polyethylene glycol, supercritical CO_2 [89], and perfluorinated solvents [90]. Fluorous solvents have many positive attributes, but they are environmentally persistent which limits their greenness [91].

A principle drawback to POM-based catalysts is that they are not efficient reservoirs of oxidizing equivalents. Most POMs have molecular weights far in excess of 1,000 amu (the cation component is often "ignored" when discussing a POM, but it should not be as it can add substantially to the overall molecular

$$\text{Oxidation: SubH}_2 + [\text{PV}^{\text{V}}_2\text{Mo}_{10}\text{O}_{40}]^{-5} \rightarrow \text{Sub}_{\text{ox}} + 2\text{H}^+ + [\text{PV}^{\text{IV}}_2\text{Mo}_{10}\text{O}_{40}]^{-7}$$

$$\text{POM Regeneration:} [\text{PV}^{\text{IV}}_2\text{Mo}_{10}\text{O}_{40}]^{-7} + 2\text{H}^+ + \tfrac{1}{2}\text{O}_2 \rightarrow \text{H}_2\text{O} + [\text{PV}^{\text{V}}_2\text{Mo}_{10}\text{O}_{40}]^{-5}$$

Scheme 1 Catalytic cycle for substrate oxidation by a polyoxometalate. The polyoxometalate catalyst is regenerated by a proton coupled two-electron transfer.

weight of the catalyst), and they contain many metal centers in high oxidation states, yet each POM only will participate in a 1 or 2 electron transfer with the substrate [92]. The POM in Scheme 1 has 12 transition metal ions, but only the two vanadium ions take part in the electron-transfer process. Even though in many cases high substrate to POM molar ratios are used in the catalytic process and impressive turnover numbers and lifetimes have been reported [93], the Mass Intensity (MI) index (MI = the total mass involved during the reaction/the mass of the final product) is low [94]. On the laboratory or preparative scale, the MI is not a significant issue, but it can become a problem as the synthetic process is scaled. Still, POMs are an excellent choice for many applications because they are oxidatively robust and often easily recovered.

Epoxidation Catalysts

Epoxides are key synthetic intermediates for a large number of specialty chemicals and pharmaceuticals and a catalytic method for generating them with high enantioselectivity opens doors for many more synthetic manipulations. A search of the Royal Society of Chemistry database using the term *epoxidation* and limited to the journal *Green Chemistry*, returned more than 120 different entries (1999–2010). Most are catalytic processes, utilize H_2O_2 as the oxidant, and they cover a wide range of types of materials and methods. Each report generally focuses on the epoxidation of a narrow set of compounds, and it is not clear how widely applicable they will become. Nevertheless, the choices in catalysts fall clearly on the green side. A non-exhaustive list includes titanium on silica, enzymes, polyoxometalates, and supported transition metal complexes. In this contribution, two of the most well-known epoxidation catalysts are examined.

Sharpless Epoxidation

A discussion of oxidation catalysts is not complete without describing one of the most important breakthroughs in stereoselective synthesis. In 1997, Professor K. Barry Sharpless won the Nobel Prize in Chemistry for stereoselective oxidation reactions (Sharpless epoxidation, asymmetric dihydroxylation, and oxyamination) [16].

The Sharpless approach is given in Fig. 9.3. It is an excellent example of a synthetically powerful reaction for a focused oxidation, namely conversion of allylic alcohols to epoxides. Its tolerance, which can be interpreted as being unreactive, toward other oxidizable groups in the molecule like acetals, alkynes, alcohols, aldehydes, amides, esters, ethers, olefins, pyridines, sulfones, and others is a testament to the reason why there are so many specialized oxidation catalysts. The catalyst is comprised of a titanium alkoxide and a chiral dialkyl tartrate to direct the point of attack by the oxidant on the allylic alcohol. The catalyst has one of the same building blocks found in TS-1, a titanium alkoxide, which places a boundary on its greenness. From a green process perspective, the solvent for the Sharpless epoxidation needs to be nonaqueous, CH_2Cl_2 is generally the best, the reaction is very sensitive to water so it is typically run in the presence of molecular sieves, and temperatures below room temperature are used. Nevertheless the utility of the catalyst is outstanding since high enantioselectivities (>90%) and yields (>80%) are obtained for a range of substituted allylic alcohols. From an economic perspective, all of the starting chemicals for the catalyst are commercially available and reasonably priced, and other than the solvent, there should not be issues with disposal of the spent catalyst [95]. There does not appear to have been concerted efforts to enhance the greenness of the reaction [96], but its power for certain segments of the specialty chemical market is so high that it is likely to receive attention at some point.

Schiff-Base Epoxidation Catalysis

One of the most successful catalytic oxidation systems for pharmaceutical and specialty chemical applications is the so-called Jacobsen-catalyst. In 1990, Professor Eric Jacobsen [97] and Professor Tsutomu Katsuki [98, 99] reported that chiral manganese(III) Schiff-base complexes like those shown in Fig. 9.16 could epoxidize olefins with high enantioselectivity. The foundations for this chemistry were laid out in studies designed around iron porphyrin complexes and achiral manganese(III) Schiff-base compounds as models for the cytochrome P-450 enzyme system and the oxygen evolving complex (OEC) of photosystem II (see below) [74, 100–102], These catalyst now are commercially available from a variety of sources.

Hugo Schiff reported the Schiff-base ligands in 1864. They coordinate virtually every transition metal ion, and they bind some main-group metals as well. In general, Schiff-base complexes are simple to prepare from readily available starting materials, yields are high, and the ability exists to manipulate structural and electronic properties in straightforward and predictable manner. Synthesis of the Jacobsen catalyst as described in Fig. 9.16 is typical for Schiff-base complexes where diamine and benzaldehyde derivatives are combined to make the ligand under mild conditions and then the metal is inserted using a simple salt. Manganese is an essential element for biological function, and according to the USEPA, it is one of the less toxic transition metals. Thus, manganese Schiff-base complexes fit reasonably well into the criteria for being green catalysts. However, as the structure of the Schiff-base ligand becomes more complicated, the number of synthetic steps

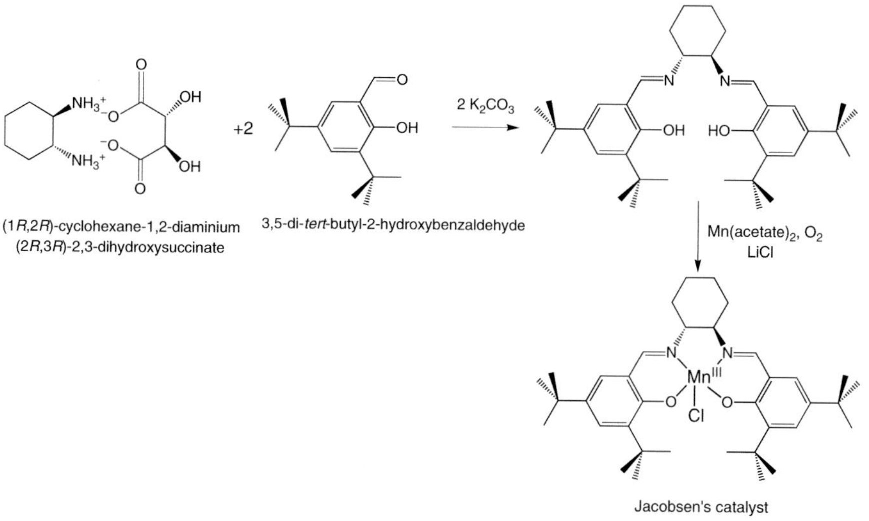

(1R,2R)-cyclohexane-1,2-diaminium   3,5-di-tert-butyl-2-hydroxybenzaldehyde
(2R,3R)-2,3-dihydroxysuccinate

Mn(acetate)₂, O₂
LiCl

Jacobsen's catalyst

Fig. 9.16   The chiral Jacobsen's Schiff-base catalyst used for epoxidation of chiral olefins

toward precursors increases which can have a significant negative impact on the overall green nature of the catalyst.

The Jacobsen-type epoxidation catalysts initially were used in nonaqueous solvents with oxidants such as iodosylbenzene rather than the favored green ones, O_2 or H_2O_2. However, over time, they have evolved tremendously to the point where they are commercially available, recoverable by their attachment to surfaces and polymers [103–105], reasonably stable during the oxidation, usable in aqueous environments, and capable of activating H_2O_2 and O_2 [106, 107]. They have achieved an impressive record for catalytic oxidations.

It was recognized early on in the epoxidation chemistry that O_2 was the oxidant of choice rather than the activated oxidants. In general, manganese(III) complexes do not react with O_2, but manganese(II) compounds readily activate oxygen. However, olefin epoxidation could be effected electrocatalytically by both manganese porphyrins [108] and achiral Schiff-base complexes [109] through the oxygen activation scheme shown in Fig. 9.17. Catalyst lifetimes for both systems were not long. A turnover number of only 2.4 was determined for the Schiff-base complex. Once again, oxidative degradation of the ligand was a factor in catalyst deactivation pathway and inactive oxo-bridged dimers also were observed. Nevertheless, these studies were extended to the use of the Jacobsen-type complexes for enantioselective electrocatalytic epoxidation of olefins [110]. Immobilization of the catalyst onto an electrode surface was a significant advancement as turnover numbers approaching 1,250 and an enantiomeric excess greater than 75% were achieved using this immobilized catalyst [111]. The turnover number is still modest, but the result clearly shows the advantage of limiting dimer formation, but self-oxidation is probably the limiting factor.

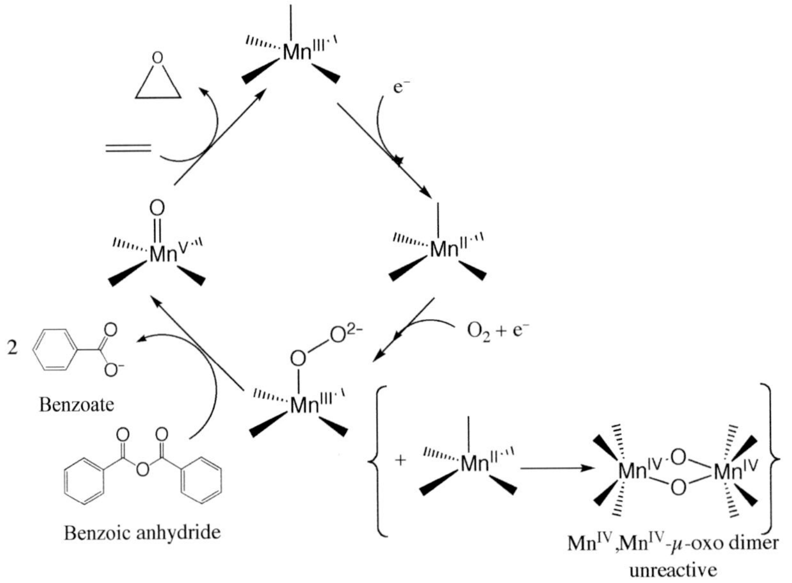

Fig. 9.17 Proposed pathway for electrocatalytic epoxidation reactions. The manganese is either in porphyrin or Schiff-base ligands. Ligand degradation and Mn^{IV}, Mn^{IV}-μ-oxo dimer formation are deactivation pathways in this chemistry

The electrocatalytic approach suffers from the need to use a sacrificial reagent (benzoic anhydride in Fig. 9.17) to remove one oxygen atom in order to generate the epoxidation catalyst. The sacrificial reagent significantly reduces the overall efficiency of the process and negatively impacts its E-factor. However, it is possible that with judicious choice of sacrificial reagent two useful products could be derived using the electrocatalytic concept. This enhanced efficiency combined with chemically modified electrodes could lead to the development of a useful and important green epoxidation catalyst.

Organonocatalysis

The previous discussion centered largely on the use of transition metal-containing compounds as oxidation catalysts for making new chemicals. The subject now shifts to organocatalysis which is *"catalysis with small organic molecules, where an inorganic element is not part of the active principle"* [112]. Professor David MacMillan in a Commentary in the journal *Nature* noted that the use of organic molecules as catalysts was mentioned "sporadically" prior to the late 1990s. Now it is recognized as a new field of research and development in organic synthesis [113]. Since the naming of the field in 2000, there have been well in excess of 2,000 publications relating to organocatalysis. This is clearly a testament to the versatility

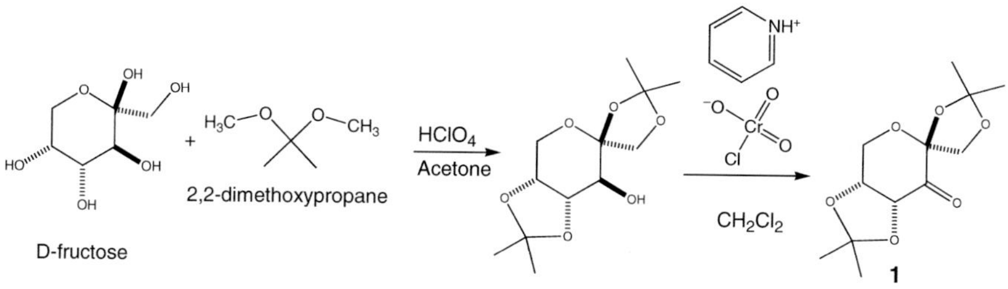

Fig. 9.18 Synthesis of a highly enantioselective epoxidation catalyst

of the field, the recognition of its diversity of synthesis, and the great potential that it has to the pharmaceutical, specialty chemical and bulk chemical markets.

The use of metal-free small organic molecules to catalyze organic transformations is thought to have broad savings in terms of reaction cost, time, and energy. Among the perceived advantages of organocatalysis over the more traditional metal-based systems are the general insensitivity of organic molecules to oxygen and moisture, the variety of organic molecules available from simple synthetic or natural sources as single enantiomers and, especially, that molecules derived from natural sources are more likely to be nontoxic and environmentally friendly. Thus, organocatalysts have many of the attributes that need to be incorporated in green oxidation catalysts. As the pace of organocatalysis accelerates, the toxicity and ecotoxicity of these chemical systems must be evaluated prior to their large-scale adoption. These new chemicals may be discharged in wastewater streams in the course of their application, and their treatment as wastes needs careful evaluation.

The strength of organocatalysis appears to be in enantioselective syntheses. It has broad utility in forming asymmetric carbon–carbon and carbon–heteroatom bonds. It was noted in a 2008 review of the subject that organocatalysts have been used across a broad spectrum of reactions, including epoxidation, aldol condensation, amination of aldehydes, Diels–Alder, and 1,3-dipolar cycloadditions, Mannich and Michael reactions, hydride transfer, nitroalkane addition to enones, and α-halogenation [114].

The use of organocatalysts for oxidation reactions was reviewed recently, and the details of the catalysts and their reactions are best left for the interested reader to pursue [115]. Here, an example illustrating the evolution of an organocatalytic reaction from a nongreen process into one that is relatively green is described. This transformation was achieved by attending to some aspects of the reaction that did not conform to green chemistry principles in addition to focusing on product output. There is still room for improvement in the chemistry, but it can serve as a model for other catalytic reactions.

In 1996, Professor Yian Shi reported that a fructose-derived ketone was an effective enantioselective catalyst for the epoxidation of *trans-* and trisubstituted olefins [116]. Compound **1**, in Fig. 9.18 is readily synthesized by a two-step process from naturally occurring D-fructose and readily available and inexpensive reagents and solvents. Thus, according to one of the common usages of the term green

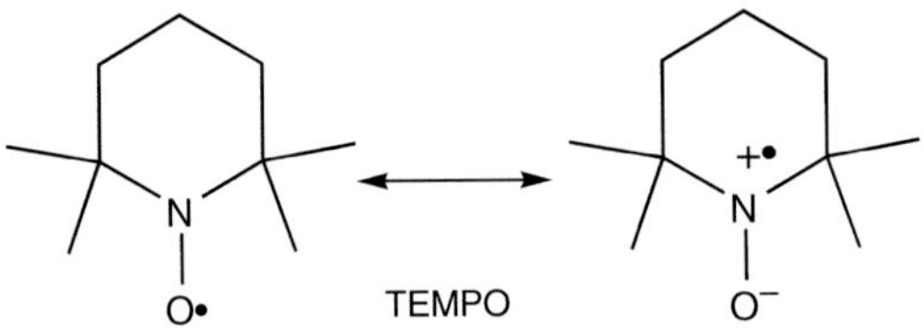

Fig. 9.19 The stable organic
radical TEMPO showing the
delocalization of the unpaired
electron over the nitrogen-
oxygen bond

TEMPO

catalyst, which is a catalyst derived from "readily available starting materials," then compound **1** is a green catalyst. However, the catalyst synthesis method is illustrative of the balance that is made between the value of the catalytic reaction and the synthetic route used to prepare the catalyst. In the first step, 70% perchloric acid is used. There is always the potential for an explosion when perchloric acid and organic materials are combined. Often it is impurities that are the problem. Then, in the second step, a stoichiometric oxidant, pyridinium chlorochromate, rather than a catalytic one, is used to oxidize the alcohol to the ketone. Chromium wastes need to be handled with care, and the mass of the spent chromium reagent is significant, so this reaction has a low E-factor.

In the 1996 report by Shi, olefins were oxidized to the corresponding epoxide in high yield, generally >75%, and with high enantioselectivities, typically >85%. However, the reaction conditions would not be considered green. First, Oxone® $(2KHSO_5 \cdot KHSO_4 \cdot K_2SO_4)$ was the oxidant. Oxone is considered to be environmentally friendly compared to Caro's acid, the traditional source of the oxidant $[HSO_5]$, but only 5% of the mass of Oxone is the active oxidant. The mass intensity and E-factors are low. Second, buffers were needed to carefully control solution pH. Third, $Na_2 \cdot EDTA$ was used to chelate any metal ions in the solution that might cause rapid Oxone decomposition. Clearly, hydrogen peroxide is the desired substitute for Oxone because of its higher active oxygen content and its reduction product is water. Subsequent studies with compound **1** showed that indeed H_2O_2 could be used as the oxidant [117]. Replacement of Oxone with H_2O_2 had additional benefits such as the quantities of solvent and salts were reduced substantially (buffer and EDTA were still used) and oxidant addition was streamlined. Most importantly, epoxidation results, yield and enantioselectivity, are comparable to those using Oxone [115].

TEMPO

TEMPO and its derivatives occupy a unique position among the oxidation organocatalysts because they have received widespread academic and industrial acceptance. TEMPO was first reported 50 years ago [118]. It is one of a class of stable radicals referred to nonconjugated nitroxyl radicals. Their stability as radicals arises from delocalization of the unpaired electron over the nitrogen-oxygen bond as shown in Fig. 9.19, and their robustness as catalysts is due to the lack of a hydrogen atom on the carbon adjacent to the nitrogen, the α-hydrogen atom [119]. The synthesis of TEMPO is challenging and costly, but there are green methods for its preparation under development [120].

Over the years, TEMPO and its derivatives have been applied as spin labels in biology, free-radical scavengers, stoichiometric oxidants, and organocatalysts. Among the most compelling suggestions that TEMPO might be a green oxidation catalyst was the application in 1996 for a USEPA Presidential Green Chemistry Challenge Award by Pharmacia and Upjohn. The collaboration resulted in a new, "green" synthesis for bisnoraldehyde (BNA), a precursor for corticosteroids and progesterone. Pharmacia and Upjohn state in their application:

> This new synthetic route to progesterone is founded on both the development of a new fermentation process which improves the utilization of a renewable, naturally derived feedstock from 15 to 100 percent, and the development of a chemical oxidation process that offers high selectivity and reduced waste streams. The fermentation employs a genetically modified bacterium to convert soya sterols directly to a new synthetic intermediate, bisnoralcohol. The new chemical process oxidizes bisnoralcohol (BA) to bisnoraldehyde {4-hydroxy-TEMPO as the catalyst to activate sodium hypochlorite (bleach)}, a key intermediate for the registered, commercial manufacture of progesterone.

Elimination of the carcinogenic organohalogen-solvent ethylene dichloride and production of the same amount of product as the previous route with 89% less nonrecoverable organic solvent waste and 79% less aqueous waste are additional green benefits of the new process. Industrially, organohalogen solvents like ethylene dichloride have been a favorite because they have many perceived beneficial properties such as robustness, recyclability, solvating ability, and others. However, they are highly regulated because of their health and safety issues and replacements are being sought.

Specialty chemical and pharmaceutical companies have described a number of commercially viable catalytic oxidations using TEMPO, TEMPO derivatives, and other nitroxyl radical organocatalysts [118, 121]. Along with adopting the TEMPO processes, these companies have replaced hypochlorite by oxidants generated electrochemically, used O_2 along with active metal catalysts (Ru^{2+}, Mn^{2+}/Co^{2+} and Cu^+) as the oxidant source, and reduced TEMPO cost by making immobilized systems. This is a clear indication that these industries and the ones that support them with new discoveries of TEMPO applications continue their rapid alignment with green chemistry principles. The enlightened outlook of these companies needs to be translated throughout the synthetic organic chemistry community where green synthesis development needs to gain a stronger foothold.

Catalysts for the Environment

The vast majority of oxidation catalysts have been designed for use in the specialty chemical or pharmaceutical areas. However, catalysts for the oxidative degradation of organic pollutants are equally significant, but they have received less attention, especially from chemists. The disposal and treatment of waterborne organic pollutants in an environmentally sound manner and at a reasonable cost is of great importance. These pollutants are from many different sources. They may be waste products from the process that makes a good that is used every day to the actual good itself once it has served its useful life. Or, they can be chemicals or products that were thought at the time to be harmless but ultimately are found to damage the environment (PCBs and CFCs are excellent examples). The environmental movement got its start with the realization by Rachel Carson that some pesticides, in particular DDT, were doing far more harm than good, and that these chemicals needed to be banned and removed from the environment. Her book *Silent Spring*, published in 1962 [122], has been named by Discover Magazine as one of the 25 greatest science books of all time.

It has often been stated that water is the only true green solvent. However, upon performing any reaction with organic chemicals in water or even contacting water with an organic substance a treatment option must be considered before discharging the water. Removal of organic (and inorganic) chemicals, now pollutants, from process waste streams is the business of hundreds of companies around the world worth multiple billions of dollars, each applying technologies that try to remove these pollutants so that the water can be safely returned to the environment possibly to be used downstream as drinking water. Thus, water is only a green solvent if it can be treated in a biological treatment plant alone [123].

Most water treatment scenarios use a biological process as the first-line attempt to remove organic pollutants from wastewaters. However, since many organic pollutants are nonbiodegradable or toxic to bacteria a more technically advanced solution needs to be applied. Physical methods like adsorption or filtration are often used but these do not solve the problem but rather shift it. For example, carbon filters remove a host of organic and inorganic pollutants by absorption or adsorption processes. Once the binding capacity of the carbon is reached, the carbon must be disposed of responsibly or regenerated which is often a thermal treatment. In contrast, a chemical oxidation may be the better choice for decontaminating the biologically recalcitrant compound because the oxidation process potentially can convert it to CO_2 and H_2O in a process referred to as mineralization or it might convert it into benign products [124]. The benign oxidation products may then undergo a subsequent biological treatment. It is important that the application of any such treatment results in cleaned water that may be directly discharged to a receiving stream or to a municipal wastewater-treatment facility without having any negative impact. The most common oxidation treatments applied to cleaning water are Fenton's reagent, wet-air oxidations (WAOs), catalytic wet-air oxidations (CWAOs), and non-Fenton processes.

Advanced Oxidation Processes (AOPs)

Fenton Chemistry

When discussing an AOP it is often assumed that the hydroxyl radical (HO$^{\bullet}$) is formed during the process. The hydroxyl radical is considered to be one of the most powerful oxidizing agents. It functions primarily by abstracting a hydrogen atom from a target organic molecule to make water (the driving force for the reaction) and a carbon centered radical. Carbon-centered radicals typically are not stable and the molecule literally disintegrates, of course helped along by more hydroxyl radical chemistry. Other more controlled reactions can be observed for the hydroxyl radical such as addition to C = C bonds or aromatic rings [125] and electron-transfer reactions [126]. Among the oldest and most thoroughly studied AOP is the so-called Fenton's reagent. More than a century ago, Henry J. H. Fenton described the oxidation of tartaric acid in the presence of iron salts and H$_2$O$_2$ [127]. The abstract reads:

> Experiments were made with standard solutions of tartaric acid, ferrous sulfate, and hydrogen peroxide to study the nature of the oxidation of tartaric acid in the presence of iron. Measured volumes of these solutions were mixed in the order named, made alkaline with soda, diluted to the same volume in tall glasses, and the depths of violet color compared. With equal quantities of tartaric acid and a fixed proportion of iron, it was observed that the depth of color increased with increasing quantities of hydrogen peroxide up to a certain limit, beyond which the color diminished as the peroxide increased, and beyond a certain limit disappeared together. From these results, it seems that the action of iron is "catalytic", a very small quantity of iron being sufficient to determine the oxidation in this direction of an almost unlimited amount of tartaric acid.

This article has been cited more than 500 times. Fenton's contribution, which appeared roughly 50 years after Perkin's, might be considered the disclosure of one of the first green oxidation catalysts as it uses the nontoxic element iron to activate the environmentally benign oxidant H$_2$O$_2$. Fenton-based technologies generally are viewed as economical and convenient but only after biological and physical treatment options are exhausted. There are excellent reviews of what is commonly referred to as "Fenton chemistry" and only some of the highlights are discussed here as they relate to the green or nongreen aspects of the chemistry [128].

Fenton chemistry is broadly applied. One can find it used in the remediation of contaminated soils [129] removal of colored effluents from the dye industry, oxidative degradation of hazardous chemicals like 2,4,6-trinitrotoluene and chlorinated organics [19]. It is often the first line of defense against a wastewater containing recalcitrant organic chemicals [130].

Fenton's reagent is catalytic with respect to iron according to Eqs. 5 and 6.

$$Fe^{2+} + H_2O_2 \rightarrow Fe^{3+} + OH^{\bullet} + OH^{-} \qquad (9.5)$$

$$Fe^{3+} + H_2O_2 \rightarrow Fe^{2+} + OOH^{\bullet} + H^{+} \qquad (9.6)$$

Scheme 2 Reactions of iron, hydrogen peroxide, and hydrogen peroxide derived species in the Fenton reaction

$$H_2O_2 + Fe^{2+} \longrightarrow Fe^{3+} + HO^{\bullet} + OH^-$$

$$H_2O_2 + Fe^{3+} \longrightarrow Fe^{2+} + HO_2^{\bullet} + H^+$$

$$HO^{\bullet} + H_2O_2 \longrightarrow HO_2^{\bullet} + H_2O$$

$$HO^{\bullet} + Fe^{2+} \longrightarrow Fe^{3+} + OH^-$$

$$Fe^{3+} + HO_2^{\bullet} \longrightarrow Fe^{2+} + O_2 + H^+$$

$$Fe^{2+} + HO_2^{\bullet} + H^+ \longrightarrow Fe^{3+} + H_2O_2$$

$$HO_2^{\bullet} + HO_2^{\bullet} \longrightarrow H_2O_2 + O_2$$

It can be seen that hydrogen peroxide acts to oxidize Fe^{2+} to Fe^{3+} and form the highly reactive hydroxyl radical ($OH^{\bullet}$), but it also reduces Fe^{3+} to Fe^{2+} forming the hydroperoxyl radical ($OOH^{\bullet}$) which is only a slow oxidizing agent in solution [131]. At a maximum, 50% of the H_2O_2 is used in productive oxidation reactions.

The chemistry is actually far more complicated than shown above and a more accurate set of reactions is given in Scheme 2. The significance of Scheme 2 is that the Fenton process is an inefficient one in terms of hydrogen peroxide consumption.

In addition to inefficient use of the oxidant, Fenton-based processes have other drawbacks such as the need to do reactions under acidic conditions (pH 3) and the formation of iron sludges that must be removed. The pH is an issue because at near neutral pH precipitation of the iron as $Fe_2O_3 \bullet nH_2O$, rust, occurs which inhibits the Fe(III)/Fe(II) cycle [132]. This has been addressed to some extent by adding chelating ligands like aminopolycarboxylates, polyhydroxy aromatics, N-heterocyclic carboxylates, and others to try to prevent formation of the sludge. While some lifetime increase is observed, these ligands are themselves oxidized and thus of limited value [133–135]. Despite these drawbacks, Fenton's reagent remains an important "green" oxidation catalyst for removing pollutants.

Catalytic Wet Air Oxidation (CWAO)

Catalytic wet-air oxidation is a process in which molecular oxygen and a catalyst are used at relatively high temperatures and pressures to oxidize organic pollutants present in water [136]. The goal of a CWAO process is to convert all of the pollutant into CO_2 and H_2O or to degrade the pollutant to a level that it can be treated biologically. Lenntech, a supplier of CWAO equipment states: "The effect of the catalyst is to provide a higher degree of COD removal than is obtained by WAO at comparable conditions (over 99% removal can be achieved), or to reduce the residence time." This is a nearly perfect statement of the function of a catalyst. Faster reactions and deeper oxidations reduce risk while allowing greater through-put, which should result in a more economic process.

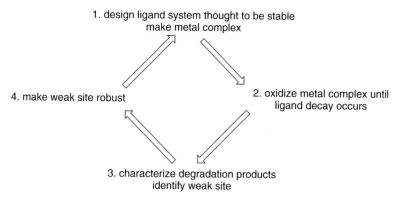

Fig. 9.20 Ligand design strategy to secure oxidatively robust catalysts

CWAO processes are applied to water-containing high concentrations of organic pollutants – chemical oxygen demand (COD) levels greater that 10,000 mg/L, in part because the quantity of alternative oxidants like H_2O_2 is too high and the waste is too recalcitrant for standard physical, biological, or chemical treatments. The majority of catalysts used in CWAO processes are oxides of Cu, Co, Mn, Cr, V, Ti, Bi, and Zn, and supported precious metals [136]. A drawback to CWAO treatments is leaching of the metal from the oxide or support into the water being treated. The metal must then be removed from the water before it can be discharged. This adds cost to the process and can make it uneconomical. There clearly are some metals in the list, particularly chromium, that have more potential for harm than others should they be discharged into our waterways. Thus judicious choice of metal can make these materials green oxidation catalysts. The benefits of using such catalysts to treat recalcitrant organic pollutants appear to be clear since in all likelihood these pollutants would have entered our water system.

Non-Fenton AOP

In 1999, Professor Terrence Collins received the Presidential Green Chemistry Challenge Award for the development and potential for wide applicability of TAML® catalysts. Collins focused on using nontoxic elements and oxidatively robust organic building blocks to surround a catalytically active iron center [137]. An iterative design strategy was developed that centered on finding and replacing weak or oxidatively sensitive points in each earlier version of the catalyst [62]. This catalyst development strategy involved the four-step process depicted in Fig. 9.20 and also resulted in a set of design parameters that define oxidatively stable catalysts. Another critical feature during the design phase of the catalysts was toxicity screening. In all instances, the catalysts were found to have toxicity well below their

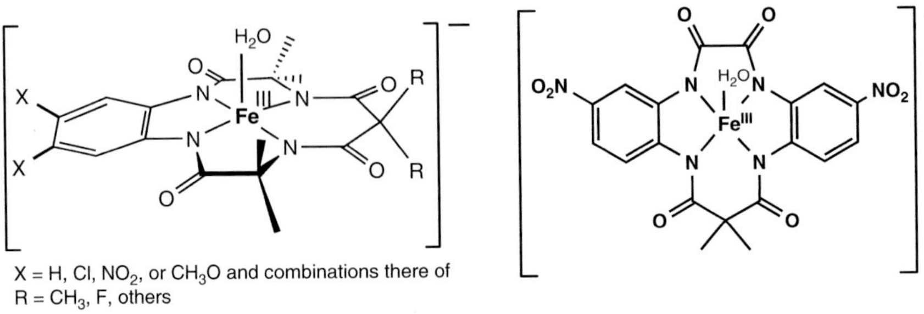

X = H, Cl, NO$_2$, or CH$_3$O and combinations there of
R = CH$_3$, F, others

Fig. 9.21 First- (*left*) and second- (*right*) generation TAML catalysts

application level so that they could be safely discharged to the environment. The ligand design strategy of using nontoxic elements and testing for catalyst and oxidation product toxicity provide a model pathway for the development of green oxidation catalysts.

The design process produced an expanding family of iron-based TAML® catalysts. Structures of first and second generation TAML catalysts are shown in Fig. 9.21 [138]. The catalysts are extremely active, achieving many thousands of reactions before being deactivated. They do this at quantities measured in the single parts per million levels or lower, which is a testament to the high stability and high catalytic activity of these molecules. Unlike most other oxidation catalysts, the TAML catalysts function in water over a pH range from acidic to highly basic (pH >14). All of these factors allow for manipulating catalyst and/or reaction conditions to achieve desired selectivities and reaction rates. Excellent recent reviews that detail kinetic and mechanistic details of the catalysts are available [139, 140].

The first generation TAML catalysts are noteworthy for the ability to manage their reactivities and lifetimes by manipulating the X and R substituents on the macrocycle. Shown in Fig. 9.22 is a summary of the deep understanding surrounding the catalysts [141]. Controlling catalyst lifetime is a particularly significant attribute as an oxidation catalyst that is intended for environmental work should not be persistent nor constantly active in the environment [142].

The TAML catalysts forte is the oxidative degradation of organic pollutants in water. The degradation does not appear to occur via a hydroxyl radical pathway [143]. Wastewater treatment has become one of the many interesting potential application areas for the TAML catalysts as they degrade chlorinated pollutants [144], phenols, dyes [145, 146], pulp-mill effluent [147], pesticides [148], pharmaceuticals, bacterial spores, [149] and endocrine disrupting compounds (EDCs) like the active ingredient in the birth-control pill – ethinylestradiol [150]. Furthermore, the oxidation products using H$_2$O$_2$ have proven to be less toxic than the target pollutant.

GreenOx Catalysts, Inc. is commercially developing TAML catalysts. They are being applied in a number of different application areas. Recombinant Innovation uses a TAML®/H$_2$O$_2$ AOP as a tankside add to completely degrade chlorinated biocides in spent metal working fluids (MWFs) in order to lower disposal costs. Approximately, 75–100 million liters per year of MWFs contain the biocide. Until

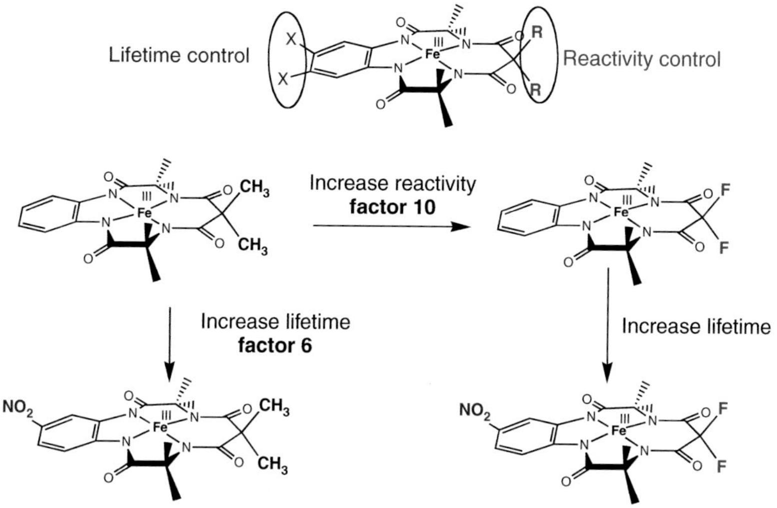

Fig. 9.22 Representation of the relationship between lifetime control and reactivity control for first-generation TAML catalysts. Note that all the catalysts have one negative charge and, in water, have two water molecules attached to the iron. These are omitted for clarity

development of the TAML®/H$_2$O$_2$ AOP, no simple, cost-effective technology existed for a tankside biocide removal. VeruTEK Technologies has developed an AOP using TAML® catalysts with their VeruSOL® technology for *in situ* remediation of contaminated soils. The TAML® catalysts substantially speed up treatment times thereby decreasing cost and risk and they provide a more complete contaminant destruction process. This combination of green chemistry and green engineering provides an excellent opportunity for demonstrating how real-world problems can be solved by green technology approaches. Catalyst concentrations in these fields of use are less than 10 ppm and can be as low as 1 ppm. These examples attest to the power of TAML®-based AOPs.

Energy

Supplying mankinds voracious and increasing appetite for energy in a sustainable manner is arguably the most important scientific and technical challenge. All fossil fuel–based technologies eventually will need replacement. The splitting of water to oxygen and hydrogen by photochemical means (artificial photosynthesis) has far-reaching potential for sustainability as it is the basis of the solar fuel-cell technology shown schematically in Fig. 9.23. The oxidation of water to O$_2$ is a complex four-electron process that also requires the release of four protons [151, 152]. Photosynthetic organisms like plants and algae, as well as certain cyanobacteria oxidize water by means of a gorgeous chemical process carried out in the oxygen-evolving

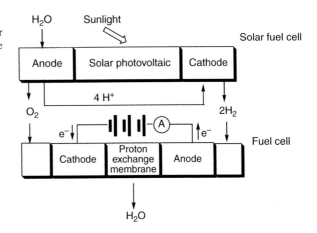

Fig. 9.23 The solar fuel cell uses sunlight to oxidize water to O_2 at the anode and reduce the released protons at the cathode to hydrogen. In the fuel cell, the hydrogen and oxygen are recombined across a membrane, which generates a flow of electrons producing electrical energy

complex (OEC) of photosystem II (PSII) [153, 154]. The O_2 is a by-product of splitting water into the electrons and protons that are ultimately used to store energy and reduce CO_2 to carbohydrates. Remarkably, only a single class of oxygen-evolving enzymes is found in all photosynthetic organisms, and they all have an inorganic core with the formula $Mn_4Ca_1O_xCl_{1-2}(HCO_3)_y$.

While the oxidation of water to O_2 now is a simple reaction in nature, it is a difficult problem physically and chemically to replicate in the laboratory. Practically, the oxidation catalyst must function efficiently for long periods of time (years), integrate with the other components of the entire charge-separating network, and be comprised of materials that are as green as possible. Efficiency means that the catalyst must remove the four electrons from two water molecules at or near the thermodynamic potential of 1.23 V. Any increase in voltage above the thermodynamic potential, which is referred to as the overvoltage, means a loss of the energy derived from the sun.

Significant advances in approaches toward development of water oxidation catalysts have occurred over the last few years. Professor Daniel Nocera has reported that nickel or cobalt ions in the presence of a variety of buffer systems produce oxygen-evolving catalyst films that operate at near 100% current efficiency [155, 156]. The catalysts are robust and are likely to integrate with other systems of the cell. From a green catalyst perspective, nickel is the better choice as it has lower toxicity [157] and is more abundant than cobalt. Furthermore, cobalt is a relatively diffuse element in the earth's crust and its main sources are mines in the Democratic Republic of the Congo and Zambia.

Professor Royce Murray is using iridium oxide nanoparticles electro-flocculated onto electrode surfaces as the electrocatalysts [158]. These electrocatalysts are very efficient for O_2 evolution. The mesoporous nature of the films encourages facile water and ion permeation. In addition, the number of active iridium sites is high, the oxidations are relatively fast one electron/one proton steps, and the overpotential is only about 0.25 V. This is an excellent conceptual approach but it is based on iridium, which is one of nature's rarest elements. Furthermore, there are ongoing debates regarding nanoparticle materials in the environmental as their toxicity is still unresolved issue.

Finally, Professor Craig Hill has expanded his research efforts around polyoxometalates as oxidation catalysts for organic molecules to the water splitting problem [159]. Hill has found that cobalt containing polyoxometalates are excellent homogeneous water-oxidation catalysts. The catalysts themselves are comprised of cobalt, tungsten, phosphorous, and oxygen. The reliance on cobalt has the same issues as the solid state form described by Professor Nocera. In addition, polyoxometalates have very high molecular weights so the substrate to catalyst mass ratio is very low. With water as the substrate, low substrate to catalyst mass ratios will inevitably be the case.

Future Directions

Outlook for Green Oxidation Catalysts

The future for green oxidation catalysis is bright. Catalysis has traditionally been in the hands of "catalysis experts" and they have provided a tremendous foundation upon which to build. Now, creative thinkers are meshing the information that has been gathered over the decades about catalyzed and uncatalyzed reactions into new paradigms.

- Rapid growth is happening in interdisciplinary approaches to solving compli- cated syntheses with catalysis as a major contributor to the process. One example is coupling organocatalysts with biocatalysts into clever reaction schemes that produce compounds that are particularly difficult to synthesize [160]. Both organocatalysis [113] and biocatalysis [161] are relatively new fields themselves having blossomed in the last decade. The rapid adoption of these technologies is a clear indication that interdisciplinary approaches are playing an increasing role in the field of catalysis, which only bodes well for making even more rapid and significant contributions.
- The elimination of even a single isolation step in a multistep synthesis saves on time, materials, and cost. Cascade and tandem catalysis are powerful methodologies, where one or more catalyst is combined with multiple reactants to assemble complex molecules without recovering intermediate products [162, 163]. These single reaction vessel reactions are challenging ones to design because all reagents in the mixture must be compatible, but the potential benefits are great.
- Organic–inorganic hybrid catalysts are the marriage of solid catalysts, which have the advantage of generally straightforward recovery and recyclability with soluble catalysts, which have a greater breadth in the variety of reactions that they can catalyze. Many different approaches, some well established and others new, are being taken to the preparation of these materials. These include

adsorbing the soluble catalyst in pores of the support, assembling the soluble catalyst within a cavity of the solid catalyst so that it cannot escape (analogous to a "ship-in-bottle"), making a covalent bond between the solid material and the soluble catalyst, and finally directly synthesizing the hybrid material [164–166]. For example, the power of the oxidation organocatalyst TEMPO is great, but a drawback is its cost. Current efforts are to covalently link it to a silica substrate like MCM-41. MCM-41 has a large surface area and a regular mesoporous structure which allows for designing a well-defined hybrid material. Preparation of the TEMPO/MCM-41 material is a relatively complicated synthetic process itself and illustrates one of the issues that must be considered in deciding whether the catalyst itself is green [167].

- Process and Green Engineering issues are also making rapid advancements in catalysis. Environmentally benign or green solvents for all types of chemical reactions, including catalysis are an intense research topic [168]. Solvents play a number of roles in catalytic processes including dissolving one or more reactant, controlling reaction temperature, isolating the desired chemical by extraction, distillation, or crystallization and other functions. They are typically used in much larger volumes than the reactants themselves and therefore pose challenges such as disposal (cost and methodology), volatilization into the atmosphere, CO_2 generation as a result of energy needed to heat and cool, and others. Solvents of the future [169] include water, ionic liquids [170], fluorous liquids [171], supercritical fluids, gas-expanded liquids [172], liquid polymers, and switchable surfactants [173]. The breadth of application space for each one of these solvents systems is still in its early stages and new solvents are likely to be developed in the coming years. It should be emphasized that fluorous solvents have many positive attributes, but they tend to be persistent in the environment (nonbiodegradable) which makes them nongreen [91]. Bi-phasic solvent systems, phase-transfer catalysis, and even solvent-free processes are considered as making major contributions to advances in green solvent systems.

- Computer-aided molecular design of environmentally benign catalysts is an expanding research area. Computer modeling of existing molecules is a highly refined process. In the case of catalyst design, the methodology is still in its early stages, but it holds the promise of greatly decreasing the time and effort required to design catalysts and to improve catalyst properties [174]. The catalysts are essentially broken into pieces with each piece evaluated for its role in the catalytic process and also its potential for environmental effects. By eliminating "negative" pathways in the catalyst design, more fruitful outcomes will result [138, 142, 175].

- High-throughput experimentation (HTE) is a tool that has been applied to catalysis during the past decade [176]. HTE utilizes an array of microreactions in which parameter space around a catalytic reaction is probed essentially simultaneously. These microreactions, which generally are robot controlled, produce a vast amount of experimental data that needs to be carefully sorted. As with any other investigation mode into a process that is ultimately destined for commercial application, it is important to investigate those parameters that are critical for scaling to laboratory, pilot-plant and commercial plant. A number of companies have been born that provide HTE services.

Clearly, this is a remarkable time to be involved in catalysis research as the field reinvents itself from one heavily focused on converting petroleum-based feedstreams for commodity chemicals into one that looks at the entire picture of the products and processes that are used for our short- and long-term needs [20, 177, 178]. It also brings into focus a "new" feedstock, renewable resources, for chemists to manipulate into value-added chemicals [20, 177]. However, the chemical knowledge base for manipulating renewable resources into high-value chemicals is limited and catalytic processes are even less well developed [179]. This knowledge gap exists because unlike petroleum-based feedstocks, those derived from renewable resources are highly functionalized molecules and thus require new methods for their manipulation. Of course, all of this must be done economically [178]. Nevertheless, the rapid increase in global demand for raw materials in all sectors of society means that better and more intelligent uses of alternative and sustainable resources are needed. Green oxidation catalysts will play a vital role in the challenge to convert these complicated feedstocks [180] into tomorrow's products.

Bibliography

1. Walker JD, Enache M, Dearden JD (2006) Quantitative cationic activity relationships for predicting toxicity of metal ions from physicochemical properties and natural occurrence levels. QSAR Comb Sci 26:522–527
2. Walker JD, Enache M, Dearden JC (2003) Quantitative cationic-activity relationships for predicting toxicity of metals. Environ toxicology chemistry SETAC 22:1916–1935
3. USEPA (2007) Framework for Metals Risk Assessment. http://www.epa.gov/raf/metalsframework/pdfs/metals-risk-assessment-final.pdf. Accessed 17 May 2010
4. Min BK, Friend CM (2007) Heterogeneous gold-based catalysis for green chemistry: low-temperature CO oxidation and propene oxidation. Chem Rev 107:2709–2724
5. Anastas PT, Kirchhoff MM, Williamson TC (2001) Catalysis as a foundational pillar of green chemistry. Appl Catal A Gen 221:3–13
6. Anastas PT, Zimmerman JB (2003) Peer reviewed: design through the 12 principles of green engineering. Environ Sci Technol 37:94–101
7. World Commission on Environment and Development (1987) Our common future. Oxford University Press, Oxford
8. Anastas PT, Williamson TC (1996) Green chemistry: an overview. ACS Symp Ser 626:1–17
9. Anastas PT, Warner JC (1998) Green: chemistry theory and practice. Oxford University Press, Oxford
10. Centi G, Perathoner S (2003) Catalysis and sustainable (green) chemistry. Catal Today 77:287–297
11. Wsniak J (2000) Jöns Jacob Berzelius a guide to the perplexed chemist. Chem Educ 5:343–350
12. Berzelius JJ (1836) Jahres-Bericht über die Fortschritte der Physichen Wissenschaften. 15:237–245 H. Laupp, Tübingen
13. Roberts M (2000) Birth of the catalytic concept (1800–1900). Catal Lett 67:1–4

14. Brégeault J-M (2003) Transition-metal complexes for liquid-phase catalytic oxidation: some aspects of industrial reactions and of emerging technologies. J Chem Soc Dalton Trans 17:3289–3302
15. Smith MB, March J (2007) March's advanced organic chemistry: reactions, mechanisms, and structure, 6th edn. Wiley, Hoboken
16. Katsuki T, Sharpless KB (1980) The first practical method for asymmetric epoxidation. J Am Chem Soc 102:5974–5976
17. Sheehan RJ (2005) Terephthalic acid, dimethyl terephthalate, and isophthalic acid. In: Book ullmann's encyclopedia of industrial chemistry. Wiley, Weinheim, pp 1–13
18. Cavani F, Teles JH (2009) Sustainability in catalytic oxidation: an alternative approach or a structural evolution? ChemSusChem 2:508–534
19. Gogate P, Pandit AB (2004) A review of imperative technologies for wastewater treatment I: oxidation technologies at ambient conditions. Adv Enviro Res 8:501–551
20. van Haveren J, Scott EL, Sanders J (2008) Bulk chemicals from biomass. Biofuels Bioprod Biorefin 2:41–57
21. Werpy T, Holladay J, White J, Peterson G, Bozell JJ, Aden A, Manheim A (2004) Top value added chemicals from biomass, vol 1.), Results of screening for potential candidates from sugars and synthetic gas U.S. Department of Energy, NREL/TP-510-35523
22. Cavani F, Ballarini N, Luciani S (2009) Catalysis for society:towards improved process efficiency in catalytic selective oxidations. Top Catal 52:935–947
23. Cavani F, Ballarini N (2009) Recent achievements and challenges for a greener chemical industry. In: Mizuno N (ed) Modern heterogenous oxidation catalysis: design, reactions and characterization. Wiley, Weinheim, pp 289–331
24. Valko M, Morris H, Cronin MTD (2005) Metals, toxicity and oxidative stress. Curr Med Chem 12:1161–1208
25. Teng Y, Jiao x, Wang J, Xu W, Yang J (2009) Environmentally geochemical characteristics of vanadium in the topsoil in the Panzhihua mining area, Sichuan Province. China Chin J Geochem 28:105–111
26. Hermans I, Spier ES, Neuenschwander U, Turrá N, Baiker A (2009) Selective oxidation catalysis: opportunities and challenges. Top Catal 52:1162–1174
27. Clerici MG, Bellussi G, Romano U (1991) Synthesis of propylene oxide from propylene and hydrogen peroxide catalyzed by titanium silicalite. J Catal 129:159–167
28. Clerici MG (2009) Titanium silicalite-1. In: Jackson SD, Hargreaves JSJ (eds) Metal oxide catalysis, vol 2. Wiley, Weinheim, pp 705–754
29. Nijhuis TA, Makkee M, Moulijn JA, Weckhuysen, BM (2006) The production of propene oxide: A catalytic processes and recent developments. Ind Eng Chem Res 45:3447–3459
30. Nijhuis TA, Huizinga BJ, Makkee M, Moulijn JA (1999) Direct epoxidation of propene using gold dispersed on TS-1 and other titanium-containing supports. Ind Eng Chem Res 38(3):884
31. Stare J, Henson NJ, Eckert J (2009) Mechanistic aspects of propene epoxidation by hydrogen peroxide. Catalytic role of water molecules, external electric field, and zeolite framework of TS-1. J Chem Info Mod 49:833–846
32. Li YG, Lee YM, Porter JF (2002) The synthesis and characterization of titanium silicalite-1. J Mater Sci 37:1959–1965
33. Baccile N, Babonneau F, Thomas B, Coradin T (2009) Introducing ecodesign in silica sol’gel materials. J Mater Chem 19:8537–8559
34. Sheldon RA (1992) Organic synthesis – past, present and future. Chem Ind Lond 23:903–906
35. Sheldon RA (1997) Catalysis: the key to waste minimization. J Chem Technol Biotechnol 68:381–388
36. Sheldon RA (2007) The E factor: fifteen years on. Green Chem 9:1273–1283
37. Calvo-Flores FG (2009) Sustainable chemistry metrics. ChemSusChem 2:905–919
38. Eckelman MJ, Zimmerman JB, Anastas PT (2008) Toward green nano: e-factor analysis of several nanomaterial syntheses. J Ind Ecol 12:316–328

39. Dahl JeA, Maddux BLS, Hutchison JE (2007) Toward greener nanosynthesis. Chem Rev 107:2228–2269
40. Hermans I, Peeters J, Jacobs PA (2008) Autoxidation chemistry: bridging the gap between homogeneous radical chemistry and (heterogeneous) catalysis. Top Catal 48:41–48
41. Thomas JM, Raja R (2006) The advantages and future potential of single-site heterogeneous catalysts. Top Catal 40:3–17
42. Niu W, Draths KM, Frost JW (2002) Benzene-free synthesis of adipic acid. Biotechnol Progr 18:201–211
43. Sato K, Aoki M, Noyori R (1998) A "Green" route to adipic acid: direct oxidation of cyclohexenes with 30 percent hydrogen peroxide. Science 281:1646–1647
44. Strigul N, Koutsospyros A, Christodoulatos C (2010) Tungsten speciation and toxicity: acute toxicity of mono – and poly-tungstates to fish. Ecotoxicol Environ Saf 73:164–171
45. Smith BJ, Patrick VA (2000) Quantitative determination of sodium metatungstate speciation by 183W NMR spectroscopy. Aust J Chem 53:965–970
46. Deng Y, Ma Z, Wang K, Chen J (1999) Clean synthesis of adipic acid by direct oxidation of cyclohexene with H_2O_2 over peroxytungstate-organic complex catalysts. Green Chem 1:275–276
47. Buonomenna MG, Golemme G, De Santo MP, Drioli E (2010) Direct oxidation of cyclohexene with inert polymeric membrane reactor. Org Proc Res Dev 14:252–258
48. York APE, Xiao T, Green MLH (2003) Brief overview of the partial oxidation of methane to synthesis gas. Top Catal 22:345–358
49. Enger BC, Lødeng R, Holmen A (2008) A review of catalytic partial oxidation of methane to synthesis gas with emphasis on reaction mechanisms over transition metal catalysts. Appl Catal A Gen 346:1–27
50. Arakawa H, Aresta M, Armor JN, Barteau MA, Beckman EJ, Bell AT, Bercaw JE, Creutz C, Dinjus E, Dixon DA, Domen K, DuBois DL, Eckert J, Fujita E, Gibson DH, Goddard WA, Goodman DW, Keller J, Kubas GJ, Kung HH, Lyons JE, Manzer LE, Marks TJ, Morokuma K, Nicholas KM, Periana R, Que L, Rostrup-Nielson J, Sachtler WMH, Schmidt LD, Sen A, Somorjai GA, Stair PC, Stults BR, Tumas W (2001) Catalysis research of relevance to carbon management: progress, challenges, and opportunities. Chem Rev 101:953–996
51. Kirillova MV, Kozlov YN, Shul'pina LS, Lyakin OY, Kirillov AM, Talsi EP, Pombeiro AJL, Shul'pin GB (2009) Remarkably fast oxidation of alkanes by hydrogen peroxide catalyzed by a tetracopper(II) triethanolaminate complex: promoting effects of acid co-catalysts and water, kinetic and mechanistic features. J Catal 268:26–38
52. Beckers J, Rothenberg G (2010) Sustainable selective oxidations using ceria-based materials. Green Chem 12:939
53. Czuprat O, Werth S, Schirrmeister S, Schiestel T, Caro J (2009) Olefin production by a multistep oxidative dehydrogenation in a perovskite hollow-fiber membrane reactor. ChemCatChem 1:401–405
54. Madeira LM, Portela MF (2002) Catalytic oxidative dehydrogenation of n-butane. Catal Rev 44:247–286
55. Grabowski R (2006) Kinetics of oxidative dehydrogenation of C_2-C_3 alkanes on oxide catalysts. Catal Rev 48:199–268
56. Meth-Cohn O, Smith M (1994) What did W. H. Perkin actually make when he oxidized aniline to obtain mauveine? Perkin Trans 1:5–7
57. Sousa MM, Melo MJ, Parola AJ, Morris PJT, Rzepa HS, de Melo JSS (2008) A study in mauve: unveiling Perkin's dye in historic samples. Chem Euro J 14:8507–8513
58. He L, Wang L-C, Sun H, Ni J, Cao Y, He H-Y, Fan K-N (2009) Efficient and selective room-temperature gold-catalyzed reduction of nitro compounds with CO and H_2O as the hydrogen source. Angew Chem Int Ed 48:9538–9541
59. Li S-C, Diebold U (2010) Reactivity of TiO_2 rutile and anatase surfaces toward nitroaromatics. J Am Chem Soc 132:64–66

60. Mann PJG, Saunders BC (1935) Peroxidase action. I. The oxidation of aniline. Proc R Soc Lond B Biol Sci 119:47–60
61. Sheldon RA, Kochi JK (1981) Metal-catalyzed oxidations of organic compounds. Academic, New York
62. Collins TJ (1994) Designing ligands for oxidizing complexes. Acc Chem Res 27:279–285
63. Schmid A, Dordick J, Hauer B, Kiener A, Wubbolts M, Witholt B (2001) Industrial biocatalysis today and tomorrow. Nature 409:258
64. Schoemaker HE, Mink D, Wubbolts MG (2003) Dispelling the myths – biocatalysis in industrial synthesis. Science 299:1694–1697
65. Yuryev R, Liese A (2010) Biocatalysis: the outcast. ChemCatChem 2:103–107
66. Matsuda T, Yamanaka R, Nakamura K (2009) Recent progress in biocatalysis for asymmetric oxidation and reduction. Tetrahedron Asymmetr 20:513–557
67. van de Velde F, Lourenço ND, Bakker M, van Rantwijk F, Sheldon RA (2000) Improved operational stability of peroxidases by coimmobilization with glucose oxidase. Biotechnol Bioeng 69:286–291
68. Kuhn D, Kholiq MA, Heinzle E, Bühler B, Schmid A (2010) Intensification and economic and ecological assessment of a biocatalytic oxyfunctionalization process. Green Chem 12:815–827
69. Que L Jr, Tolman WB (2008) Biologically inspired oxidation catalysis. Nature 455:333–340
70. Costas M, Chen K, Que L Jr (2000) Biomimetic nonheme iron catalysts for alkane hydroxylation. Coord Chem Rev 200–202:517–544
71. Piera J, Bäckvall J-E (2008) Catalytic oxidation of organic substrates by molecular oxygen and hydrogen peroxide by multistep electron transfer – a biomimetic approach. Angew Chem Int Ed 47:3506–3523
72. Meunier B, de Visser SP, Shaik S (2004) Mechanism of oxidation reactions catalyzed by cytochrome P450 enzymes. Chem Rev 104:3947–3980
73. McLain JL, Lee J, Groves JT (2000) Biomimetic oxygenations related to cytochrome P450: metal-oxo and metal-peroxo intermediates. In: Meunier B (ed) Biomimetic oxidations catalyzed by transition metal complexes. Imperial College Press, London, pp 91–169
74. Groves JT, Nemo TE, Myers RS (1979) Hydroxylation and epoxidation catalyzed by iron-porphine complexes. oxygen transfer from iodosylbenzene. J Am Chem Soc 101:1032–1033
75. Leroy J, Bondon A (2007) β-Fluorinated porphyrins and related compounds: an overview. Eur J Org Chem 2008:417–433
76. Cai Y, Liu Y, Lu Y, Gao G, He M (2008) Ionic manganese porphyrins with S-containing counter anions: mimicking cytochrome P450 activity for alkene epoxidation. Catal Lett 124:334–339
77. Nagarajan S, Nagarajan R, Bruno F, Samuelson LA, Kumar J (2009) A stable biomimetic redox catalyst obtained by the enzyme catalyzed amidation of iron porphyrin. Green Chem 11:334–338
78. Liu Y, Zhang H-J, Lu Y, Cai Y-Q, Liu X-L (2007) Mild oxidation of styrene and its derivatives catalyzed by ionic manganese porphyrin embedded in a similar structured ionic liquid. Green Chem 9:1114–1119
79. Lente G, Espenson JH (2005) Oxidation of 2, 4, 6-trichlorophenol by hydrogen peroxide Comparison of different iron-based catalysts. Green Chem 7:28–34
80. Stahl SS (2004) Palladium oxidase catalysis: selective oxidation of organic chemicals by direct dioxygen-coupled turnover. Angew Chem Int Ed 43:3400–3420
81. Allardyce CS, Dyson PJ (2001) Ruthenium in medicine: current clinical uses and future prospects. Platinum Met Rev 45:62–69
82. Pope MT, Muller A (2002) Polyoxometalate chemistry from topology via self assembly to applications. Kluwer, New York
83. Jeannin YP (1998) The nomenclature of polyoxometalates: how to connect a name and a structure. Chem Rev 98:51–76
84. Hill CL, Delannoy L, Duncan DC, Weinstock IA, Renneke RF, Reiner RS, Atalla RH, Han JW, Hillesheim DA, Cao R, Anderson TM, Okun NM, Musaev DG, Geletii YV

(2007) Complex catalysts from self-repairing ensembles to highly reactive air-based oxidation systems. Comptes rendus Chim 10:305–312

85. Kozhevnikov IV (1998) Catalysis by heteropoly acids and multicomponent polyoxometalates in liquid-phase reactions. Chem Rev 98:171–198

86. Long D-L, Burkholder E, Cronin L (2007) Polyoxometalate clusters, nanostructures and materials: from self assembly to designer materials and devices. Chem Soc Rev 36:105–121

87. Trubitsyna T, Kholdeeva O (2008) Kinetics and mechanism of the oxidation of 2, 3, 6-trimethylphenol with hydrogen peroxide in the presence of Ti-monosubstituted polyoxometalates. Kinet Catal 49:371–378

88. Neumann R (2010) Activation of molecular oxygen, polyoxometalates, and liquid-phase catalytic oxidation. Inorg Chem 49:3594–3601

89. Maayan G, Ganchegui B, Leitner W, Neumann R (2006) Selective aerobic oxidation in supercritical carbon dioxide catalyzed by the $H_5PV_2Mo_{10}O_{40}$ polyoxometalate. Chem Commun 21:2230–2232

90. Vazylyev M, Sloboda-Rozner D, Haimov A, Maayan G, Neumann R (2005) Strategies for oxidation catalyzed by polyoxometalates at the interface of homogeneous and heterogeneous catalysis. Top Catal 34:93–99

91. Clark JH, Tavener SJ (2007) Alternative solvents: shades of green. Org Proc Res Dev 11:149–155

92. Collins TJ (2001) Papermaking: green chemistry through the mill. Nature 414:161–162

93. Weiner H, Finke RG (1999) An all-inorganic, polyoxometalate-based catechol dioxygenase that exhibits >100 000 catalytic turnovers. J Am Chem Soc 121:9831–9842

94. Curzons AD, Constable DJ, Mortimer DN, Cunningham VL (2001) So you think your process is green, how do you know? Using principles of sustainability to determine what is green – a corporate perspective. Green Chem 3:1–6

95. Höft E (1993) Enantioselective epoxidation with peroxidic oxygen. Top Curr Chem 164:63–77

96. Canali L, Karjalainen JK, Sherrington DC, Hormi O (1997) Efficient polymer-supported sharpless alkene epoxidation catalyst. Chem Commun 1997:123–124

97. Zhang W, Loebach JL, Wilson SR, Jacobsen EN (1990) Enantioselective epoxidation of unfunctionalized olefins catalyzed by (Salen)manganese complexes. J Am Chem Soc 112:2801–2803

98. Irie R, Noda K, Ito Y, Matsumoto N, Katsuki T (1990) Catalytic asymmetric epoxidation of unfunctionalized olefins. Tetrahedron Lett 31:7345–7348

99. Katsuki T (1995) Catalytic asymmetric oxidations using optically-active (salen)manganese (III) complexes as catalysts. Coord Chem Rev 140:189–214

100. Srinivasan K, Michaud P, Kochi JK (1986) Epoxidation of olefins with cationic (salen)Mn^{III} complexes. The modulation of catalytic activity by substituents. J Am Chem Soc 108: 2309–2320

101. Groves JT, Myers RS (1983) Catalytic asymmetric epoxidations with chiral iron porphyrins. J Am Chem Soc 105:5791–5796

102. Limburg J, Brudvig GW, Crabtree RH (2000) Modeling the oxygen-evolving complex in photosystem II. In: Meunier B (ed) Biomimetic oxidations catalyzed by transition metal complexes. Imperial College Press, London, pp 509–541

103. Zou X, Fu X, Li Y, Tu X, Fu S, Luo Y, Wu X (2010) Highly enantioselective epoxidation of unfunctionalized olefins catalyzed by chiral Jacobsen's catalyst immobilized on phenoxy-modified zirconium poly(styrene-phenylvinylphosphonate)phosphate. Adv Synth Catal 352:163–170

104. Das P, Silva AR, Carvalho AP, Pires J, Freire C (2009) Enantioselective epoxidation of alkenes by Jacobsen catalyst anchored onto aminopropyl-functionalised laponite, MCM-41 and FSM-16. Catal Lett 129:367–375

105. Fraile JM, García JI, Mayoral JA (2009) Noncovalent immobilization of enantioselective catalysts. Chem Rev 109:360–417

106. Grigoropoulou G, Clark JH, Elings JA (2003) Recent developments on the epoxidation of alkenes using hydrogen peroxide as an oxidant. Green Chem 5:1–7
107. Bhattacharjee S, Anderson JA (2004) Epoxidation by layered double hydroxide-hosted catalysts. Catalyst synthesis and use in the epoxidation of R-(+)-Limonene and (-)-α-Pinene using molecular oxygen. Catal Lett 95:119–125
108. Creager SE, Raybuck SA, Murray RW (1986) An efficient electrocatalytic model cytochrome P-450 epoxidation cycle. J Am Chem Soc 108:4225–4227
109. Horwitz CP, Creager SE, Murray RW (1990) Electrocatalytic olefin epoxidation using manganese Schiff-base complexes and dioxygen. Inorg Chem 29:1006–1011
110. Guo P, Wong K-Y (1999) Enantioselective electrocatalytic epoxidation of olefins by chiral manganese Schiff-base complexes. Electrochem Commun 1:559–563
111. Fatibello-Filho O, Dockal ER, Marcolino-Junior LH, Teixeira MFS (2007) Electrochemical modified electrodes based on metal-salen complexes. Anal Lett 40:1825–1852
112. List B (2007) Introduction: organocatalysis. Chem Rev 107:5413–5415
113. MacMillan DWC (2008) The advent and development of organocatalysis. Nature 455:304–308
114. Dondoni A, Massi A (2008) Asymmetric organocatalysis: from infancy to adolescence. Angew Chem Int Ed 47:4638–4661
115. Wong OA, Shi Y (2008) Organocatalytic oxidation. Asymmetric epoxidation of olefins catalyzed by chiral ketones and iminium salts. Chem Rev 108:3958–3987
116. Tu Y, Wang Z-X, Shi Y (1996) An efficient asymmetric epoxidation method for trans-olefins mediated by a fructose-derived ketone. J Am Chem Soc 118:9806–9807
117. Shu L, Shi Y (1999) Asymmetric epoxidation using hydrogen peroxide (H_2O_2) as primary oxidant. Tetrahedron Lett 40:8721–8724
118. Sheldon RA, Arends IWCE (2004) Organocatalytic oxidations mediated by nitroxyl radicals. Adv Synth Catal 346:1051–1071
119. Bowman DF, Gillan T, Ingold KU (1971) Kinetic applications of electron paramagnetic resonance spectroscopy. III. Self-reactions of dialkyl nitroxide radicals. J Am Chem Soc 93:6555–6561
120. Phukan P, Khisti RS, Sudalai A (2006) Green protocol for the synthesis of N-oxides from secondary amines using vanadium silicate molecular sieve catalyst. J Mol Catal A Chem 248:109–112
121. Ciriminna R, Pagliaro M (2010) Industrial oxidations with organocatalyst TEMPO and its derivatives. Org Proc Res Dev 14:245–251
122. Carson R (1962) Silent spring. Houghton Mifflin, Boston
123. Blackmond DG, Armstrong A, Coombe V, Wells A (2007) Water in organocatalytic processes: debunking the myths. Angew Chem Int Ed 46:3798–3800
124. Mantzavinos D, Psillakis E (2004) Enhancement of biodegradability of industrial wastewaters by chemical oxidation pre-treatment. J Chem Tech Biotech 79:431–454
125. Walling C (1975) Fenton's reagent revisited. Acc Chem Res 8:125–131
126. Richter HW, Waddell WH (1983) Mechanism of the oxidation of dopamine by the hydroxyl radical in aqueous solution. J Am Chem Soc 105:5434–5440
127. Fenton HJH (1894) Oxidation of tartaric acid in presence of iron. J chem Soc Trans 65:899–910
128. Masarwa A, Rachmilovich-Calis S, Meyerstein N, Meyerstein D (2005) Oxidation of organic substrates in aerated aqueous solutions by the fenton reagent. Coord Chem Rev 249: 1937–1943
129. Watts RJ, Teel AL (2005) Chemistry of modified fenton's reagent (catalyzed H_2O_2 propagations-CHP) for in situ soil and groundwater remediation. J Enviro Eng 131:612–622
130. Millioli VS, Freire DDC, Cammarota MC (2002) Testing the efficiency of Fenton's reagent in treatment of petroleum-contaminated sand. Engenharia Térmica, Edição Especial 44–47

131. Nadtochenko V, Kiwi J (1998) Photoinduced mineralization of xylidine by the Fenton reagent. 2. Implications of the precursors formed in the dark. Environ Sci Technol 32:3282–3285

132. Yu R-F, Chen H-W, Liu K-Y, Cheng W-P, Hsieh P-H (2010) Control of the fenton process for textile wastewater treatment using artificial neural networks. J Chem Technol Biotechnol 85:267–278

133. Sun Y, Pignatello JJ (1992) Chemical treatment of pesticide wastes. Evaluation of Fe(III) chelates for catalytic hydrogen peroxide oxidation of 2, 4-D at circumeutral pH. J Agric Food Chem 40:322–327

134. Sun Y, Pignatello JJ (1993) Activation of hydrogen peroxide by iron(III) chelates for abiotic degradation of herbicides and insecticides in water. J Agric Food Chem 41:308–312

135. Lewis S, Lynch A, Bachas L, Hampson S, Ormsbee L, Bhattacharyya D (2009) Chelate-modified fenton reaction for the degradation of trichloroethylene in aqueous and two-phase systems. Enviro Eng Sci 26:849–859

136. Cybulski A (2007) Catalytic wet air oxidation: are monolithic catalysts and reactors feasible? Ind Engr Chem Res 46:4007–4033

137. Collins TJ, Gordon-Wylie SW, Bartos MJ, Horwitz CP, Woomer CG, Williams SA, Patterson RE, Vuocolo LD, Paterno SA, Strazisar SA, Peraino DK and Dudash CA (1998) The design of green oxidants. In: Anastas P, Warner JC (eds) Greem Chemistry, pp. 46–71

138. Ellis WC, Tran CT, Denardo MA, Fischer A, Ryabov AD, Collins TJ (2009) Design of more powerful iron-TAML peroxidase enzyme mimics. J Am Chem Soc 131:18052–18053

139. Collins TJ, Khetan SK, Ryabov AD (2009) Chemistry and applications of iron-TAML catalysts in green oxidation processes based on hydrogen peroxide. In: Anastas PT, Crabtree R (eds) Handbook of green chemistry, vol 1. Wiley, Weinheim, pp 39–77

140. Ryabov AD, Collins TJ (2009) Mechanistic considerations on the reactivity of green FeIII-TAML activators of peroxides. Adv Inorg Chem 61:471–521

141. Chanda A, Ryabov AD, Mondal S, Alexandrova L, Ghosh A, Hangun-Balkir Y, Horwitz CP, Collins TJ (2006) Activity-stability parameterization of homogeneous green oxidation catalysts. Chem Euro J 12:9336–9345

142. Polshin V, Popescu D-L, Fischer A, Chanda A, Horner DC, Beach ES, Henry J, Qian Y-L, Horwitz CP, Lente G, Fabian I, Muenck E, Bominaar EL, Ryabov AD, Collins TJ (2008) Attaining control by design over the hydrolytic stability of Fe-TAML oxidation catalysts. J Am Chem Soc 130:4497–4506

143. Georgi A, Schierz A, Trommler U, Horwitz CP, Collins TJ, Kopinke F (2007) Humic acid modified Fenton reagent for enhancement of the working pH range. Appl Catal B Enviro 72:26–36

144. Gupta SS, Stadler M, Noser CA, Ghosh A, Steinhoff B, Lenoir D, Horwitz CP, Schramm K-W, Collins TJ (2002) Rapid total destruction of chlorophenols by activated hydrogen peroxide. Science 296:326–328

145. Horwitz CP, Fooksman DR, Vuocolo LD, Gordon-Wylie SW, Cox NJ, Collins TJ (1998) Ligand design approach for securing robust oxidation catalysts. J Am Chem Soc 120:4867–4868

146. Chahbane N, Popescu D-L, Mitchell DA, Chanda A, Lenoir D, Ryabov AD, Schramm K-W, Collins TJ (2007) Fe^{III}-TAML-catalyzed green oxidative degradation of the azo dye Orange II by H$_2$O$_2$ and organic peroxides: products, toxicity, kinetics, and mechanisms. Green Chem 9:49–57

147. Horwitz CP, Collins TJ, Spatz J, Smith HJ, Wright LJ, Stuthridge TR, Wingate KG, McGrouther K (2006) Iron-TAML catalysts in the pulp and paper industry. ACS Symp Ser 921:156–169

148. Chanda A, Khetan SK, Banerjee D, Ghosh A, Collins TJ (2006) Total degradation of fenitrothion and other organophosphorus pesticides by catalytic oxidation employing Fe-TAML peroxide activators. J Am Chem Soc 128:12058–12059

149. Banerjee D, Markley AL, Yano T, Ghosh A, Berget PB, Minkley E Jr, Khetan SK, Collins TJ (2006) "Green" oxidation catalysis for rapid deactivation of bacterial spores. Angew Chem Int Ed 45:3974–3977

150. Shappell NW, Vrabel MA, Madsen PJ, Harrington G, Billey LO, Hakk H, Larson GL, Beach E, Horwitz CP, Ro K, Hunt PG, Collins TJ (2008) Destruction of estrogens using Fe-TAML/peroxide catalysis. Environ Sci Technol 42:1296–1300

151. Lewis NS, Nocera DG (2006) Powering the planet: chemical challenges in solar energy utilization. Proc Nat Acad Sci 103:15729–15735

152. Nocera DG (2009) Chemistry of personalized solar energy. Inorg Chem 48:10001–10017

153. Dismukes GC, van Willigen RT (2006) Manganese: the oxygen-evolving complex & models. Encyclopedia of Inorganic Chemistry 5:1–15

154. Raymond J, Blankenship RE (2008) The origin of the oxygen-evolving complex. Coord Chem Rev 252:377–383

155. Dincă M, Surendranath Y, Nocera DG (2010) Nickel-borate oxygen-evolving catalyst that functions under benign conditions. Proc Nat Acad Sci 107:10337–10341

156. Surendranath Y, Dinca M, Nocera DG (2009) Electrolyte-dependent electrosynthesis and activity of cobalt-based water oxidation catalysts. J Am Chem Soc 131:2615–2620

157. Cross DP, Ramachandran G, Wattenberg EV (2001) Mixtures of nickel and cobalt chlorides induce synergistic cytotoxic effects: implications for inhalation exposure modeling. Ann Occup Hyg 45:409–418

158. Nakagawa T, Beasley CA, Murray RW (2009) Efficient electro-oxidation of water near its reversible potential by a Mesoporous IrOx Nanoparticle Film. J Phys Chem C 113:12958–12961

159. Yin Q, Tan JM, Besson C, Geletii YV, Musaev DG, Kuznetsov AE, Luo Z, Hardcastle KI, Hill CL (2010) A fast soluble carbon-free molecular water oxidation catalyst based on abundant metals. Science 328:342–345

160. Baer K, Kraußr M, Burda E, Hummel W, Berkessel A, Gröger H (2009) Sequential and modular synthesis of chiral 1, 3-diols with two stereogenic centers: access to all four stereoisomers by combination of organo- and biocatalysis. Angew Chem Int Ed 48:9355–9358

161. Turner NJ (2009) Directed evolution drives the next generation of biocatalysts. Nat Chem Biol 5:567–573

162. Fogg DE, dos Santos EN (2004) Tandem catalysis: a taxonomy and illustrative review. Coord Chem Rev 248:2365–2379

163. Nicolaou KC, Chen JS (2009) The art of total synthesis through cascade reactions. Chem Soc Rev 38:2993–3009

164. Forster PM, Cheetham AK (2003) Hybrid inorganic-organic solids: an emerging class of nanoporous catalysts. Top Catal 24:79–86

165. Kaneda K, Mizugaki T (2009) Development of concerto metal catalysts using apatite compounds for green organic syntheses. Energy Enviro Sci 2:655–673

166. Wight AP, Davis ME (2002) Design and preparation of organic-inorganic hybrid catalysts. Chem Rev 102:3589–3614

167. Brunel D, Fajula F, Nagy JB, Deroide B, Verhoef MJ, Veum L, Peters JA, van Bekkum H (2001) Comparison of two MCM-41 grafted TEMPO catalysts in selective alcohol oxidation. Appl Catal A Gen 213:73–82

168. Capello C, Fischer U, Hungerbühler K (2007) What is a green solvent? a comprehensive framework for the environmental assessment of solvents. Green Chem 9:927–934

169. Olivier-Bourbigou H, Magna L, Morvan D (2009) Ionic liquids and catalysis: recent progress from knowledge to applications. Appl Catal A Gen 373:1–56

170. Deetlefs M, Seddon KR (2010) Assessing the greenness of some typical laboratory ionic liquid preparations. Green Chem 12:17–30

171. Zhang W (2009) Green chemistry aspects of fluorous techniques-opportunities and challenges for small-scale organic synthesis. Green Chem 11:911–920

172. Akien GR, Poliakoff M (2009) A critical look at reactions in class I and II gas-expanded liquids using CO_2 and other gases. Green Chem 11:1083–1100
173. Liu Y, Jessop PG, Cunningham M, Eckert CA, Liotta CL (2006) Switchable surfactants. Science 313:958–960
174. Chavali S, Lin B, Miller DC, Camarda KV (2004) Environmentally-benign transition metal catalyst design using optimization techniques. Comput Chem Eng 28:605–611
175. Guidoni L, Spiegel K, Zumstein M, Röthlisberger U (2004) Green oxidation catalysts: computational design of high-efficiency models of galactose oxidase. Angew Chem Int Ed 43:3286–3289
176. Schüth F, Busch O, Hoffmann C, Johann T, Kiener C, Demuth D, Klein J, Schunk S, Strehlau W, Zech T (2002) High-throughput experimentation in oxidation catalysis. Top Catal 21:55–66
177. Corma A, Iborra S, Velty A (2007) Chemical routes for the transformation of biomass into chemicals. Chem Rev 107:2411–2502
178. Christensen CH, Rass-Hansen J, Marsden CC, Taarning E, Egeblad K (2008) The renewable chemicals industry. ChemSusChem 1:283–289
179. Bozell JJ, Petersen GR (2010) Technology development for the production of biobased products from biorefinery carbohydrates – the US Department of Energy's "Top 10" revisited. Green Chem 12:539–554
180. Scott EL, Sanders JPM, Steinbüchel A (2010) Perspectives on Chemicals from Renewable Resources. In: Sustainable biotechnology, pp 195–210

Chapter 10
Supercritical Carbon Dioxide (CO_2) as Green Solvent

Tianbin Wu and Buxing Han

Glossary

Chemical structure of carbon dioxide	Carbon dioxide (CO_2) has a structure O=C=O.
Green solvent	A green solvent should have some basic properties, such as low toxic, chemically stable, readily available, and easily recyclable.
Ionic liquids	Ionic liquids are salts that are liquid at ambient conditions.
Microemulsion	A microemulsion is a thermodynamically stable dispersion formed from immiscible substances with the aid of surfactants.
Supercritical fluid	A substance is called as a supercritical fluid (SCF) when the temperature and pressure are higher than its critical values.

Definition of the Subject

A substance is called as a supercritical fluid (SCF) when the temperature and pressure are higher than its critical values. Therefore, CO_2 becomes supercritical when its temperature and pressure are higher than 31.1°C (critical temperature, Tc) and pressure 7.38 MPa (critical pressure, P_c). SCFs have many unique properties, such as strong solvation power for different solutes, large diffusion coefficient

This chapter was originally published as part of the Encyclopedia of Sustainability Science and Technology edited by Robert A. Meyers. DOI:10.1007/978-1-4419-0851-3

T. Wu (✉) • B. Han
Institute of Chemistry, Chinese Academy of Sciences, Beijing 100190, China
e-mail: wtb@iccas.ac.cn; hanbx@iccas.ac.cn

P.T. Anastas and J.B. Zimmerman (eds.), *Innovations in Green Chemistry and Green Engineering*, DOI 10.1007/978-1-4614-5817-3_10,
© Springer Science+Business Media New York 2013

comparing with liquids, zero surface tension, and their physical properties can be tuned continuously by varying the pressure and temperature because the isothermal compressibility of SCFs is very large, especially in the critical region. These unique properties of SCFs lead to great potential for the development of innovative technologies. Besides these common advantages of SCFs, supercritical CO_2 (scCO$_2$) has some other advantages, such as nontoxic, nonflammable, chemically stable, readily available, cheap, and easily recyclable, and it has easily accessible critical parameters. Therefore, scCO$_2$ can be used as green solvent in different fields.

Introduction

It is well known that chemical industry has made great contribution to mankind. However, many volatile, toxic organic solvents are used in chemical processes, which results in environment and safety problems. In recent years, sustainable development has become one of the most important topics in the world, and great effort has been devoted to green chemistry. Green chemistry can be simply defined as "design of chemical products and processes that reduce or eliminate the use and generation of hazardous substances." Effective utilization of green solvents is one of the main topics of green chemistry.

SCFs are typical green solvents. In general, physicochemical properties of SCFs are intermediate to those of the liquid and gaseous states. The ability of a SCF to dissolve low-volatile substances was found more than century ago [1]. However, SCFs did not receive enough attention before 1960s. Systemic study of the properties of SCFs and their applications began in 1970s. It is now well known that SCFs possess many unusual properties, such as strong solvation power for many substances, excellent diffusivity, and near zero surface tension. Moreover, many properties of SCFs can be tuned continuously by varying pressure and/or temperature. Especially, a small change in pressure and temperature remarkably alters the properties of a SCF in the vicinity of its critical point, such as density, viscosity, diffusivity, dielectric constant, and solvent power. The tunable nature of SCFs is favorable to developing new technologies. ScCO$_2$ has the common advantages of SCFs. In addition, it has more advantages because CO_2 is cheap, nontoxic, nonflammable, readily available, and has easily accessible critical point.

Effective utilization of scCO$_2$ and other SCFs as greener media has attracted much attention in recent years. On one hand, they can be used to replace hazardous organic solvents in many chemical processes. On the other hand, the efficiency of some processes can be optimized by using the special properties of SCFs, even some new technologies with obvious advantages can be developed. Both fundamental researches and developing high technologies have achieved significantly over the past decades [2, 3]. In this entry, the basic properties of scCO$_2$ and its applications in extraction and fractionation, chemical reactions, polymeric synthesis, material science, supercritical chromatography, painting, dyeing and cleaning, emulsions related with CO_2, and so on are discussed. It should be emphasized that in each field only some examples are discussed because too many excellent publications are available in the open literature.

The Properties of scCO$_2$

In the immediate vicinity of the critical point, the density of scCO$_2$ is about 0.4 g/mL and increases with increasing pressure and decreases as temperature rises. The diffusion coefficient of scCO$_2$ is much larger than that of liquid solvents. The increase of temperature and decrease of pressure lead to the enlargement of self-diffusion coefficient of scCO$_2$. In general, the viscosity of scCO$_2$ is at least an order of magnitude lower than that of liquid solvents and rapidly increases with pressure near the critical point. ScCO$_2$ shows essentially the properties of nonpolar solvent. It is therefore more suitable for dissolution of weak polar substances. In general, scCO$_2$ is a good solvent for many nonpolar molecules with low molecular weights. It is however a very poor solvent for most high molecular weight polymers under readily achievable conditions. Nevertheless, the addition of some cosolvents (e.g. methanol, ethanol, acetone) to scCO$_2$ can improve the solubility of the polar substances in scCO$_2$ significantly. Its surface tension reduces to zero when temperature exceeds its critical temperature (31.1°C), which makes scCO$_2$ easily penetrate a microporous solid structure. Moreover, scCO$_2$ can swell many polymers. These properties are beneficial for the preparation of porous polymer materials and metal nanoparticles stabilized in porous polymer materials.

Applications of scCO$_2$

Extraction and Fractionation Using scCO$_2$ as Solvent

Supercritical fluid extraction (SFE) is the process of separating one or some components from others using SCFs as the extracting solvent. SFE is an alternative to traditional extraction methods using organic solvents. Extraction is suitable for both solid matrixes and liquids. SFE has some obvious advantages. For example, the process is simple; mild operation temperature; the extraction rate and phase separation can be much faster than those of conventional extraction methods; the extraction efficiency can be controlled through changes of the system pressure and/or temperature because the solubility of a solute in SCFs depends on both vapor pressure of the solute and the solute–solvent interaction; toxic organic solvents can be avoided and the quantity of residual solvent in the products is negligible.

Moreover, the addition of small amount of cosolvents to scCO$_2$ can improve the extraction efficiency greatly in many cases. SFE has been studied extensively in food, flavoring, pharmaceutical, petroleum, and coal industries, and some techniques have been commercialized. Some other applications such as catalyst regeneration and soil protection have also been studied.

SFE using scCO$_2$ as the solvent has unusual advantages for the applications in food, flavor, and pharmaceutical industries due to low operation temperature and no solvent residue in the products. Application of SFE in food industry started from 1970s, and this technique has been industrialized [4]. The earliest industrial application of SFE is extracting caffeine from coffee beans using scCO$_2$, and it is shown that the technique has some prominent advantages compared with organic solvent extraction [4]. Another successful application of SFE is the extraction of hops in beer industry [5]. Nonpolar CO$_2$ is an ineffective solvent for substances of high polarity. Therefore, polar cosolvents are sometimes added to enhance the overall polarity of the fluid phase during extraction. For example, the solubility of phospholipids in neat CO$_2$ is extremely low in SFE. However, addition of ethanol as cosolvent makes it possible to selectively extract phospholipids from soybeans [6]. Extraction of soybean isoflavones using scCO$_2$ modified with aqueous methanol is studied [7], and the results show that the isoflavones recovery can be very high at suitable condition.

In the past 3 decades, rapid progress has been obtained for the extraction of essential oils and flavor compounds using scCO$_2$ [8, 9]. On the basis of experimental data of scCO$_2$ extraction of essential oils from plant materials, a mathematical model for the partition of a solute between the solid matrix and the solvent has been proposed [10]. Peach almond oil can also be obtained by means of SFE with a satisfactory yield [11]. The extraction pressure, CO$_2$ flow rate, and particle size of the matrix on the extraction are evaluated, and the scale-up methodologies are studied.

SFE technique has also been used to upgrade petroleum feedstocks. The separation of petroleum residuum using SFE technique can obtain high-quality deasphalted oil that has been widely used as the raw materials of catalytic cracking and lubricants [12]. The effect of temperature and pressure on the extraction of crude oil and bitumen using scCO$_2$ is investigated [13], and it is demonstrated that higher oil yields can be obtained. In the crude oil and native bitumen extractions, heavier compounds are extracted in supercritical phase as the extraction time and/or the extraction pressure increased. Quantitative recovery of petroleum hydrocarbons from a contaminated soil [14] by SFE technique has also achieved satisfactory results.

Coals can be separated into oils, asphaltenes, and pre-asphaltenes by SFE method. The oil is divided into aliphatics, aromatics, and polar compounds. The prediction method of the solubilities of naphthalene, phenanthrene, phenol, and 2-naphthol in scCO$_2$ has been developed [15]. Study of applications of scCO$_2$ extraction in pharmaceutical industry [16], treatment of nuclear waste [17], and tobacco industry [18] has also obtained satisfactory progress. Some studies have shown that scCO$_2$ can be used to extract triglycerides from a number of natural products effectively. It is concluded that high concentration of triglycerides in the extract and high conversion of methylation lead to a high-quality biodiesel [19].

Chemical Reactions in scCO$_2$

Many chemical reactions have been conducted in scCO$_2$. Besides the solvent is greener, chemical reactions in scCO$_2$ or under supercritical condition have many other advantages. For example, reaction rates, yields, and selectivity can be adjusted by varying temperature and/or pressure; mass transfer can be improved for heterogeneous reactions; and simultaneous reaction and separation can be accomplished more easily for some reactions. In some cases, scCO$_2$ acts as both reactant and solvent.

Hydrogenation Reactions in scCO$_2$

Hydrogenation is a class of important reaction in petroleum, food, and fine chemical industries. The solubility of H$_2$ in liquids is very small. The excellent miscibility of scCO$_2$ with H$_2$ and liquid reactants can help to avoid mass-transfer limitation, which are frequently encountered in gas-liquid phase reactions. Application of scCO$_2$ to hydrogenation reactions is particularly attractive [20, 21].

It is known that allyl alcohol easily isomerizes into by-product propanal or acetone during its hydrogenation to produce 1-propanol catalyzed by Pd nanoparticles [22]. To prevent isomerization, Pd nanoparticles can be embedded in polymer to enhance the selectivity of 1-propanol and the stability of the catalysts, but polymer restricted allyl alcohol to access active sites of Pd nanoparticles, which reduce the reaction rate. It is possible improve the reaction rate effectively by using scCO$_2$ due to its higher diffusivity, near zero interfacial tension and swelling effect on polymer. The hydrogenation of allyl alcohol in scCO$_2$ over Pd nanocatalysts on silica support with cross-linked polystyrene coating has been conducted [23]. The selectivity of the reaction can be enhanced significantly by the polymer coating. The catalysts are very stable due to the insoluble nature of the cross-linked polymers, and scCO$_2$ can accelerate the reaction rate of the reaction significantly.

Selective hydrogenation of α, β-unsaturated aldehydes into the corresponding unsaturated alcohols is of importance in flavor, fragrance, and pharmaceutical industries. The selective hydrogenation of citral in scCO$_2$ has been carried out. The selectivity to the partially saturated aldehyde (citronellal) or unsaturated alcohols (geraniol and nerol) can be tuned by varying metal catalysts and CO$_2$ pressure. The monometallic Pt catalyst is highly selective to the unsaturated alcohol (geraniol and nerol), whereas the bimetallic Pt-Ru catalyst becomes selective to the partially saturated aldehyde (citronellal) under the same scCO$_2$ conditions. The selectivity for the unsaturated alcohols is larger at higher CO$_2$ pressures. It has also been shown that the selective hydrogenation of citral to produce only geraniol (*trans*) with remarkable selectivity at a very high conversion, and the selectivity for geraniol (*trans*) strongly depends on CO$_2$ and hydrogen pressure, reaction time, citral concentration, and the reaction temperature [24]. The effect of CO$_2$ pressure on product distribution for the hydrogenation of citral over Pd and Ru nanoparticles hosted in reverse micelles with scCO$_2$ as the continuous phase has been investigated [25]. The selective formation of a particular hydrogenated product can be achieved by carefully tuning the pressure of the fluid, which can alter the balance between solvation of the molecule into the fluid and its binding affinity to the metal surface.

The selective hydrogenation of nitro compounds is commonly used to manufacture amines, which are important intermediates for dyes, urethanes, agro-chemicals, and pharmaceuticals. The hydrogenation of nitrobenzene in $scCO_2$ and in ethanol catalyzed by Pd, Pt, Ru, and Rh supported on C, SiO_2, and Al_2O_3 has been carried out [26]. For all the catalysts, higher selectivity to aniline has been obtained in $scCO_2$ compared with ethanol. The yield to aniline can approach 100% at optimized condition. The hydrogenation of a series of substituted nitro compounds such as 2-,3-,4-nitroanisole, 2-,3-,4-nitrotoluene, 2,4-dinitrobenzene, and 2,4-dinitrotoluene in $scCO_2$ and ethanol has been conducted with carbon supported platinum catalyst [27]. The solubility of these compounds in $scCO_2$ has also been examined at different conditions. The solubility of the nitro compounds increases with increasing CO_2 pressure, but decreases in the presence of hydrogen. Although the total conversion obtained in $scCO_2$ is similar to that in ethanol, the selectivity to amino products is higher in the former reaction medium.

The hydrogenation of phenol is an important approach for preparation of cyclohexanone. Pd is a commonly used catalyst. However, the activity of the reaction is low under mild condition, and the product cyclohexanone can be further hydrogenated to by-product cyclohexanol easily. Therefore, the attainment of high selectivity at elevated conversion with a satisfactory rate is very difficult. Recent work has illustrated that commercial supported Pd catalysts and solid Lewis acids have excellent synergy in the hydrogenation of phenol to cyclohexanone, and the conversion and selectivity of the reaction approach 100% simultaneously under mild condition [28]. Moreover, it is further found that in $scCO_2$ reaction is faster and the reaction efficiency depends on the phase behavior of the reaction system, and the separation process is very easy.

Asymmetric hydrogenation plays a critical role in both the pharmaceutical and agrochemical industries. Hydrogenation of α, β-unsaturated carboxylic acids, such as tiglic acid, has been performed in $scCO_2$ with Ru-based catalyst. The enantiomeric excess (ee) of the product in $scCO_2$ is comparable to that in methanol and greater than that in hexane. $ScCO_2$ has also been used as the medium for the asymmetric hydrogenation of α-enamides using a cationic Rh complex as the catalyst [29]. The reactions can proceed homogeneously at controlled temperature and pressure. The ee is fair to excellent and generally comparable to that obtained in conventional solvents. The asymmetric hydrogenation of C=O [30] and C=N [31] double bonds have also been performed in $scCO_2$. It is demonstrated that $scCO_2$ is an attractive alternative reaction medium for the enantioselective hydrogenation. $ScCO_2$ provides advantages over the organic solvents and enhances the enantioselectivity and catalytic efficiency.

Continuous flow $scCO_2$ has been shown to be a viable medium for conducting continuous asymmetric hydrogenation in the presence of catalysts [32]. The hydrogenation of complex pharmaceutical intermediate, rac-sertraline imine has been optimized in a continuous flow process utilizing a palladium/calcium carbonate catalyst in $scCO_2$ [33]. Superior levels of selectivity can be obtained in the flow system. It is suggested that the excellent heat transfer properties of the $scCO_2$ help to maintain exceptional levels of chemoselectivity even at elevated temperature. The application of $scCO_2$ in continuous, fixed bed reactors has allowed the successful development of a variety of industrially viable synthetic transformations. The multireaction, supercritical flow reactor was commissioned in 2002. The development of this project from laboratory to plant scale is highlighted, particularly in the context of the hydrogenation of isophorone [34].

Oxidation Reactions in $scCO_2$

Oxidation reactions are among the most important processes for introducing functional groups into hydrocarbons. $ScCO_2$ is advantageous because of its inertness towards oxidation, thus providing additional safety and avoiding side products from solvent oxidation. The selective oxidation of hydrocarbons with molecular oxygen is of special interest for synthetic chemistry from an economic as well as an ecological point of view. $ScCO_2$ allows the combination of high concentration of substrates and oxygen with the additional safety. It has been reported that the presence of CO_2 can improve yields, reaction rates, and/or selectivities in some selective alkane and alkene oxidation processes.

The selective oxidation of cyclohexane into cyclohexanol and cyclohexanone is of considerable importance in chemical industry for producing Nylon-6 polymers. The selective oxidation reaction of cyclohexane to produce cyclohexanol and cyclohexanone is carried out in CO_2 at 398.2 K using oxygen as an oxidant, and the reaction mixture is controlled to be in the two-phase region, very close to the critical point, and in the supercritical region of the reaction system. The effect of a small amount of butyric acid cosolvent on the reaction in $scCO_2$ is also studied [35]. The conversion and selectivity of the reaction in $scCO_2$ change considerably with the phase behavior or the apparent density of the reaction system. Addition of a small amount of butyric acid cosolvent to $scCO_2$ enhances the conversion significantly, and the selectivity also changes considerably. The by-products of the reaction in $scCO_2$ with and without the cosolvent are much less than that of the reaction in liquid solvents or in the absence of solvents. The use of supported metal oxides leads to higher yields for target products cyclohexanone and cyclohexanol compared to the immobilized complexes. A series of CoAPO-5 with different Co contents have been synthesized and used to catalyze selective oxidation of cyclohexane in $scCO_2$ [36]. The CoAPO-5 catalysts are effective catalysts for the reaction and the total selectivity of objective products are high in $scCO_2$.

The oxidation of cyclooctane to cyclooctanone with molecular oxygen and acetaldehyde as a coreductant proceeds efficiently in the presence of compressed CO_2 [37]. CO_2 is more effective than other inert diluting gases at the same conditions. The selective oxidation of other alkanes has also been studied, and the results are promising [38]. For example, the oxidation of cycloalkanes or alkylarenes with molecular oxygen and acetaldehyde as sacrificial coreductant occurs efficiently in scCO$_2$ under mild multiphase conditions without catalyst. In comparison with other inert gases, the yields of alkane oxygenates in CO$_2$ is higher at identical reaction conditions.

Wacker reaction is an efficient way to produce methyl ketones via the Pd(II)-catalyzed oxidation of terminal olefins. It has been shown that the Wacker reaction of oct-1-ene can be proceeded smoothly in scCO$_2$ or methanol/scCO$_2$ mixed SCF [39]. The selectivity of product octa-2-one of the reaction proceeded in scCO$_2$ is obviously higher than that in methanol at the same reaction time. However, the reaction rate in scCO$_2$ is lower than that in methanol due to the insoluble nature of PdCl$_2$ and CuCl$_2$ in scCO$_2$. Nevertheless, addition of cosolvent methanol in scCO$_2$ can accelerate the reaction rate and reduce the isomerization of oct-1-ene. PdCl$_2$/polystyrene-supported benzoquinone can also be employed as the catalyst for the acetalization of terminal olefins with electron-withdrawing groups [40]. In scCO$_2$, the reaction can proceed with high yield and selectivity, and the catalyst can be easily recycled.

The selective oxidation of styrene to acetophenone over Pd-Au bimetallic catalyst using H$_2$O$_2$ as the oxidant in scCO$_2$ medium has been studied [41]. The bimetallic catalyst supported on Al$_2$O$_3$ is very effective for the reaction. The presence of CO$_2$ improves the oxidation of styrene to acetophenone and inhibits the formation of the by-products. An electron-donating group at the *para*-position of styrene increases the conversion of styrene, but an electron-withdrawing group at the *meta*-position of styrene reduces both the conversion and the selectivity to acetophenone.

Poly(ethylene glycol) (PEG) is an inexpensive, nonvolatile, and environmentally benign solvent, which is another kind of green reaction medium. Combination of PEG and scCO$_2$ has some obvious advantages for chemical reactions. The PdCl$_2$-catalyzed aerobic oxidation of styrene in a biphasic PEG/scCO$_2$ system results in two possible products, benzaldehyde and the Wacker oxidation product acetophenone [42]. The selectivity of the reaction can be switched by the cocatalyst CuCl, from 92% in favor of acetophenone in the presence of CuCl to 85% favoring benzaldehyde without any cocatalyst. The PEG in the reaction system can effectively stabilize the catalysts.

Alkyl 3,3-dialkoxypropanoates are important intermediates in organic synthesis and have been used to synthesize a variety of compounds. The oxidation of methyl acrylate catalyzed by Pd(II)/CuCl$_2$ can be carried out in scCO$_2$ to give the dimethyl acetal as major product when an excess amount of methanol is used [43]. A conversion of 99.4% and selectivity of 96.6% can be obtained under the suitable condition. The reaction medium and the electron property of alkene substituent group obviously influence the regioselectivity of oxidation for methyl acrylate.

The aerobic oxidation of alcohols to aldehydes and ketones is a fundamental chemical transformation for the production of a variety of important intermediates and fine chemical products. Aerobic oxidation of benzyl alcohol in scCO$_2$ catalyzed with perruthenate immobilized on polymer supports has been conducted [44]. The catalyst is very active and highly selective in scCO$_2$. The reaction rate in scCO$_2$

depends strongly on pressure and reaches maximum at about 14 MPa, which can be explained by the effects of pressure on the phase behavior of reaction system, diffusivity of the components, and solvent power of scCO$_2$. The catalyst can be reused directly after extraction of the products using scCO$_2$. PEG/CO$_2$ biphasic system has also been used for aerobic oxidation of alcohols. For example, a highly efficient PEG/CO$_2$ biphasic catalytic system for the aerobic oxidation for allylic alcohols, primary and secondary benzylic alcohols, cyclic alcohol, and 1-butanol has been reported [45], which is highly active, selective, and stable for a broad range of substrates. In the system, the PEG matrix effectively stabilizes and immobilizes the catalytically active Pd nanoparticles, whereas the unique solubility and mass-transfer properties of scCO$_2$ allow continuous processing at mild conditions, even for the low-volatility substrates. This system well exemplifies the benefits resulting from the combination of scCO$_2$ and PEG for the reactions. Study of aerobic oxidation of benzhydrol, 1-phenylethanol, and cyclohexanol to corresponding ketones in PEG/scCO$_2$ biphasic system using unsupported and supported CoCl$_2$·6H$_2$O as the catalysts [46] has indicated that the CoCl$_2$·6H$_2$O, Co(II)/Al$_2$O$_3$, and Co(II)/ZnO are all active and selective for the reactions, and the yields of the desired products can be optimized by the pressure of scCO$_2$. Co(II)/ZnO is most stable and can be reused. Recently, a miniature catalytic reactor has been developed, which can be used for the continuous oxidation of primary and secondary alcohols with molecular oxygen in scCO$_2$. Satisfactory yields can be achieved at the optimized conditions [47].

Photo-oxidation in scCO$_2$ has become an interesting topic. A continuous photocatalytic reactor for performing reactions of ^{1}O$_2$ in scCO$_2$ has been developed. The reactor has demonstrated the potential of using light-emitting diodes (LED) for performing synthetic photochemistry in a continuous milliliter-scale reactor. This has been accomplished by the combination of high-power LED technology with a high-pressure scCO$_2$ reactor system capable of supporting high concentrations of O$_2$ with negligible mass-transfer limitations [48].

A scCO$_2$-soluble photosensitiser has been used to perform ^{1}O$_2$ reactions of α-terpinene, 2,3-dimethyl-2-butene, and 1-methyl-cyclohexene in scCO$_2$ [49]. The three reactions show zero-order kinetics and quantitative conversions with reasonable reaction rate. Some insoluble photosensitisers have also been evaluated with highly fluorinated surfactants and dimethyl carbonate cosolvent. It is clear that the applicability of ^{1}O$_2$ reactions in scCO$_2$ has been significantly enhanced by the use of these surfactants and cosolvent.

Carbon–Carbon Bond Forming Reactions

The efficient generation of a carbon–carbon bond is the essence of synthetic chemistry. The efficient formation of carbon–carbon bonds is still a synthetic challenge in organic chemistry although this topic has been the focus for many years.

The Diels–Alder reactions are among the most important synthetic reactions for the construction of polycyclic ring compounds. The Diels–Alder reaction in scCO$_2$ is also an interesting topic [50, 51]. Investigation of the Diels–Alder reaction between p-benzoquinone and cyclopentandiene in CO$_2$ illustrates that the reaction effectively occurs throughout the liquid and supercritical range with no discontinuity and that the rate of the reaction is about 20% larger than those obtained in diethyl ether [52]. It has been shown that for the Diels–Alder reaction in scCO$_2$ using scandium tris(heptadecafluorooctanesulfonate) as a Lewis acid catalyst [53], the catalyst activity is improved by increasing the length of the perfluoroalkyl chain and its solubility. This catalyst has also been used in the aza-Diels–Alder reaction of Danishefsky's diene with the imine in scCO$_2$, and 99% yield of the corresponding aza-Diels–Alder adduct can be obtained.

PdCl$_2$ is a commonly used catalyst for the cyclotrimerization of alkynes to substituted benzene derivatives. Cyclotrimerization of alkynes to substituted benzene derivatives in scCO$_2$ can smoothly proceed [54]. The yields and regioselectivities are high as CuCl$_2$ and methanol are added. CuCl$_2$ can not only accelerate the rate of cyclotrimerization of alkynes, but also is an absolutely indispensable factor for the cyclotrimerization of unsymmetrical alkynes.

The Glaser coupling is a synthesis of symmetric or cyclic bisacetylenes via a coupling reaction of terminal alkynes. There is a competition between coupling and carbonylation reactions when carbonylation reaction is operated in scCO$_2$. The Glaser coupling can be carried out effectively in scCO$_2$ using a solid base instead of amines [55]. The coupling reaction gives low conversion and yield in pure scCO$_2$, in which CuCl$_2$ and NaOAc cannot dissolve. However, the presence of methanol remarkably enhanced the rate of the reaction. The higher CO$_2$ pressure is favorable to the enhancing the reaction. In scCO$_2$, the coupling can also take place without base with higher conversion and yield than those in the presence of pyridine.

Suzuki and Heck reactions are among the most versatile families of reactions. The catalysts soluble in scCO$_2$ are crucial for high reaction efficiency. Polyfluoroalkylphosphine ligand has been prepared to enhance the solubility of Pd(II) catalysts for coupling reactions of phenyl iodide [56]. While PdCl$_2$[P(C$_6$H$_5$)$_3$]$_2$ and Pd(O$_2$-CCH$_3$)$_2$[P(C$_6$H$_5$)$_3$]$_2$ are insoluble in scCO$_2$, their analogues containing P(C$_6$H$_5$)$_{3-n}$(CH$_2$CH$_2$C$_6$F$_{13}$)n ($n = 1$ or 2) ligands are partially soluble. In the Suzuki reaction of phenyl boronicacid catalyzed by Pd[((C$_6$F$_{13}$CH$_2$CH$_2$)$_2$PC$_6$H$_5$]$_2$Cl$_2$, the yield of biphenyl in scCO$_2$ is higher than that in benzene medium. Study has also indicated that the homocoupling of iodoarenes catalyzed by Pd(OCOCF$_3$)$_2$/P(2-furyl)$_3$ occurs effectively in scCO$_2$ [57], which provides an attractive alternative to conventional procedures. In addition, there is preferential solvation effect for the reaction in scCO$_2$.

Study of Heck coupling reaction of phenyl iodide [58] demonstrates that the coupling reaction can proceed in scCO$_2$ with rates and selectivites comparable to those in toluene. The fluorinated phosphines, particularly tris[3,5-bis (trifluoromethyl)phenyl]phosphine, result in high conversions due to the ability of these ligands to enhance the solubility of the metal complexes in scCO$_2$. CO$_2$-philic ligands can expand the utility of scCO$_2$ for homogeneous catalysis efficiently.

Hydroformylation of olefins is an important industrial reaction used to transform alkenes into aldehydes using carbon monoxide and hydrogen. The reaction of propene with a Co catalyst $Co_2(CO)_8$ occurs at a slightly lower rate in $scCO_2$ than in hydrocarbon solvents such as methylcyclohexane and heptane [59], and the selectivity to the desired linear aldehyde, butanal is higher than that in the organic solvent. The linear-to-branch ratio is slightly influenced by the reaction pressure and temperature [60]. The linear product selectivity increases as pressure increases. One possible explanation for the change in the selectivity is a steric effect of a CO_2 molecule coordinated to the central metal as observed in the CoH $(CO)_3[P(C_4H_9)_3]$ catalyst system. Kinetic study of the reaction has demonstrated that the activation energy obtained in $scCO_2$ is comparable to or somewhat lower than those in organic solvents. More detailed study for the effect of different solvents on hydroformylation of olefins suggests that $scCO_2$ has some obvious advantages. Continuous process for the selective hydroformylation of higher olefins in $scCO_2$ is investigated [61, 62]. The catalyst shows high selectivity and activity over several hours and decrease in performance does not occur over several days.

Olefin metathesis refers to the mutual alkylidene exchange reaction of alkenes. $[Ru(H_2O)_6](OTs)_2$ (Ts=p-toluenesulfonyl) can catalyze the ring-opening metathesis polymerization (ROMP) of norbornene in $scCO_2$ [63]. The product can be isolated by just venting the CO_2. The chemical yield and molecular weight of the polymers are comparable to those in conventional solvents. Norbornene is quite soluble in $scCO_2$ under the reaction conditions, although the Ru catalyst is not completely soluble. However, the catalyst is sufficiently soluble for the completion of the reaction. In the presence of methanol cosolvent, the catalyst dissolved. The polymers obtained in $scCO_2$ have high *cis* olefin content, but adding methanol in $scCO_2$ results in a significant decrease in the *cis* content, indicating that the structure of the polymer is controllable by adjusting the polarity of the CO_2 medium with methanol. The isolable metal-carbene complexes are also highly effective for the ROMP of norbornene in $scCO_2$ [64]. Although the Ru-carbene complex retains catalytic activity even under aqueous emulsion conditions, $scCO_2$ is another environmentally benign solvent besides H_2O and offers the possibility of developing new polymer syntheses without solvent waste. It has also been reported that $scCO_2$ is a versatile reaction medium for ROMP and ring-closing olefin metathesis reactions using ruthenium and molybdenum catalysts. The unique properties of $scCO_2$ provide significant advantages beyond simple solvent replacement. This pertains to highly convenient workup procedures both for polymeric and low molecular weight products, to catalyst immobilization, to reaction tuning by density control, and to applications of $scCO_2$ as a protective medium [65].

Catalysis of Enzyme in $scCO_2$

Enzymes have been used widely as biocatalysts. However, the possibility of using enzymes as heterogeneous catalysts in supercritical media opens up more possibilities for chemical synthesis. Enzyme catalyzed reactions have been studied widely since the mile 1980s [66, 67]. Among enzymes active in scCO$_2$, lipase has been most frequently used because it catalyzes various types of reactions with wide substrate specificity. Thus, lipase-catalyzed reactions in scCO$_2$ have been carried out for different reactions, such as hydrolysis of triacylglycerol [68], alcoholysis of palm kernel oil [69], glycerolysis of soybean oil [70], acylation of glucose with lauric acid [71], transesterifications of milk fat with canola oil [72], and synthesis of poly-L-lactide [73].

The effect of compressed CO$_2$ on the specific activity of chloroperoxidase to catalyze the chlorination of 1,3-dihydroxybenzene in cetyltrimethylammonium chloride/H$_2$O/octane/pentanol reverse micellar solution has been studied [74]. The results show that the specific activity of the enzyme can be enhanced significantly by compressed CO$_2$, and the specific activity can be tuned continuously by changing pressure.

The main advantages of using scCO$_2$ as solvent for biocatalyzed reactions are the tunability of solvent properties and simple downstream processing features that can be readily combined with other unit operations. Although many enzymes are stable in scCO$_2$, attention should be paid to identify the correct reaction conditions for each substrate/enzyme/SCF system. One of the persistent problems is the instability and deactivation of enzymes under pressure and temperature [75].

Some Other Reactions in scCO$_2$

Many other reactions in scCO$_2$ have also been studied, and promising results have been obtained. For example, the reaction between 1,n-terminal diols (n = 3 or 6) with simple alcohols (MeOH, EtOH, and n-PrOH) in scCO$_2$ over an acid catalyst leads to two possible products, a mono and a bis-ether [76]. The selectivity of the reaction with 1,6-hexanediol and MeOH can be switched from 1:20 in favor of the bis-ether to 9:1 in favor of the desymmetrized mono-ether with the increase of CO$_2$ pressure. It is demonstrated that the switch in selectivity is associated with the phase state of the reaction mixture.

High-yielding and greener routes for the continuous synthesis of methyl ethers in scCO$_2$ have been developed [77]. Ethers have been efficiently produced using a methodology which eliminates the use of toxic alkylating agents and reduces the waste generation that is characteristic of traditional etherification processes. The use of acidic heterogeneous catalysts can achieve etherification when using scCO$_2$ as a reaction medium.

The esterification of acetic acid and ethanol in scCO$_2$ has been studied [78]. The phase behavior and the isothermal compressibility of reaction system are also determined under the reaction conditions. The conversion increases with increasing pressure in the two-phase region and reaches a maximum in the critical region of the reaction system where the system just becomes one phase. Then the conversion decreases with pressure after the pressure is higher than the critical value. The study on transesterification between ethyl acetate and n-butanol in scCO$_2$ demonstrates that the equilibrium conversion is very sensitive to pressure as the reaction mixture approaches the critical point, bubble point, and dew point of the reaction mixtures in the critical region, whereas the effect of pressure on equilibrium conversion is not significant outside the critical region [79]. All these results indicate that the equilibrium conversions of the reversible reactions can be tuned effectively by pressure in the critical regions of the reaction mixtures. Continuous dehydration of alcohols in scCO$_2$ shows that CO$_2$ can provide significant advantages in these heterogeneous, acid-catalyzed reactions over a wide range of temperatures and pressures [80]. Furthermore, using scCO$_2$ as the solvent for ether formation will encourage the phase separation of H$_2$O, which appears to reduce back reaction via rehydration of the product.

Dimethyl carbonate(DMC) is a very useful chemical. Synthesis of DMC using scCO$_2$ and methanol is studied, and the phase behavior and critical density of the reaction system are also determined [81]. The reaction is carried out at various pressures that correspond to conditions in the two-phase region, the critical region as well as the single-phase supercritical region. The original ratios of the reactants CO$_2$:CH$_3$OH are 8:2 and 7:3, and the corresponding reaction temperatures were 353.2 K and 393.2 K, respectively, which are slightly higher than the critical temperatures of the reaction systems. The results indicate that the phase behavior affects the equilibrium conversion of methanol significantly and the conversion reaches a maximum in the critical regions of the reaction system. At 353.2 K, the equilibrium conversion in the critical region is about 7%, and can be about three times as large as those in other phase regions. At 393.15 K, the equilibrium conversion in the critical region is also much higher and can be twice as large as those in other phase regions.

Chemical Reactions in scCO$_2$/Ionic Liquid Systems

Ionic liquids (ILs) are salts that are liquid at ambient conditions. They are a relatively new type of solvents, which have attracted much interest as an alternative for volatile organic compounds. ILs have some unique properties, such as extremely low vapor pressure, excellent solvent power for both organic and inorganic compounds, high thermal and chemical stability. In addition, functions ILs can be designed by varying the cations and anions to afford certain specific properties. Combination of CO$_2$ and ILs is an interesting topic. The very distinct properties of scCO$_2$ and ILs make them good combination to form biphasic solvent systems with special properties. High-pressure CO$_2$ is soluble in ILs, while ILs are almost insoluble in scCO$_2$ [82]. Therefore, at suitable conditions the organic compounds can be extracted from the ILs without cross-contamination. With these unique properties, CO$_2$/IL biphasic systems can be used not only in separation, but also in biphasic reactions. It has been shown that scCO$_2$/IL biphasic system can shift chemical equilibrium of revisable reactions and improve conversion and selectivity of chemical reactions [83].

The oxidation of aromatic alcohol into corresponding aldehyde is an important transformation because aromatic aldehdyes are versatile intermediates for the production of pharmaceuticals, plastic aitives, and perfumes. However, a major drawback of such oxidation reactions is their lack of selectivity owing to the easy over-oxidation of aldehydes into carboxylic acids. Electro-oxidation reaction of benzyl alcohol in $scCO_2$/ILs is investigated using 1-butyl-3-methylimidazolium tetrafluoroborate ([Bmim][BF$_4$]) and 1-butyl-3-methylimidazolium hexafluorophosphate ([Bmim][PF$_6$]) as the ILs and electrolytes [84]. It is demonstrated that benzyl alcohol can be efficiently electro-oxidized to benzaldehyde. [Bmim][BF$_4$] is more effective medium for the electro-oxidation of benzyl alcohol. The product can be easily recovered from the IL by using $scCO_2$ extraction after the electrolysis, and the IL can be reused. The Faradic efficiency (FE) and selectivity of benzaldehyde increase with the pressure of CO_2 when the pressure is lower than about 9.3 MPa, while the FE decreases as the pressure is increased further. This phenomenon can be explained reasonably on the basis of solubility difference of the reactant and product.

Enhancing equilibrium conversion of reversible reactions has been a topic of great importance for a long time. The effect of CO_2 on the phase behavior of the reaction system and equilibrium conversion for esterification of acetic acid and ethanol in IL 1-butyl-3-methylimidazolium hydrogen sulfate ([bmim][HSO$_4$]) is studied up to 15 MPa [85]. There is only one phase in the reaction system in the absence of CO_2. The reaction system undergoes two-phase→three-phases→ two-phase transitions with increasing pressure. The pressure of CO_2 or the phase behavior of the system affected the equilibrium conversion of the reaction markedly. As the pressure is less than 3.5 MPa, there are two phases in the system, and the equilibrium conversion increases as pressure is increased. In the pressure range of 3.5–9.5 MPa, there exist three phases, and the equilibrium conversion increases more rapidly with increasing pressure. As the pressure is higher than 9.5 MPa, the reaction system enters another two-phase region and the equilibrium conversion is nearly independent of pressure. The total equilibrium conversion was 64% without CO_2 and can be as high as 80% as pressure is higher than 9.0 MPa. The apparent equilibrium constants (Kx) in different phases are also determined, showing that the Kx in the mile phase or top phase is much greater than that in the bottom phase.

Oxidation of 1-hexene by molecular oxygen is conducted in [bmim] [PF$_6$], $scCO_2$, $scCO_2$/[bmim] [PF$_6$] biphasic system, and in the absence of solvent [86]. The difference in conversion at various reaction times in $scCO_2$ and in the CO_2/IL mixture is not considerable. However, in the $scCO_2$/IL biphasic system, the selectivity to the desired product is much higher than that in $scCO_2$. Moreover, the selectivity in the biphasic system increases slightly with reaction time, while the selectivity decreases slowly with reaction time in $scCO_2$. The conversion is very high and is nearly independent of pressure. However, the selectivity increases with pressure significantly, especially in the low-pressure range. A reasonable explanation is that the solvent power of CO_2 increases with increasing pressure. Therefore, less reactant exists in the IL-rich phase at higher pressure, which favors reduction of the isomerization of the reactant. Meanwhile, the solubility of CO_2 in the IL increases with pressure, and so the diffusivity of the solvent is improved more significantly at the higher pressures, which may also enhance the selectivity. In addition, the catalyst used is more stable in the $scCO_2$/IL biphasic system than in $scCO_2$.

A continuous flow system for the hydroformylation of relatively low-volatility alkenes in scCO$_2$/IL biphasic systems has been reported [87]. The catalyst is dissolved in an IL, while the substrate and gaseous reagents dissolved in scCO$_2$ are transported into the reactor, which simultaneously acts as a transport vector for aldehyde products. Decompression of the fluid mixture yields products which are free of both reaction solvent and catalyst. The nature of the ILs is very important in achieving high rates, with 1-alkyl-3-methylimidazolium bis(trifluoromethane-sulfonyl)amides giving the best activity if the alkyl chain is at least C-8. High catalyst turnover frequencies have been obtained, with the better rates at higher substrate flow rates. The leaching of the catalyst into the product stream can be very low. Under certain process conditions, scCO$_2$/IL biphasic system can be operated continuously for several weeks without any visible sign of catalyst degradation.

Polymerization in scCO$_2$

ScCO$_2$ has received considerable attention as an environmentally benign reaction medium for polymerization. The early research on polymerization in CO$_2$ can be dated back to the 1960s [88]. However, less attention was received on this because viscous solid or liquid polymer with low molecular weight were produced, which are practically useless. In early 1990s, homogeneous solution polymerization of 1,1-dihydroperfluorooctyl methyl acrylate monomers in scCO$_2$ was reported using azo-bisisobutyronitrile as the initiator [89]. The polymerization in scCO$_2$ has become an interesting topic since then [90–92]. There are some unique aspects of polymerization in scCO$_2$, such as the solvent is green; the molecular weight and the morphologies of the products can be tailored by pressure; the resulting polymer can be isolated from the reaction medium by simple depressurization, resulting in a dry, solvent-free product. These make it possible for scCO$_2$ to replace hazardous volatile organic solvents in some industrial applications.

The solubility of polymer in scCO$_2$ has a decisive effect on the polymerization. While CO$_2$ is a good solvent for many low molecular weight monomers, it is a very poor solvent for most polymers. The only polymers to have good solubility in pure scCO$_2$ under mild conditions are amorphous fluoropolymers and silicones. This has allowed the synthesis of high molecular weight fluoropolymers by homogeneous solution polymerization in scCO$_2$. Except for the synthesis of fluoropolymers, most polymerizations in scCO$_2$ have been carried out by hetero-geneous polymerization, mainly including suspension, emulsion, dispersion, and precipitation polymerizations. The fore three polymerizations need the addition of special fluoro-surfactants that can dissolve in scCO$_2$. Although some nonfluoro-nonionic-surfactants and nonfluoro-ionic-surfactants have been reported, they cannot meet the requirement satisfactorily.

Although the solubility of most polymers in scCO$_2$ is extremely low, the solubil-ity of scCO$_2$ in many polymers is fairly large. This can lead to a dramatic decrease in the glass transition temperature (Tg) of these materials (i.e., plasticisation), even at modest pressures [93]. It has been shown by various methods that CO_2 is a good plasticizing agent for a number of polymeric materials [94]. In heterogeneous polymerization, plasticization can facilitate diffusion of monomer and initiator into the polymer phase to increase the rate of polymerization.

Preparation of Materials

SCF technology has penetrated into various fields of materials science, including generation of superfine particles, microporous materials, composite materials, and semiconductor devices [95]. The material preparation using $scCO_2$ technology has some unique advantages, and some special materials that are difficult to be prepared by other techniques can be fabricated due to the special properties of SCFs. Application of $scCO_2$ in material science has bright future.

Preparation of Fine Particles and Porous Materials

There are two kinds of methods to prepare superfine particles using $scCO_2$, i.e., physical and chemical methods. Physical method mainly includes rapid expansion of SCF solutions (RESS) and supercritical anti-solvent (SAS) process, and some new methods have been derived on the basis of RESS and SAS.

The technique of RESS was developed in the late 1980s [96]. The most attractive advantage of RESS is that it can produce particles with relatively narrow size distribution, which can be particularly important for drug delivery systems. The main disadvantage of RESS is that it can only be applied to produce materials which are soluble in $scCO_2$. Since the solubility of most materials in $scCO_2$ is relatively low, a large amount of CO_2 is used to prepare small amount of particles. Therefore, this technique is relatively costly and is usually used in laboratory or producing high-value-added products. Up to now, many particles have been fabricated by RESS, such as fine particles of composite polymer microparticles [97], semiconductor particles [98], and pharmaceutical particles [99].

Study of SAS technique began in early 1990s [100], which has been developed rapidly. This technique can be used to process the substances that are not soluble in the supercritical medium. In the process, the materials of interesting are dissolved into an organic liquid that is miscible with $scCO_2$ under appropriate process conditions. When in contact with liquid solution, the SCF dissolves into the liquid, a large and fast expansion of the liquid phase occurs, and small particles of solute can be precipitated [101]. The SAS technique has been further developed using compressed CO_2 as antisolvent to recover the nanoparticles and nanocomposites synthesized in reverse micelles, such as protein [102] and ZnS [103], nanoparticles, and Ag/PS nanocomposites [104].

Another method to prepare superfine particles using scCO$_2$ as medium involves chemical reaction. Inorganic materials including metals, semiconductors, nitrides, and oxides are prepared by thermal decomposition reaction or redox reaction in scCO$_2$. ScCO$_2$ can be a good solvent for many organometallic compounds and the ligands, but a poor solvent for the inorganic compounds. Therefore, precursor reduction occurs at the solution/solid interface at significantly lower temperature and higher reagent concentration than those of vapor-phase techniques like chemical vapor deposition. In addition, the presence of scCO$_2$ facilitates desorption of the ligands on the metal surfaces, which are produced from the decomposition of organometallic compounds [105].

Inorganic nanomaterials in scCO$_2$ can be prepared by the reduction of organometallic compounds with H$_2$ due to the high miscibility of scCO$_2$ with H$_2$ [106]. Preparation of core-shell magnetic materials from the precursor bis(hexafluoroacetylacetonate)-copper(II) solubilized in the scCO$_2$/ethanol mixture has been studied [107]. The results show that Ni/Cu core-shell structure particles increase the coercive field due to the existence of Cu shell layer.

The polymerization of tetraalkoxysilanes tetramethoxysilane and 1,4-bis (triethoxysilyl)benzene in scCO$_2$ is carried out by using formic acid as the condensation reagent, and silica aerogels and 1,4-phenylene-bridged polysilsesquioxane aerogels are fabricated, respectively [108]. ScCO$_2$ appears to be an excellent solvent for preparing highly porous architectures with both meso- and macroporous structure, and seems to be a better solvent than ethanol for preparing phenylene-bridged polysilsesquioxane aerogels. In addition, the process can be used to generate monolithic aerogels in a single step from their monomeric precursors. It has also been reported that silica aerogel particles can be obtained in scCO$_2$ using acetic acid as the condensation agent for silicon alkoxides [109]. It is shown that acetic acid is a mild and controllable agent for sol-gel route preparation of silica aerogel particles. By changing the ratio of silicon alkoxide, acetic acid, and water, the polymerization rate is tunable and precipitation can be prevented. Submicron particle sizes can be obtained when the sol-gel solution is destabilized by pressure reduction and particles as small as 100 nm can be formed using the improved RESS process.

The synthesis of hollow silica spheres with mesoporous wall structures via CO$_2$-in-water emulsion templating in the presence of nonionic PEO-PPO-PEO block copolymer surfactants as mesostructure-directing templates has been investigated under various conditions [110]. The successful synthesis of hollow spheres strongly depends on the pressure and density of CO$_2$. The pore size and morphology of the silica hollow spheres can be controlled by varying the pressure of CO$_2$.

Well-ordered mesoporous silicate films have been prepared by infusion and selective condensation of silicon alkoxides within microphase-separated block copolymer templates dilated with scCO$_2$ [111]. Confinement of metal oxide deposition to specific subdomains of the preorganized template yields high-fidelity, three-dimensional replication of the copolymer morphology, enabling the preparation of structures with multiscale order in a process that closely resembles biomineralization. Ordered mesoporous silicate films synthesized have very low dielectric constants and excellent mechanical properties. The films survive the chemical–mechanical polishing step required for device manufacturing.

Polymer foams created using scCO$_2$ as a processing solvent have attracted much interest in recent years, and some progress has been achieved. For example, the foaming of fluorinated ethylene propylene copolymer films using scCO$_2$ is investigated [112]. The films prepared show a homogeneous cellular structure with thin compact polymer layers on both sides. The foaming behavior of the polymer films is dependence on the exposure time of the polymer to the scCO$_2$, the processing temperature and pressure as well as on the pressure drop rate during the foaming process. The resulting density and the foam morphology depend on the pressure during the saturation process. The size of the cells generated depends mainly on the foaming temperature, whereas their number per volume is determined mainly by the applied pressure. Furthermore, the pressure drop rate influences the size of the cells in a way that significantly fewer but larger bubbles are formed for longer times of depressurizing. The properties of scCO$_2$ make it suited to replace organic solvents that are being phased out for environmental reasons. A number of processes and products are already being used in industry [113].

Preparation of Composite Materials and Devices

As mentioned above, SCFs can easily carry dissolved matters into any accessible place, which provides a new approach for the preparation of composite materials. Various composite materials using scCO$_2$ as a medium have been prepared and studied extensively, including polymer composites, organic/inorganic hybrid materials and inorganic/inorganic composites.

The ability of scCO$_2$ to swell polymers is beneficial for the preparation of polymer composites. ScCO$_2$ can carry dissolved small molecules into swollen polymer matrix to prepare polymer-based composites. A number of polymeric composites have been fabricated using this route. For example, styrene monomer and initiator can be infused into polymer substrates such as poly(chlorotrifluoroethylene), poly(4-methyl-1-pentne), poly(ethylene), nylon 66, poly (oxymethylene), and bisphenol A polycarbonate, and the corresponding polymer blends are formed after free-radical polymerization [114].

Because many organometallic precursors are soluble in scCO$_2$, metal/polymer materials can also be prepared by impregnate metal precursors into polymer matrix using scCO$_2$ as the solvent, followed by chemical or thermal reduction of the precursors to the base metals [115, 116]. Similarly, the swelling property of scCO$_2$ has also been used to fabricate Eu$_2$O$_3$/polystyrene composites via decomposition of europium nitrate hexahydrate in scCO$_2$/ethanol solution in the presence of dispersed polymeric hollow spheres [117]. In addition to adhering to the outer surface, the Eu$_2$O$_3$ nanoparticles impregnate the shell and the cavity of the polystyrene substrate.

Carbon nanotubes (CNTs) have received much attention owing to their special electrical, mechanical, and optical properties. SCF techniques show intriguing advantages for the synthesis of various CNT-based composites. The composites can be fabricated by deposition of metal and metal oxide on CNTs using scCO$_2$ as the solvent [118, 119]. Because the deposition process involves only H$_2$ and precursors, it is possible to apply the same method to fabricate other metal/CNT composites, or to coat CNTs with metal films or nanoparticles. From the various results presented, it is clear that chemical fluid deposition in scCO$_2$ provides an effective and clean way to decorate CNTs. For instance, coating of Eu$_2$O$_3$ on CNTs in scCO$_2$-ethanol mixture can be accomplished [120]. Similarly, Poly(2,4-hexadiyne-1,6-diol) (poly(HDiD)) can be coated on the outer walls of CNTs with the aid of scCO$_2$, resulting in poly(HDiD)/CNT nanocomposites, which possess optical properties originated from poly(HDiD) [121]. Coating CNTs with solvent resistant polymer in scCO$_2$ has also been achieved [122], which permits the selective deposition of high molecular weight fluorinated graft poly(methyl vinyl ether-alt-maleic anhydride) polymer onto CNTs and forms quasi one-dimensional nanostructures with conducting cores and insulating surfaces.

Some special microporous, mesoporous, and layer composites have also been prepared in scCO$_2$, such as size-controlled Pt nanoparticles in micro- and mesoporous silica. ScCO$_2$ is an excellent solvent for the introduction of nanoparticles into small spaces on a nanometer scale due to its special properties [123].

Zr-TiO$_2$ nanotubes have been synthesized via a surfactant-free sol-gel route in scCO$_2$. The morphology of the Zr-TiO$_2$ nanotubes can be tailored by changing the concentration of the starting materials or the acid-to-metal-alkoxide ratio. The formed Zr-TiO$_2$ nanotubes have high surface area, small crystallite size, and high thermal stability, and are highly desirable as catalysts, support materials, semiconductors, and electrodes in dye-sensitized solar cells. This synthesis procedure is simple and reaction condition is mild, and provides a high yield and high-quality nanotubes [124].

High-purity conformal metal oxide films can be deposited onto planar and etched silicon wafers by surface-selective precursor hydrolysis in scCO$_2$ using a cold wall reactor [125]. Continuous films of cerium, hafnium, titanium, mobium, tantalum, zirconium, and bismuth oxides with different thick are grown using CO$_2$ soluble precursors. The as-deposited films are pure single-phase oxide in the cases of HfO$_2$, ZrO$_2$, and TiO$_2$ and composed of oxides of mixed oxidation states in the cases of cerium, tantalum, niobium, and bismuth oxides.

Carbon black supported single Pd, Pt and bimetallic Pd-Pt nanoparticles have been prepared utilizing scCO$_2$ deposition method with hydrogen as the reducing agent [126]. Increasing reduction temperature and metal loading caused an increase in Pd particle size. The Pt nanoparticles are homogeneously distributed with a size range of 2–6 nm. Binary metal nanoparticles can also be produced by simultaneous adsorption and subsequent reduction of the precursors. Addition of Pt increases the homogeneity and reduces the particle size on the support compared to single Pd nanoparticles. Pt-rich nanoparticles have diameters of around 4 nm, whereas Pd-rich nanoparticles are larger with diameters of around 10 nm.

Supercritical Chromatography

SCFs have been used as chromatographic mobile phases for several decades [127, 128]. Supercritical fluid chromatography (SFC) using scCO$_2$ as a mobile phase has been applied to analyze a wide variety of materials [129], including chiral separation [130, 131]. It has also been proven that SFC has high performance to control the enantiomeric purity of the final drug substance [132]. SFC has recently been utilized for the separation of furocoumarins of essential oils [133] and evaluation of a condensation nucleation light scattering detector for the analysis of synthetic polymer [134]. SFC combines some advantages of both liquid chromatography and gas chromatography, including high efficiency, fast separation, low-temperature analysis, and applicability to wide variety of detectors. It permits separation and determination of some compounds that are not conveniently handled by either liquid chromatography or gas chromatography.

Painting and Dyeing

ScCO$_2$ is well suited to spray painting applications because of its excellent solvent power, low toxicity, and nonflammability. Spray painting with scCO$_2$ can reduce volatile organic compounds (VOCs) emissions and increases transfer efficiency [135].

Due to the environmental benefit of avoiding wastewater, an alternative dyeing process in SCFs has been proposed [136]. ScCO$_2$ is widely adopted as a suitable solvent, which has good plasticizing and swelling action on many textiles [137]. Some promising results have been obtained in the dyeing of some synthetic polymers with disperse dyes in scCO$_2$ [138]. Similar to SFE, the presence of small amount of polar or nonpolar cosolvents can remarkably increase the solubility of different dyes in scCO$_2$ and allows good dyeing results at less severe working conditions [139].

Cleaning Using scCO$_2$

Due to special physical properties scCO$_2$, some significant achievements have been made using scCO$_2$ as cleaning solvent in the fields of microelectronics [140], metallic surfaces [141], and printing and packaging industries [142]. Additionally, water-in-CO$_2$ microemulsion formed from surfactants in nonpolar CO$_2$ can effectively remove residues. The study for removing post-etch residues from patterned porous low-k dielectrics using water-in-CO$_2$ microemulsions has demonstrated that scCO$_2$ has no effect on the thickness and refractive index, indicating that collapse or voiding of the pores does not occur and post-etch residue can be removed with a solvent containing water, CO_2, and a hydrocarbon surfactant [143]. Combination of scCO_2 and hydrocarbon surfactants has also been utilized to remove water from photoresists without pattern collapse due to capillary forces. Moreover, the addition of highly branched hydrocarbon surfactants to CO_2 reduces the amount of solvent required for drying and lowers the interfacial tension [144].

Emulsions Related with CO_2

Surfactants have the ability to self-assemble into morphologically different structures, such as micelles, reverse micelles, vesicles, and liquid crystals. The functions and properties of surfactant systems depend strongly on their microstructures. Therefore, surfactant assemblies have wide applications in different fields of chemical engineering, material science, biology, environmental science, food industry, detergency, and enhanced oil recovery because desired functions can be achieved by controlling the microstructures. To tune the properties of surfactant assemblies is of great importance from both scientific and practical application viewpoints. Formation of water-in-CO_2 microemulsions and tuning the properties of surfactant aggregates using compressed CO_2 have been interesting areas in recent years.

Microemulsions with scCO_2 as the continuous phase have been studied extensively, which show some special properties comparing with the conventional water-in-oil microemulsions. For most of the CO_2-continuous microemulsions, water is used as the dispersed phase, forming the water-in-CO_2 microemulsions. The applications of this kind of microemulsions in nanoparticle synthesis and chemical reactions have been studied extensively. The design of surfactants compatible with CO_2 is crucial for the formation of stable water-in-CO_2 microemulsions. The most effective compounds that have been reported to stabilize water-in-CO_2 microemulsions are the partially or fully fluorinated surfactants. On consideration of the environmental and economical factors, some efforts have been made for the hydrocarbon surfactants and hybrid fluorocarbon–hydrocarbon surfactants [145, 146]. Recently, the formation of IL-in-CO_2 microemulsions was reported, of which the nanosized IL droplets are dispersed in scCO_2 with the aid of a perfluorinated surfactant [147]. Due to the unique features of scCO_2 and ILs, this kind of IL-in-CO_2 microemulsions may find various potential applications.

Controlling the properties of surfactant solutions using compressed CO_2 is another interesting topic, which exhibits unique properties in the synthesis of nanomaterials and chemical reactions because the properties of the microemulsions can be tuned continuously by pressure and the operation pressure is relatively low. Some copolymers can form reverse micelles in organic solvents, and different methods to induce micellization, such as changing temperature and adding salts, have been studied. The formation of copolymer reverse micelles induced by CO_2 has been investigated. It shows that compressed CO_2 can induce the formation of reverse micelles of the PEO-PPO-PEO copolymers [148]. The pressure at which reverse micelles begin to form is defined as critical micelle pressure. The unique advantage of this kind of reverse micelles is that the formation and breaking of reverse micelles can be repeated easily by controlling the pressure, and CO_2 can be removed completely after depressurization.

Recent study demonstrates that compressed CO_2 can switch the phase transition between lamellar liquid crystal and micellar solution of sodium bis(2-ethylhexyl) sulphosuccinate (AOT)/water system [149]. In the low pressure range, the viscosity of the system is very high and gradually decreases with increasing pressure of CO_2. As the pressure reaches an optimum value, the viscosity of the surfactant system reduces dramatically and changes into a transparent solution. Different techniques proved that this phenomenon corresponds to the phase transition from lamellar liquid crystal to micellar solution. The phase transition is reversible and can be repeated by controlling the pressure. This method to induce phase transition has some unique futures. For instance, CO_2 can induce the phase transition at ambient temperature, whereas the phase transition occurs at or above $140°C$ in the absence of CO_2 for the surfactant system studied. Second, the CO_2-induced phase transition is reversible and can be easily accomplished by the controlling pressure of CO_2. As the CO_2 can be easily removed by release of pressure, the post-processing steps are much easier. Gold and silica nanostructures in the AOT/water lamellar liquid crystal and AOT/water/CO_2 micelles are synthesized using the hydrophilic precursor of $HAuCl_4$ and hydrophobic precursor of tetraethyl orthosilicate, respectively. It is shown that the morphologies of the materials depend on pressure of CO_2. Similarly, compressed CO_2 can also induce micelle-to-vesicle transition (MVT) of dodecyltrimethylammonium bromide/sodium dodecyl sulfate mixed surfactants in aqueous solution reversibly [150]. It is deduced that the MVT induced by CO_2 is not originated from the change of the property of water. The main reason may be that CO_2 can insert into hydrocarbon chain region of the interfacial films, which increases the structural packing parameter. Silica particles with micelle template and vesicle template have prepared by controlling the pressure of CO_2. Switching the MVT by CO_2 may find different applications in material science and chemical reactions.

Future Directions

$ScCO_2$ as a green solvent has been used in different fields, and will be used more widely in extraction and fractionation, chemical reactions, material science, microelectronics, and so on. It also provides many opportunities for developing new technologies using its unusual properties. However, it must be realized that there are some disadvantages of $scCO_2$ in applications. For example, high pressure is always required when using $scCO_2$, and therefore the energy problem, safety aspects, capital investment for equipments must be carefully considered. Nevertheless, it is no doubt that $scCO_2$ will play an important role in the development of new and green technologies, especially those in which the usual properties of scCO$_2$ can be used effectively. Attention should also be paid to combine supercritical technology and other technologies, and therefore makes the processes more efficient. Further studies are surely needed to require true interdisciplinary cooperation involving academic and industrial researchers. With growing environmental concerns, deep realization of the properties of scCO$_2$, and the requirement of sustainable development, scCO$_2$-based technology will be widely applied in many fields.

Bibliography

Primary Literature

1. Hannay JB, Hogarth J (1879) On the solubility of solids in gases. Proc R Soc Lond 29:324–326
2. Poliakoff M, Licence P (2007) Sustainable technology: green chemistry. Nature 450:810–812
3. Eckert CA, Knutson BL, Debenedetti PG (1996) Supercritical fluids as solvents for chemical and materials processing. Nature 383:313–318
4. Zosel K (1978) Separation with supercritical gases: practical applications. Angew Chem Int Ed 17:702–709
5. Daoud IS, Kusinski S (1993) Liquid CO$_2$ and ethanol extraction of hops. 3. Effect of hop deterioration on utilization and beer quality. J Inst Brew 99:147–152
6. Montanari L, Fantozzi P, Snyder JM, King JW (1999) Selective extraction of phospholipids from soybeans with supercritical carbon dioxide and ethanol. J Supercrit Fluids 14:87–93
7. Zuo YB, Zeng AW, Yuan XG, Yu KT (2008) Extraction of soybean isoflavones from soybean meal with aqueous methanol modified supercritical carbon dioxide. J Food Eng 89:384–389
8. Hawthorne SB, Krieger MS, Miller DJ (1988) Analysis of flavor and fragrance compounds using supercritical fluid extraction coupled with gas chromatography. Anal Chem 60:472–477
9. Abbasi H, Rezaei K, Rashidi L (2008) Extraction of essential oils from the seeds of pomegranate using organic solvents and supercritical CO$_2$. J Am Oil Chem Soc 85:83–89
10. Araus K, Uquiche E, del Valle JM (2009) Matrix effects in supercritical CO$_2$ extraction of essential oils from plant material. J Food Eng 92:438–447
11. Mezzomo N, Martinez J, Ferreira SRS (2009) Supercritical fluid extraction of peach (*Prunus persica*) almond oil: kinetics, mathematical modeling and scale-up. J Supercrit Fluids 51:10–16
12. Nelson SR, Roodman RG (1985) Rose-the energy-efficient bottom of the barrel alternative. Chem Eng Prog 81:63–68
13. Deo MD, Hwang J, Hanson FV (1992) Supercritical fluid extraction of a crude oil, bitumen-derived liquid and bitumen by carbon dioxide and propane. Fuel 71:1519–1526
14. Low GKC, Duffy GJ (1995) Supercritical fluid extraction of petroleum hydrocarbons from contaminated soils. Trends Anal Chem 14:218–225
15. Bartle KD, Clifford AA, Shilstone GF (1989) Prediction of solubilities for tar extraction by supercritical carbon dioxide. J Supercrit Fluids 2:30–34
16. Bristow S, Shekunov BY, York P (2001) Solubility analysis of drug compounds in supercritical carbon dioxide using static and dynamic extraction systems. Ind Eng Chem Res 40:1732–1739

17. Lin YH, Smart NG, Wai CM (1995) Supercritical fluid extraction of uranium and thorium from nitric acid solutions with organophosphorus reagents. Environ Sci Technol 29:2706–2708

18. Rincon J, De Lucas A, Garcia MA, Garcia A, Alvarez A, Carnicer A (1998) Preliminary study on the supercritical carbon dioxide extraction of nicotine from tobacco wastes. Sep Sci Technol 33:411–423

19. Chen WH, Chen CH, Chang CMJ, Chiu YH, Hsiang D (2009) Supercritical carbon dioxide extraction of triglycerides from *Jatropha curcas* L. seeds. J Supercrit Fluids 51:174–180

20. Hitzler MG, Poliakoff M (1997) Continuous hydrogenation of organic compounds in supercritical fluids. Chem Commun 1667–1668

21. Baiker A (1999) Supercritical fluids in heterogeneous catalysis. Chem Rev 99:453–474

22. Jiang YJ, Gao QM (2006) Heterogeneous hydrogenation catalyses over recyclable Pd(0) nanoparticle catalysts stabilized by PAMAM-SBA-15 organic–inorganic hybrid composites. J Am Chem Soc 128:716–717

23. Wu TB, Jiang T, Hu BJ, Han BX, He JL, Zhou XS (2009) Cross-linked polymer coated Pd nanocatalysts on SiO_2 support: very selective and stable catalysts for hydrogenation in supercritical CO_2. Green Chem 11:798–803

24. Chatterjee M, Zhao FY, Ikushima Y (2004) Hydrogenation of citral using monometallic Pt and bimetallic Pt-Ru catalysts on a mesoporous support in supercritical carbon dioxide medium. Adv Synth Catal 346:459–466

25. Meric P, Yu KMK, Kong ATS, Tsang SC (2006) Pressure-dependent product distribution of citral hydrogenation over micelle-hosted Pd and Ru nanoparticles in supercritical carbon dioxide. J Catal 237:330–336

26. Zhao FY, Zhang R, Chatterjee M, Ikushima Y, Arai M (2004) Hydrogenation of nitrobenzene with supported transition metal catalysts in supercritical carbon dioxide. Adv Synth Catal 346:661–668

27. Zhao FY, Fujitaa S, Sun JM, Ikushima Y, Arai M (2004) Hydrogenation of nitro compounds with supported platinum catalyst in supercritical carbon dioxide. Catal Today 98:523–528

28. Liu HZ, Jiang T, Han BX, Liang SG, Zhou YX (2009) Selective phenol hydrogenation to cyclohexanone over a dualsupported Pd-Lewis acid catalyst. Science 326:1250–1252

29. Burk MJ, Feng S, Gross MF, Tumas W (1995) Asymmetric catalytic hydrogenation reactions in supercritical carbon dioxide. J Am Chem Soc 117:8277–8278

30. Wang SN, Kienzle F (2000) The syntheses of pharmaceutical intermediates in supercritical fluids. Ind Eng Chem Res 39:4487–4490

31. Kainz S, Brinkmann A, Leitner W, Pfaltz A (1999) Iridium-catalyzed enantioselective hydrogenation of imines in supercritical carbon dioxide. J Am Chem Soc 121:6421–6429

32. Stephenson P, Licence P, Ross SK, Poliakoff M (2004) Continuous catalytic asymmetric hydrogenation in supercritical CO_2. Green Chem 6:521–523

33. Pr C, Poliakoff M, Wells A (2007) Continuous flow hydrogenation of a pharmaceutical intermediate, [4-(3, 4-Dichlorophenyl)-3, 4-dihydro-2H-naphthalenyidene]methylamine, in supercritical carbon dioxide. Adv Synth Catal 349:2655–2659

34. Licence P, Ke J, Sokolova M, Ross SK, Poliakoff M (2003) Chemical reactions in supercritical carbon dioxide: from laboratory to commercial plant. Green Chem 5:99–104

35. Hou ZS, Han BX, Gao L, Liu ZM, Yang GY (2002) Selective oxidation of cyclohexane in compressed CO_2 and in liquid solvents over MnAPO-5 molecular sieve. Green Chem 4:426–430

36. Zhang RZ, Qin ZF, Dong M, Wang GF, Wang JG (2005) Selective oxidation of cyclohexane in supercritical carbon dioxide over CoAPO-5 molecular sieves. Catal Today 110:351–356

37. Theyssen N, Leitner W (2002) Selective oxidation of cyclooctane to cyclootanone with molecular oxygen in the presence of compressed carbon dioxide. Chem Commun 410–411. doi: 10.1039/B111212K

38. Theyssen N, Hou ZS, Leitner W (2006) Selective oxidation of alkanes with molecular oxygen and acetaldehyde in compressed (supercritical) carbon dioxide as reaction Medium. Chem Eur J 12:3401–3409

39. Jiang HF, Jia LQ, Li JH (2000) Wacker reaction in supercritical carbon dioxide. Green Chem 2:161–164

40. Wang ZY, Jiang HF, Qi CR, Wang YG, Dong YS, Liu HL (2005) PS-BQ: an efficient polymer-supported cocatalyst for the Wacker reaction in supercritical carbon dioxide. Green Chem 7:582–585

41. Wang XG, Venkataramanan NS, Kawanami H, Ikushima Y (2007) Selective oxidation of styrene to acetophenone over supported Au-Pd catalyst with hydrogen peroxide in supercritical carbon dioxide. Green Chem 9:1352–1355

42. Wang JQ, Cai F, Wang E, He LN (2007) Supercritical carbon dioxide and poly(ethylene glycol): an environmentally benign biphasic solvent system for aerobic oxidation of styrene. Green Chem 9:882–887

43. Jia LQ, Jiang HF, Li JH (1999) Palladium(II)-catalyzed oxidation of acrylate esters to acetals in supercritical carbon dioxide. Chem Commun 985–986. doi: 10.1039/A901935I

44. Xie Y, Zhang ZF, Hu SQ, Song JL, Li WJ, Han BX (2008) Aerobic oxidation of benzyl alcohol in supercritical CO_2 catalyzed by perruthenate immobilized on polymer supported ionic liquid. Green Chem 10:278–282

45. Hou ZS, Theyssen N, Brinkmann A, Leitner W (2005) Biphasic aerobic oxidation of alcohols catalyzed by poly(ethylene glycol)-stabilized palladium nanoparticles in supercritical carbon dioxide. Angew Chem Int Ed 44:1346–1349

46. He JL, Wu TB, Jiang T, Zhou XS, Hu BJ, Han BX (2008) Aerobic oxidation of secondary alcohols to ketones catalyzed by cobalt(II)/ZnO in poly(ethylene glycol)/CO_2 system. Catal Commun 9:2239–2243

47. Chapman AO, Akien GR, Arrowsmith NJ, Licence P, Poliakoff M (2010) Continuous heterogeneous catalytic oxidation of primary and secondary alcohols in scCO$_2$. Green Chem 12:310–315

48. Bourne RA, Han X, Poliakoff M, George MW (2009) Cleaner continuous photo-oxidation using singlet oxygen in supercritical carbon dioxide. Angew Chem Int Ed 48:5322–5325

49. Han X, Bourne RA, Poliakoff M, George MW (2009) Strategies for cleaner oxidations using photochemically generated singlet oxygen in supercritical carbon dioxide. Green Chem 11:1787–1792

50. Clifford AA, Pople K, Gaskill WJ, Bartle KD, Rayner CM (1997) Reaction control and potential tuning in a supercritical fluid. Chem Commun 595–596. doi: 10.1039/A608006I

51. Qian J, Timko MT, Allen AJ, Rusell CJ, Winnik B, Buckley B, Steinfeld JI, Tester JW (2004) Solvophobic acceleration of Diels-Alder reactions in supercritical carbon dioxide. J Am Chem Soc 126:5465–5474

52. Isaacs NS, Keating N (1992) The rates of a Diels-Alder reaction in liquid and supercritical carbon dioxide. J Chem Soc Chem Commun 876–877. doi: 10.1039/C39920000876

53. Matsuo J, Tsuchiya T, Odashina K, Kobayashi S (2000) Lewis acid catalysis in supercritical carbon dioxide. use of scandium tris(heptadecafluorooctanesulfonate) as a lewis acid catalyst in Diels-Alder and Aza Diels-Alder reactions. Chem Lett 29:178–179

54. Cheng JS, Jiang HF (2004) Palladium-catalyzed regioselective cyclotrimerization of acetylenes in supercritical carbon dioxide. Eur J Org Chem 643–646. doi: 10.1002/ejoc.200300299

55. Li JH, Jiang HF (1999) Glaser coupling reaction in supercritical carbon dioxide. Chem Commun 2369–2370. doi: 10.1039/A908014G

56. Carroll MA, Holmes AB (1998) Palladium-catalysed carbon-carbon bond formation in supercritical carbon dioxide. Chem Commun 1395–1396. doi: 10.1039/A802235F

57. Shezad N, Clifford AA, Rayner CM (2002) Pd-catalysed coupling reactions in supercritical carbon dioxide and under solventless conditions. Green Chem 4:64–67

58. Morita DK, David SA, Tumas W, Pesiri DR, Glaze WH (1998) Palladium-catalyzed cross-coupling reactions in supercritical carbon dioxide. Chem Commun 1397–1398. doi: 10.1039/A802621A

59. Rathke JW, Klingler RJ, Krause TR (1991) Propylene hydroformylation in supercritical carbon dioxide. Organometallics 10:1350–1355

60. Guo Y, Akgerman A (1997) Hydroformylation of propylene in supercritical carbon dioxide. Ind Eng Chem Res 36:4581–4585

61. Webb PB, Kunene TE, Cole-Hamilton DJ (2005) Continuous flow homogeneous hydroformylation of alkenes using supercritical fluids. Green Chem 7:373–379

62. Meehan NJ, Poliakoff M, Sandee AJ, Reek JNH, Kamer PCJ, van Leeuwen PWNM (2000) Continuous, selective hydroformylation in supercritical carbon dioxide using an immobilised homogeneous catalyst. Chem Commun 1497–1498. doi: 10.1039/B002526G

63. Mistele CD, Thorp HH, DeSimone JM (1996) Ring-opening metathesis polymerizations in carbon dioxide. J Macromol Sci Pure Appl Chem 33:953–960

64. Fürstner A, Koch D, Langemann K, Leitner W, Six C (1997) Olefin metathesis in compressed carbon dioxide. Angew Chem Int Ed Engl 36:2466–2469

65. Fürstner A, Ackermann L, Beck K, Hori H, Koch D, Langemann K, Liebl M, Six C, Leitner W (2001) Olefin metathesis in supercritical carbon dioxide. J Am Chem Soc 123:9000–9006

66. Randolph TW, Blanch HW, Prausnitz JM, Wilke CR (1985) Enzymatic catalysis in a supercritical fluid. Biotechnol Lett 7:325–328

67. Hobbs HR, Thomas NR (2007) Biocatalysis in supercritical fluids, in fluorous solvents, and under solvent-free conditions. Chem Rev 107:2786–2820

68. Hampson JW, Foglia TA (1999) Effect of moisture content on immobilized lipase-catalyzed triacylglycerol hydrolysis under supercritical carbon dioxide flow in a tubular fixed-bed reactor. J Am Oil Chem Soc 76:777–781

69. Oiveira JV, Oiveira D (2000) Kinetics of the enzymatic alcoholysis of palm kernel oil in supercritical CO_2. Ind Eng Chem Res 39:4450–4454

70. Jackson MA, King JW (1997) Lipase-catalyzed glycerolysis of soybean oil in supercritical carbon dioxide. J Am Oil Chem Soc 74:103–106

71. Tsitsimkou C, Stamatis H, Sereti V, Daflos H, Kolisis FN (1998) Acylation of glucose catalysed by lipases in supercritical carbon dioxide. J Chem Technol Biotechnol 71:309–314

72. Yu ZR, Rizvi SSH, Zolloweg JA (1992) Enzymic esterification of fatty acid mixtures from milk fat and anhydrous milk fat with canola oil in supercritical carbon dioxide. Biotechnol Prog 8:508–513

73. Garcia-Arrazola R, López-Guerrero DA, Gimeno M, Bárzana E (2009) Lipase-catalyzed synthesis of poly-L-lactide using supercritical carbon dioxide. J Supercrit Fluids 51:197–201

74. Chen J, Zhang JL, Han BX, Li JC, Li ZH, Feng XY (2006) Effect of compressed CO_2 on the chloroperoxidase catalyzed halogenation of 1, 3-dihydroxybenzene in reverse micelles. Phys Chem Chem Phys 8:877–881

75. Knez E (2009) Enzymatic reactions in dense gases. J Supercrit Fluids 47:357–372

76. Licence P, Gray WK, Sokolova M, Poliakoff M (2005) Selective monoprotection of 1, n-terminal diols in supercritical carbon dioxide: a striking example of solvent tunable desymmetrization. J Am Chem Soc 127:293–298

77. Gooden PN, Bourne RA, Parrott AJ, Bevinakatti HS, Irvine DJ, Poliakoff M (2010) Continuous acid-catalyzed methylations in supercritical carbon dioxide: comparison of methanol, dimethyl ether and dimethyl carbonate as methylating agents. Org Process Res Dev 14:411–416

78. Hou ZS, Han BX, Zhang XG, Zhang HF, Liu ZM (2001) Pressure tuning of reaction equilibrium of esterification of acetic acid with ethanol in compressed CO_2. J Phys Chem B 105:4510–4513

79. Gao L, Wu WZ, Hou ZS, Jiang T, Han BX, Liu J, Liu ZM (2003) Transesterification between ethyl acetate and *n*-butanol in compressed CO$_2$ in the critical region of the reaction system. J Phys Chem B 107:13093–13099

80. Gray WK, Smail FR, Hitzler MG, Ross SK, Poliakoff M (1999) The continuous acid-catalyzed dehydration of alcohols in supercritical fluids: a new approach to the cleaner synthesis of acetals, ketals, and ethers with high selectivity. J Am Chem Soc 121:10711–10718

81. Hou ZS, Han BX, Liu ZM, Jiang T, Yang GY (2002) Synthesis of dimethyl carbonate using CO$_2$ and methanol: enhancing the conversion by controlling the phase behavior. Green Chem 4:467–471

82. Blanchard LA, Hancu D, Beckman EJ, Brennecke JF (1999) Green processing using ionic liquids and CO$_2$. Nature 399:28–29

83. Jiang T, Han BX (2009) Ionic liquid catalytic systems and chemical reactions. Curr Org Chem 13:1278–1299

84. Zhao GY, Jiang T, Wu WZ, Han BX, Liu ZM, Gao HX (2004) Electro-oxidation of benzyl alcohol in a biphasic system consisting of supercritical CO$_2$ and ionic liquids. J Phys Chem B 108:13052–13057

85. Zhang ZF, Wu WZ, Han BX, Jiang T, Wang B, Liu ZM (2005) Phase separation of the reaction system induced by CO$_2$ and conversion enhancement for the esterification of acetic acid with ethanol in ionic liquid. J Phys Chem B 109:16176–16179

86. Hou ZS, Han BX, Gao L, Jiang T, Liu ZM, Chang YH, Zhang XG, He J (2002) Wacker oxidation of 1-hexene in 1-n-butyl-3-methylimidazolium hexafluorophosphate ([bmim] [PF$_6$]), supercritical (SC) CO$_2$, and SC CO$_2$/[bmim][PF$_6$]mixed solvent. New J Chem 26:1246–1248

87. Webb PB, Sellin MF, Kunene TE, Williamson S, Slawin AMZ, Cole-Hamilton DJ (2003) Continuous flow hydroformylation of alkenes in supercritical fluid-ionic liquid biphasic systems. J Am Chem Soc 125:15577–15588

88. Biddulph RH, Plesch PH (1960) The low-temperature polymerisation of isobutene. Part IV. exploratory experiments. J Chem Soc 3913–3920. doi: 10.1039/JR9600003913

89. Desimone JM, Guan Z, Elsbernd CS (1992) Synthesis of fluoropolymers in supercritical carbon doxide. Science 257:945–947

90. O'Connor P, Zetterlund PB, Aldabbagh F (2010) Effect of monomer loading and pressure on particle formation in nitroxide-mediated precipitation polymerization in supercritical carbon dioxide. Macromolecules 43:914–919

91. Kim J, Dong LB, Kiserow DJ, Roberts GW (2009) Complex effects of the sweep fluid on solid-state polymerization: poly(bisphenol a carbonate) in supercritical carbon dioxide. Macromolecules 42:2472–2479

92. Grignard B, Jerome C, Calberg C, Jerome R, Wang WX, Howdle SM, Detrembleur C (2008) Dispersion atom transfer radical polymerization of vinyl monomers in supercritical carbon dioxide. Macromolecules 41:8575–8583

93. Chiou JS, Barlow JW, Paul DR (1985) Plasticization of glassy polymers by CO$_2$. J Appl Polym Sci 30:2633–2642

94. Wissinger RG, Paulaitis ME (1991) Molecular thermodynamic model for sorption and swelling in glassy polymer-carbon dioxide systems at elevated pressures. Ind Eng Chem Res 30:842–851

95. Romang AH, Watkins JJ (2010) Supercritical fluids for the fabrication of semiconductor devices: emerging or missed opportunities? Chem Rev 110:459–478

96. Matson DW, Fulton JL, Petersen RC, Smith RD (1987) Rapid expansion of supercritical fluid solutions: solute formation of powders, thin films, and fibers. Ind Eng Chem Res 26:2298–2306

97. Reverchon E, Adami R, Cardea S, Della Porta G (2009) Supercritical fluids processing of polymers for pharmaceutical and medical applications. J Supercrit Fluids 47:484–492

98. Sun YP, Guduru R, Lin F, Whiteside T (2000) Preparation of nanoscale semiconductors through the rapid expansion of supercritical solution (RESS) into liquid solution. Ind Eng Chem Res 39:4663–4669

99. Türk M (2009) Manufacture of submicron drug particles with enhanced dissolution behaviour by rapid expansion processes. J Supercrit Fluids 47:537–545

100. Gallagher PM, Coffey MP, Krukonis VJ, Hillstrom WW (1992) Gas anti-solvent recrystallization of RDX: Formation of ultra-fine particles of a difficult-to-comminute explosive. J Supercrit Fluids 5:130–142

101. Reverchon E, Porta GD, Trolio AD, Pace S (1998) Supercritical antisolvent precipitation of nanoparticles of superconductor precursors. Ind Eng Chem Res 37:952–958

102. Chen J, Zhang JL, Han BX, Li ZH, Li JC, Feng XY (2006) Synthesis of cross-linked enzyme aggregates (CLEAs) in CO_2-expanded micellar solutions. Colloids Surf B Biointerfaces 48:72–76

103. Zhang JL, Han BX, Liu JC, Zhang XG, Liu ZM, He J (2001) A new method to recover the nanoparticles from reverse micelles: recovery of ZnS nanoparticles synthesized in reverse micelles by compressed CO_2. Chem Commun 2724–2725. doi: 10.1039/B109802K

104. Zhang JL, Liu ZM, Han BX, Liu DX, Chen J, He J, Jiang T (2004) A novel method to synthesize polystyrene nanospheres immobilized with silver nanoparticles by using compressed CO_2. Chem Eur J 10:3531–3536

105. Watkins JJ, Blackburn JM, McCarthy TJ (1999) Chemical fluid deposition: reactive deposition of platinum metal from carbon dioxide solution. Chem Mater 11:213–215

106. Blackburn JM, Long DP, Cabañas A, Watkins JJ (2001) Deposition of conformal copper and nickel films from supercritical carbon dioxide. Science 294:141–145

107. Pessey V, Garriga R, Weill F, Chevalier B, Etourneau J, Cansell F (2000) Core-shell materials elaboration in supercritical mixture CO_2/ethanol. Ind Eng Chem Res 39: 4714–4719

108. Loy DA, Russick EM, Yamanaka SA, Baugher BM, Shea KJ (1997) Direct formation of aerogels by sol-gel polymerizations of alkoxysilanes in supercritical carbon dioxide. Chem Mater 9:2264–2268

109. Sui RH, Rizkalla AS, Charpentier PA (2004) Synthesis and formation of silica aerogel particles by a novel sol-gel route in supercritical carbon dioxide. J Phys Chem B 108: 11886–11892

110. Wang JW, Xia YD, Wang WX, Poliakoff M, Mokaya R (2006) Synthesis of mesoporous silica hollow spheres in supercritical CO_2/water systems. J Mater Chem 16:1751–1756

111. Pai RA, Humayun R, Schulberg MT, Sengupta A, Sun JN, Watkins JJ (2004) Mesoporous silicates prepared using preorganized templates in supercritical fluids. Science 303:507–510

112. Zirkel L, Jakob M, Münstedt H (2009) Foaming of thin films of a fluorinated ethylene propylene copolymer using supercritical carbon dioxide. J Supercrit Fluids 49:103–110

113. Tomasko DL, Burley A, Feng L, Yeh SK, Miyazono K, Nirmal-Kumar S, Kusaka I, Koelling K (2009) Development of CO_2 for polymer foam applications. J Supercrit Fluids 47:493–499

114. Watkins JJ, McCarthy TJ (1994) Polymerization in supercritical fluid-swollen polymers: a new route to polymer blends. Macromolecules 27:4845–4847

115. Watkins JJ, McCarthy TJ (1995) Polymer/metal nanocomposite synthesis in supercritical CO_2. Chem Mater 7:1991–1994

116. Liao WS, Pan HB, Liu HW, Chen HJ, Wai CM (2009) Kinetic study of hydrodechlorination of chlorobiphenyl with polymer-stabilized palladium nanoparticles in supercritical carbon dioxide. J Phys Chem A 113:9772–9778

117. Wang JQ, Zhang CL, Liu ZM, Ding KL, Yang ZZ (2006) A simple and efficient route to prepare inorganic compound/polymer composites in supercritical fluids. Macromol Rapid Commun 27:787–792

118. Liu ZM, Han BX (2009) Synthesis of carbon-nanotube composites using supercritical fluids and their potential applications. Adv Mater 21:825–829

119. Ye XR, Lin YH, Wang CM, Wai CM (2003) Supercritical fluid fabrication of metal nanowires and nanorods templated by multiwalled carbon nanotubes. Adv Mater 15:316–319

120. Fu L, Liu ZM, Liu YQ, Han BX, Wang JQ, Hu PA, Cao LC, Zhu DB (2004) Coating carbon nanotubes with rare earth oxide multiwalled nanotubes. Adv Mater 16:350–352

121. Dai XH, Liu ZM, Han BX, Sun ZY, Wang Y, Xu J, Guo XL, Zhao N, Chen J (2004) Carbon nanotube/poly(2,4-hexadiyne-1,6-diol) nanocomposites prepared with the aid of supercritical CO₂. Chem Commun 2190–2191. doi: 10.1039/B407605B

122. Wang JW, Khlobystov AN, Wang WX, Howdle SM, Poliakoff M (2006) Coating carbon nanotubes with polymer in supercritical carbon dioxide. Chem Commun 1670–1672

123. Wakayama H, Setoyama N, Fukushima Y (2003) Size-controlled synthesis and catalytic performance of Pt nanoparticles in micro- and mesoporous silica prepared using supercritical solvents. Adv Mater 15:742–745

124. Lucky RA, Charpentier PA (2008) A one-Step approach to the synthesis of ZrO₂-modified TiO₂ nanotubes in supercritical carbon dioxide. Adv Mater 20:1755–1759

125. O'Neil A, Watkins JJ (2007) Reactive deposition of conformal metal oxide films from supercritical carbon dioxide. Chem Mater 19:5460–5466

126. Cangül B, Zhang LC, Aindow M, Erkey C (2009) Preparation of carbon black supported Pd, Pt and Pd-Pt nanoparticles using supercritical CO₂ deposition. J Supercrit Fluids 50:82–90

127. Klesper E, Corwin AH, Turner DA (1962) High pressure gas chromatography above critical temperatures. J Org Chem 27:700–701

128. Lee ML, Markides KE (1987) Chromatography with supercritical fluids. Science 235:1342–1347

129. Taylor LT (2009) Supercritical fluid chromatography for the 21st century. J Supercrit Fluids 47:566–573

130. Zhao YQ, Pritts WA, Zhang SH (2008) Chiral separation of selected proline derivatives using a polysaccharide-type stationary phase by supercritical fluid chromatography and comparison with high-performance liquid chromatography. J Chromatogr A 1189:245–253

131. West C, Bouet A, Gillaizeau I, Coudert G, Lafosse M, Eric L (2010) Chiral separation of phospine-containing α-amino acid derivatives using two complementary cellulosic stationary phases in supercritical fluid chromatography. Chirality 22:242–251

132. Miller L, Potter M (2008) Preparative chromatographic resolution of racemates using HPLC and SFC in a pharmaceutical discovery environment. J Chromatogr B 875:230–236

133. Desmortreux C, Rothaupt M, West C, Lesellier E (2009) Improved separation of furocoumarins of essential oils by supercritical fluid chromatography. J Chromatogr A 1216:7088–7095

134. Takahashi K, Kinugasa S, Yoshihara R, Nakanishi A, Mosing RK, Takahashi R (2009) Evaluation of a condensation nucleation light scattering detector for the analysis of synthetic polymer by supercritical fluid chromatography. J Chromatogr A 1216:9008–9013

135. Munshi P, Bhaduri S (2009) Supercritical CO₂: a twenty-first century solvent for the chemical industry. Curr Sci 97:63–72

136. Liu ZT, Sun ZF, Liu ZW, Lu J, Xiong HP (2008) Benzylated modification and dyeing of ramie fiber in supercritical carbon dioxide. J Appl Polym Sci 107:1872–1878

137. Montero GA, Smith CB, Hendrix WA, Butcher DL (2000) Supercritical fluid technology in textile processing: an overview. Ind Eng Chem Res 39:4806–4812

138. van der Kraan M, Cid MVF, Woerlee GF, Veugelers WJT, Witkamp GJ (2007) Dyeing of natural and synthetic textiles in supercritical carbon dioxide with disperse reactive dyes. J Supercrit Fluids 40:470–476

139. Banchero M, Ferri A, Manna L (2009) The phase partition of disperse dyes in the dyeing of polyethylene terephthalate with a supercritical CO₂/methanol mixture. J Supercrit Fluids 48:72–78

140. Weibel GL, Ober CK (2003) An overview of supercritical CO₂ applications in microelectronics processing. Microelectron Eng 65:145–152

141. Ventosa C, Rébiscoul D, Perrut V, Ivanova V, Renault O, Passemard G (2008) Copper cleaning in supercritical CO_2 for the microprocessor interconnects. Microelectron Eng 85:1629–1638
142. Porta GD, Volpe MC, Reverchon E (2006) Supercritical cleaning of rollers for printing and packaging industry. J Supercrit Fluids 37:409–416
143. Zhang XG, Pham JQ, Martinez HJ, Wolf PJ, Green PF, Johnston KP (2003) Water-in-carbon dioxide microemulsions for removing post-etch residues from patterned porous low-k dielectrics. J Vac Sci Technol B 21:2590–2598
144. Zhang XG, Pham JQ, Ryza N, Green PF, Johnston KP (2004) Chemical-mechanical photo-resist drying in supercritical carbon dioxide with hydrocarbon surfactants. J Vac Sci Technol B 22:818–825
145. Johnston KP, da Rocha SRP (2009) Colloids in supercritical fluids over the last 20 years and future directions. J Supercrit Fluids 47:523–530
146. Zhang JL, Han BX (2009) Supercritical CO_2-continuous microemulsions and compressed CO_2-expanded reverse microemulsions. J Supercrit Fluids 47:531–536
147. Liu JH, Cheng SQ, Zhang JL, Feng XY, Fu XG, Han BX (2007) Reverse micelles in carbon dioxide with ionic-liquid domains. Angew Chem Int Ed 46:3313–3315
148. Zhang R, Liu J, He J, Han BX, Zhang XG, Liu ZM, Jiang T, Hu GH (2002) Compressed CO_2-assisted formation of reverse micelles of PEO-PPO-PEO copolymer. Macromolecules 35:7869–7871
149. Zhang JL, Han BX, Li W, Zhao YJ, Hou MQ (2008) Reversible switching of lamellar liquid crystals into micellar solutions using CO_2. Angew Chem Int Ed 47:10119–10123
150. Li W, Zhang JL, Zhao YJ, Hou MQ, Han BX, Yu CL, Ye JP (2010) Reversible switching a micelle-to-vesicle transition reversibly by compressed CO_2. Chem Eur J 16:1296–1305

Books and Reviews

Anastas PT, Farris CA (1994) Benign by design: alternative synthetic design for pollution prevention, vol 577. American Chemical Society, Washington, DC
Brunner G (2004) Supercritical fluid as solvents and reaction media. Elsevier, Amsterdam
Centi G, Perathoner S (2002) Selective oxidation-industrial, Encyclopedia of catalysis. Wiley, New York. doi:10.1002/0471227617.eoc188
Clifford AA (1998) Fundamentals of supercritical fluid. Oxford University Press, Oxford, UK
Gupta RB, Shim JJ (2007) Solubility in supercritical carbon dioxide. CRC Press, Boca Raton
Jessop PG, Leitner W (1999) Chemical synthesis using supercritical fluids. Wiley-VCH, Weinheim
McHugh MA, Krukonis VJ (1994) Supercritical fluid extraction, 2nd edn. Butterworth-Heinemann, Stoneham, MA
Noyori R (1994) Asymmetric catalysis in organic Synthesis. John Wiley & Sons, New York
Quinn EL, Jones CL (1936) Carbon dioxide. Reinhold, New York

Index

P.T. Anastas and J.B. Zimmerman (eds.), *Innovations in Green Chemistry
and Green Engineering*, DOI 10.1007/978-1-4614-5817-3,
© Springer Science+Business Media New York 2013